SUSTAINABLE TOURISM DEVELOPMENT

Futuristic Approaches

Advances in Hospitality and Tourism

SUSTAINABLE TOURISM DEVELOPMENT

Futuristic Approaches

Edited by

Anukrati Sharma, PhD

APPLE ACADEMIC PRESS

Apple Academic Press Inc. 3333 Mistwell Crescent Oakville, ON L6L 0A2 Canada	Apple Academic Press Inc. 1265 Goldenrod Circle NE Palm Bay, Florida 32905 USA

Exclusive worldwide distribution by CRC Press, a member of Taylor & Francis Group

International Standard Book Number-13: 978-1-77188-772-4 (Hardcover)
International Standard Book Number-13: 978-0-42939-799-8 (eBook)

Library and Archives Canada Cataloguing in Publication

Title: Sustainable tourism development : futuristic approaches / edited by Anukrati Sharma, PhD.
Other titles: Sustainable tourism development (Oakville, Ont.)

Names: Sharma, Anukrati, 1981- editor.

Series: Advances in hospitality and tourism book series.

Description: Series statement: Advances in hospitality and tourism book series | Includes bibliographical references and index.

Identifiers: Canadiana (print) 20190119241 | Canadiana (ebook) 20190119306 | ISBN 9781771887724 (hardcover) | ISBN 9780429397998 (ebook)

Subjects: LCSH: Sustainable tourism. | LCSH: Branding (Marketing)

Classification: LCC G156.5.S87 S87 2019 | DDC 910.68/8—dc23

CIP data on file with US Library of Congress

Apple Academic Press also publishes its books in a variety of electronic formats. Some content that appears in print may not be available in electronic format. For information about Apple Academic Press products, visit our website at **www.appleacademicpress.com** and the CRC Press website at **www.crcpress.com**

DEDICATION

Dedicated with lots of love to
my darling daughter 'Vidushi Choudhary' and
with devotion to SAINATH

ABOUT THE EDITOR

Anukrati Sharma, PhD
Associate Professor,
Faculty of Commerce and Management,
University of Kota, Rajasthan, India

Anukrati Sharma, PhD, is currently working as an Associate Professor in the Faculty of Commerce and Management, University of Kota, Kota Rajasthan, India. Dr. Sharma has worked as an internal trainer and teacher in the management arena. In 2015, she received a Research Award Fellowship by the Univeristy Grants Commission (UGC), New Delhi, India, for her project "Analysis of the Status of Tourism in Hadoti and Shekhawati Region/Circuit (Rajasthan): Opportunities, Challenges, and Future Prospectus." Seeing her work on tourism, the UGC has given her one-year extension for her project till the year 2018. Her doctorate from the University of Rajasthan is in tourism marketing, and she completed her dissertation research on "Tourism in Rajasthan: Progress & Prospects." She has two postgraduate degrees specialties—one in International Business (Masters of International Business) and another in Business Administration (Masters of Commerce). Her core subjects are tourism, strategic management, and international business management.

She has edited book titled *Tourism Events in Asia Marketing and Development, Routledge, USA* under *Advances in Events Research Series*, *Sustainable Destination Branding and Marketing: Strategies for Tourism Development* under CABI, U.K and *Maximizing Business Performance and Efficiency through Intelligent Systems*, under IGI Global.

She is serving as an editorial board member of 26 reputed national and international journals, as well as serving as a reviewer of 17 journals of national and international repute. She has attended and presented over 50 papers at international and national conferences and is a member of 17 professional bodies. She has been invited for many talks/lectures/panel discussions by different universities, such as the International Development Institute Pokhara, Nepal; University of Colombo, Sri Lanka; Kastamonu University, Turkey; Aligarh Muslim University, etc. She has been invited by the Government of Rajasthan and FICCI to deliver a talk on agritourism. Dr. Sharma continues to take training sessions at Rajasthan Police Academy, Jaipur, on topics such as change management, communication skills, gender discrimination, personality development, etc.

ADVANCES IN HOSPITALITY AND TOURISM BOOK SERIES FROM APPLE ACADEMIC PRESS, INC.

Editor-in-Chief:

Mahmood A. Khan, PhD
Professor, Department of Hospitality and Tourism Management,
Pamplin College of Business,
Virginia Polytechnic Institute and State University,
Falls Church, Virginia, USA
Email: mahmood@vt.edu

Books in the Series:

Food Safety: Researching the Hazard in Hazardous Foods
Editors: Barbara Almanza, PhD, RD, and Richard Ghiselli, PhD

Strategic Winery Tourism and Management: Building Competitive Winery Tourism and Winery Management Strategy
Editor: Kyuho Lee, PhD

Sustainability, Social Responsibility and Innovations in the Hospitality Industry
Editor: H. G. Parsa, PhD
Consulting Editor: Vivaja "Vi" Naraparreddy, PhD
Associate Editors: SooCheong (Shawn) Jang, PhD,
Marival Segarra-Oña, PhD, and Rachel J. C. Chen, PhD, CHE

Managing Sustainability in the Hospitality and Tourism Industry: Paradigms and Directions for the Future
Editor: Vinnie Jauhari, PhD

Management Science in Hospitality and Tourism: Theory, Practice, and Applications
Editors: Muzaffer Uysal, PhD, Zvi Schwartz, PhD, and
Ercan Sirakaya-Turk, PhD

Tourism in Central Asia: Issues and Challenges
Editors: Kemal Kantarci, PhD, Muzaffer Uysal, PhD, and Vincent Magnini, PhD

Poverty Alleviation through Tourism Development: A Comprehensive and Integrated Approach
Robertico Croes, PhD, and Manuel Rivera, PhD

Chinese Outbound Tourism 2.0
Editor: Xiang (Robert) Li, PhD

Hospitality Marketing and Consumer Behavior: Creating Memorable Experiences
Editor: Vinnie Jauhari, PhD

Women and Travel: Historical and Contemporary Perspectives
Editors: Catheryn Khoo-Lattimore, PhD, and Erica Wilson, PhD

Wilderness of Wildlife Tourism
Editor: Johra Kayeser Fatima, PhD

Medical Tourism and Wellness: Hospitality Bridging Healthcare (H2H)©
Editor: Frederick J. DeMicco, PhD, RD

Sustainable Viticulture: The Vines and Wines of Burgundy
Claude Chapuis

The Indian Hospitality Industry: Dynamics and Future Trends
Editors: Sandeep Munjal and Sudhanshu Bhushan

Evolving Paradigms in Tourism and Hospitality in Developing Countries: A Case Study of India
Editors: Bindi Varghese, PhD

The Hospitality and Tourism Industry in China: New Growth, Trends, and Developments
Editors: Jinlin Zhao, PhD

Labor in Tourism and Hospitality Industry: Skills, Ethics, Issues, and Rights
Abdallah M. Elshaer, PhD, and Asmaa M. Marzouk, PhD

Sustainable Tourism Development: Futuristic Approaches
Editor: Anukrati Sharma, PhD

ABOUT THE SERIES EDITOR

Mahmood A. Khan, PhD, is a Professor in the Department of Hospitality and Tourism Management, Pamplin College of Business at Virginia Tech's National Capital Region campus. He has served in teaching, research, and administrative positions for the past 35 years and worked at major U.S. universities. Dr. Khan is the author of several books and has traveled extensively for teaching and consulting on management issues and franchising. He has been invited by national and international corporations to serve as a speaker, keynote speaker, and seminar presenter on different topics related to franchising and services management. He is the author of *Restaurant Franchising: Concepts, Regulations, and Practices, Third Edition, Revised and Updated*, published by Apple Academic Press, Inc.

Dr. Khan has received the Steven Fletcher Award for his outstanding contribution to hospitality education and research. He is also a recipient of the John Wiley & Sons Award for the lifetime contribution to outstanding research and scholarship; the Donald K. Tressler Award for scholarship; and the Cesar Ritz Award for scholarly contribution. He also received the Outstanding Doctoral Faculty Award from Pamplin College of Business.

He has served on the Board of Governors of the Educational Foundation of the International Franchise Association, on the Board of Directors of the Virginia Hospitality and Tourism Association, as a Trustee of the International College of Hospitality Management, and as a Trustee on the Foundation of the Hospitality Sales and Marketing Association's International Association. He is also a member of several professional associations.

CONTENTS

CONTRIBUTORS

Harini K. Amarasinghe
University of Sri Jayewardenepura, Sri Soratha Mawatha, Nugegoda, 10250, Sri Lanka,
E-mail: harinik@sjp.ac.lk

G. Anbalagan
Assistant Regional Director, Indira Gandhi National Open University (IGNOU),
Regional Centre, Madurai, Tamil Nadu-625018, India

R. Annapoorani
Professor, Department of Economics, Avinashilingam University, Coimbatore, Tamil Nadu, India

Gabriella G. C. Bertozzi
University of São Paulo, Rua Arlindo Bettio 1000, São Paulo, 03828-000, Brazil | Technically certified in Hospitality by Fundação Escola de Comércio Álvares Penteado in 2012 and graduated in Leisure and Tourism by The University of São Paulo in 2017, E-mail: gcbertozzi@gmail.com

Shivam Prakash Bhartiya
CMTHS, HNB Garhwal Central University, Srinagar (G), Uttarakhand, 246174, India

Vaibhav Bhatt
CMTHS, HNB Garhwal Central University, Srinagar (G), Uttarakhand, 246174, India

Bruno H. M. Carvalho
University of São Paulo, Rua Arlindo Bettio 1000, São Paulo, 03828–000, Brazil | Graduated in Business Management by The European College of Management Dublin in 2012 and in Leisure and Tourism by the University of São Paulo in 2017, E-mail brunohmcarvalho@gmail.com

James Malitoni Chilembwe
Mzuzu University, Private Bag 201, Luwinga, Mzuzu, Malawi,
E-mail: chilembwe1975@gmail.com

Istemi Comlekci
Duzce University, Faculty of Business, 81620, Turkey, E-mail: istemicomlekci@duzce.edu.tr

Cynthia Correa
University of São Paulo, Rua Arlindo Bettio 1000, São Paulo, 03828–000, Brazil |
Associate Professor of the Graduate Program in Cultural Studies, University of São Paulo,
E-mail: cynthia.correa@outlook.com

Bekir Bora Dedeoğlu
Tourism Faculty, Nevsehir Hacı Bektaş Veli University, Nevsehir, 50300, Turkey,
E-mail: b.bora.dedeoglu@nevsehir.edu.tr

Manoj Edward
Associate Professor, School of Management Studies, Cochin University of Science and Technology,
Kochi, Kerala, India

Sheeba Hamid
Professor (Tourism), Department of Commerce, Aligarh Muslim University, Aligarh, India,
E-mail:sheebaranyah@gmail.com

Azizul Hassan
PhD Researcher, Cardiff Metropolitan University, UK,
E-mail: M.Hassan15@outlook.cardiffmet.ac.uk

M. E. Korstanje
University of Palermo, Argentina, Fellow at CERS University of Leeds, UK

Alok Kumar
Associate Professor, Department of Tourism Management & Hospitality Studies, IPHC,
Mekelle University, Ethiopia (Africa), Mobile: 251-985024149;
E-mail: kumaralok1975@yahoo.co.in

Narendra Kumar
Amity Institute of Travel and Tourism, Amity University, Noida, Uttar Pradesh, India,
E-mail: nkumar14@amity.edu

Arjun Kumar Limbu
Sr. Program Officer, Kathmandu Environmental Education Project, Nepal,
E-mail: arjunlimbu@outlook.com

Prithbi Bahadur Limbu
Central Department of Sociology, Tribhuvan University, Nepal, E-mail: prithbilimbu@outlook.com

James Manalel
Professor, School of Management Studies, Cochin University of Science and Technology, Kochi,
Kerala, India

S. Meera
Assistant Professor, Indian Institute of Tourism and Travel Management, Nellore, Andhra Pradesh,
India, E-mail: mirasmadhav@yahoo.com

Muammer Mesci
Duzce University, School of Akcakoca Tourism and Hotel Management, 81650, Turkey,
E-mail: muammermesci@duzce.edu.tr

Zeynep Mesci
Duzce University, School of Akcakoca Tourism and Hotel Management, 81650, Turkey,
E-mail: zeynepmesci@duzce.edu.tr

Victor Ronald Mweiwa
Malawi Institute of Tourism, P.O. Box 2673, Blantyre, Malawi

Manvinder Singh Pahwa
Associate Professor, Department of Business Administration, Manipal University Jaipur,
India VPO Dehmi Kalan, Sanganer Jaipur, (Rajasthan) India,
E-mail: manvindersingh.pahwa@jaipur.manipal.edu

Samik Ray
Ex-Faculty of Department of Folklore University of Kalyani, Ex-Faculty of Travel and Tourism
Management in MPTI (Kolkata), and WTCC School of Trade and Commerce,
Ex-Trainer and Faculty of RLG Training (Govt. of India, Department of Tourism),
presently working as RLG (Govt. of India, Department of Tourism),
Editor of 'Tourism Theory and Practice,' author of several essays on Tourism Studies and Management,
Social Science, and Literary Criticism

Hugues Seraphin
University of Winchester, UK

Anukrati Sharma
Associate Professor, Department of Commerce and Management, University of Kota, Kota (Rajasthan) India, E-mail: dr.anukratisharma@gmail.com

Swati Sharma
Amity Institute of Travel and Tourism, Amity University, Noida, Uttar Pradesh, India, E-mail: ssharma3@amity.edu

D. A. C. Suranga Silva
University of Colombo, Kumaratunga Munidasa Mawatha, Colombo, 00700, Sri Lanka, E-mail: drsuranga3@gmail.com

Shivendra Singh
Mahatma Gandhi Central University of Bihar, Mothihari, 845401, Bihar, India

Sujood
Research Scholar (Junior Research Fellow-Tourism), Department of Commerce, Aligarh Muslim University, Aligarh, India, E-mail: sjdkhancool@gmail.com

Toney K. Thomas
Taylor's University Lakeside Campus, No.1 JLN Taylor's, 47500, Subang Jaya, Malaysia, E-mail: tonythomas@taylors.edu.my

A. Vinodan
Assistant Professor Indian Institute of Tourism and Travel Management, Nellore, Andhra Pradesh, India

ABBREVIATIONS

AHP	analytic hierarchy process
ASI	Archeological Survey of India
ATCMs	Antarctica Treaty Consultative Meetings
ATS	Antarctic Treaty System
AVE	average variance extracted
CAGR	compound annual growth rate
CAPM	capital asset pricing model
CBNRM	community-based natural resource management
CBNT	community-based nature tourism
CBRM	community-based resource management
CBT	community-based tourism
CCF	conservator of forest
CDS	cognitive dissonance scale
CE	conservation effort
CES	cognitive emotional satisfaction
CI	commercial intervention
CIS	community intervention strategies
CMBS	commercial mortgages backed securities
CP	community perception
CPS	cognitive product satisfaction scale
CSRs	customer service representatives
CSS	cognitive service satisfaction
DBI	doing business index
DFW	Department of Forest and Wild Life
DMO	Destination Marketing Organizations
DOE	Directorate of Ecotourism
DOT	Department of Tourism
EC	economic condition
EDCs	eco-development committees
EEM	expectation experience matrix
EI	expert interview
EPFDC	East Pakistan Film Development Corporation
EWAM	equal weight averaging model

FAO	Food and Agricultural Organization
FDA	Forest Development Agency
FGD	focused group discussions
FTA's	foreign tourist arrivals
GFI	goodness of fit index
GNH	gross national happiness
GOK	Government of Kerala
GOP	gross operating profit
HACCP	hazard analysis and critical control point
HDI	human development index
IAATO	International Association of Antarctica Tour Operators
ICT	information and communication technologies
IMO	International Maritime Organization
INR	Indian rupees
IPA	importance-performance analysis
IUCN	International Union for Conservation of Nature
LWR	Liwonde Wildlife Reserve
MIRR	modified internal rate of return
MPT	modern portfolio theory
MSMEs	Micro, Small and Medium Enterprises
NCCP	national codex contact point
NGO	Non-Governmental Organization
NPD	new product development
NPV	net present value
NWR	Nyika Wildlife Reserve
OTAs	online travel portals
PAs	protected areas
PPP	purchasing power parity
PPT	pro-poor tourism
PRPs	prerequisite programme
REITs	real estate investment trusts
ROI	return on investment
ROS	recreational opportunity spectrum
SD	sustainable development
SEM	structural equation modeling
SI	social image
SM	social media

SMART	simple multi-attitude rating technique
SMP	social media pyramid
SMS	social media sharing
SRI	socially responsible investment
SWOT	strength weaknesses opportunities threats
TEDC	Tourism Eco-Development Committee
THT	Tamang Heritage Trail
TSA	tourism satellite account
TTCI	travel and tourism competitiveness index
UBP	unique branding proposition
UGC	user-generated content
USP	unique selling proposition
VICE	Visitor Industry Community Environment
VOLM	virtual online learning material
VR	virtual reality
VTB	vendor take back
WHO	World Health Organization
WOM	word-of-mouth
WWF	World Wildlife Fund

PREFACE

Information technology and communication are powerful forces behind the development of transformation in the tourism industry. Today, the tourist wants to be entertained, informed, and motivated before visiting a place. An attempt has been made in this book to fulfill this delightful aspect from the perspective of different planners, policy makers and stakeholders of travel, tourism and hospitality, and information technology industry. The book gives ample opportunities to the readers to find out the urge of content writing, social media and community involvement in an extensive way for successful destination branding and marketing. This book provides an insight into the new inventions which can create interest of potential tourists not only to attract them towards a destination through storytelling, etc., but also meticulously focuses upon a long-term relationship with them. The book is a blend of content marketing, destination branding and community-based tourism for reimaging and attracting tourists with logic, perception, and learning. It certainly provides such a conceptual framework for addressing the general problem, opportunities and challenges of tourism, destination branding, and marketing. The book is a conscientious effort of numerous minds. The uniqueness of this manuscript is that it is not systematically focused on theory but on the practical aspects which are relevant too. The main focus of the book is to generate and extend the practical implications of marketing, sustainable tourism and community development which the readers can find worthy especially in travel, tourism, and hospitality industry. Instead of just unfolding the formal definitions, terms, and steps, each chapter of the book will give explicit suggestions that the tourism industry people should pursue for promoting destinations. The editor felt the need for this book because Sustainable Tourism Development is a topic on which different sections of people hold diverse views. It is Therefore, important to reach a concrete decision. The book is designed in such a way that it will be beneficial not only for researchers, academicians, students but also for the government, local people and the tourists.

—Anukrati Sharma, PhD

CHAPTER 1

SUSTAINABLE TOURISM: GROWTH AND DIVERSIFICATION

MUAMMER MESCI,[1] ISTEMI COMLEKCI,[2] and ZEYNEP MESCI[3]

[1]*Duzce University, School of Akcakoca Tourism and Hotel Management, 81650, Turkey, E-mail: muammermesci@duzce.edu.tr*

[2]*Duzce University, Faculty of Business, 81620, Turkey, E-mail: istemicomlekci@duzce.edu.tr*

[3]*Duzce University, School of Akcakoca Tourism and Hotel Management, 81650, Turkey, E-mail: zeynepmesci@duzce.edu.tr*

ABSTRACT

Although mass tourism appears to be an important source of income for countries, it has been adversely affected in recent years as mass tourism has been applied unfavorably to countries and regions. This negative situation, which mass tourism has shown, is an important need for sustainable tourism development.

The innovations brought by the technology today have caused a positive contribution of tourism to the world. For example, with the development of information technology and transportation, new countries and regions have begun to be discovered, and tourism activities have increased. The unplanned application of tourism activities in a destination and the extreme increase of tourism to the region lead to the deterioration of the social, cultural, and natural resources of these regions. Here, sustainable tourism should be given importance and developed in order to overcome these negativities.

On the basis of sustainable development lies the protection and development of natural, social, and cultural resources. In addition, these resources should be managed to meet the needs of future generations. On the other hand, in sustainable tourism, while preserving the surrounding environment, the importance should be given to cultural integrity, ecological processes, biodiversity, and transport capacity.

In this study, the growth of sustainable tourism and what needs to be done for its development will be discussed. In addition, the basic dimension of sustainable tourism, environmental, social, and economic dimension will be discussed in detail.

1.1 INTRODUCTION

The concept of sustainable tourism emerged together with the concept of sustainable development on account of the fact that the other sectors and other elements attach importance to the concept of sustainability. Nevertheless, human-nature relation is more intense and continuous than other sectors, and it has become compulsory for this concept to come together with the concept of tourism.

In addition, tourism is an activity that can have a real impact on sustainable development compared to other sectors (Angelkova et al., 2012). The trend that began with people using the environmental elements and natural resources in an unplanned and extensive way, along with the industrial revolution and consuming the resources of continuous use with this planlessness, showed that these resources could not be renewed and that the destroyed nature elements could not be replaced over time. With the emergence of damages related to this subject, the opinions of the various institutions and organizations along with their views on the possibility that these sources cannot be renewed and excessive consumption will cause material and moral damages to humanity in the coming years have brought this issue to the agenda. Nevertheless, the fact that globalization and the new world order created by removing the borders raised awareness, revealed problems, and laid burdens has made the concept of sustainability spread to the world. The fact that tourism enterprises are affected by globalization and maintaining the constant dynamism within the framework of the needs of the people make an important contribution to the filling of this concept, and academic research is being carried out for possible solutions.

1.2 SUSTAINABLE TOURISM

Today, it is observed that the most important problems faced by service sectors stem from their unique characteristics. One of them is that tourists take initiatives for the products created. Another characteristic is that the standardization cannot be achieved. In this context, the concept of sustainability has become a part of tourism activities, as a result of the damage of people to the environment and natural resources within the scope of establishing tourist attractions (Halis, 2009).

The concept of sustainability is first summarized in 1987 as a brief description of the concept mentioned in the United Nations' report on the Environment and Development Commission, a harmonious integration of the economy with the economy. The concept of sustainable tourism is defined as the ability of a tourism zone to be sensitive to future developments and to protect from the adverse effects of these developments (Hunter, 1997). According to another definition, the concept of sustainable tourism emerges as a special term that comes from a broader concept of sustainable growth and means that this growth is applied to the special context of tourism. Maintaining the quality of the environment, raising the quality of life of the touristic regions, and carrying out quality visitor experience are aimed with this concept (Kaypak, 2010). Sustainable tourism can be expressed as preserving and sustaining the regional and local charms that are the source of tourism (Avcıkurt, 2015). Sustainable tourism is largely influenced by environmental quality, protection of natural and cultural heritage, and products and resources, along with other values, to protect development and competitiveness (Angelkova, et al., 2012).

When we look at the content of the concept of sustainable tourism, it is seen that environmental and cultural issues are essential. The fact that it includes subjects which can easily be destroyed but cannot be restored is important for the continuation of humanity. In other respects, the protection of natural and artificial cultural assets seems to be important in terms of being knowledgeable about historical processes and transferring them to generations. They may not be regarded to be vital in terms of other sectors, but these issues have an extremely important proposition in terms of tourism and tourism types (Er & Bardakoğlu, 2016; Karapınar & Barakazi, 2017). When we consider that many types of tourism emerged along with the concept of sustainability, it can be seen that many types of tourism can be sustained with the development and prevalence of sustainable tourism. On the other hand, sustainable tourism, where some researchers take part in alternative tourism, has laid the foundation for the emergence and development of many types of tourism (Güniz, 2011). The most important ones are culture tourism, ecotourism, cave, plateau, hunting, and golf tourism.

1.3 DIMENSIONS OF SUSTAINABLE TOURISM

Hasna investigated these dimensions from three perspectives in his research on the dimensions of sustainability (Hasna, 2007). These dimensions are listed together with the reminder that interdisciplinary variables should be considered socially, economically and ecologically. Voniov and Smith emphasized that the dimensions of sustainability should be considered within a total system framework and should not be considered independently of each other (Voinov & Smith, 2008). They have laid out the dimensions that Hasna revealed in his research. Lozano and Huisingh (2011), Holmberg (1992), and Kade (2012) have also emphasized three dimensions of sustainability. Krajnc and Glavic (Krajnc & Peter, 2005) have emphasized economic, environmental and social dimensions in their research of companies focusing on improving their performances. Considering current research, the dimensions of sustainability are generally collected under three main headings. This study will be considered environmental, social, and economic dimensions of sustainable tourism.

The most important features of sustainable tourism dimensions gathered around the three main headings are inseparable and mutually supportive. Each of these dimensions will be presented in following subheadings within their own characteristics.

1.3.1 THE ENVIRONMENTAL DIMENSION OF SUSTAINABLE TOURISM

The environment, the most important factor for survival of the living, has been polluted and destroyed rapidly in recent years because of uncontrolled human behavior. As a more important issue for the tourism industry, the environment has vital importance for many types of tourism (Katerina & Rakicevik, 2012). Especially for the types of tourism associated with nature, it is the leading cause of existence. Therefore, it is an undeniable fact that the environmental dimension and environmental quality attract tourists (Grabara & Bajdor, 2013). As a result, the environment is the most important tourism resource for the quality, diversity, and development of tourism (Kade, 2012).

An unprotected and undeveloped environment has negative consequences for all stakeholders. Tanja (2016) states that the environmental dimension, which is highly favorable for the development of tourism activities and for the protection of cultural heritage, is not considered by interest groups to respond to growing demand and achieve economic gain. The areas where the tourism enterprises are located in our country and in the world are generally locations attracting people and having high levels of attractiveness (Katerina & Rakicevik, 2012; Kavaliauskė & Kočytė, 2014). These locations are being built up by economic benefit providers without regard to any environmental factor on account of the opportunities that arise because people are attracted to the high material and spiritual atmosphere of these locations. On the other hand, there are many reasons why the relationship with the environment does not go well after it has been uncovered outside the structured image of the businesses discovered in the framework of this construction. The leading reason comes from infrastructure deficiencies (Avcıkurt, 2003; Priskin, 2003). Carrying capacity is one of the essential issues when it comes to infrastructure deficiencies.

Carrying capacity literally means that a maximum number of people in an environment can be supported without diminishing the ability to support future generations on the planet (Kormondy, 1996). According to another definition, it can be defined as maximum usage without a negative effect on resources, undesirable effect on local economy and culture without decreasing visitor satisfaction (Avcı, 2007). With regard to these definitions, it is not only limited to the environmental dimension, but also includes social and economic dimensions. In terms of environmental dimension; ignoring the protection of the environment for the development of tourism means that a sustainable development of tourism should be abandoned. If the resources are destroyed or damaged in some way, as a result of touristic use of natural resources, the ecological carrying capacity will be exceeded (Mercan, 2010).

Another factor is waste (Garrod & Fyall, 1998). New destinations are created by removing the tourism destination areas from cities. It is extremely important that these destinations be cleared of the negative effects of waste on the environment (Kahraman & Türkay, 2006). In this respect, the most important element is the classification of waste (Kavaliauskė & Kočytė, 2014). In particular, the benefits that can be gained from the recycling of environmentally sustainable wastes can be assessed both at the environmental and economic dimension. Recently, with the development of technology, the electric energy gain obtained from the wastes is regarded as important for regional energy by providing non-renewable energy production. Another important factor in waste is the removal of uncontrolled and unplanned forms. Especially, the removal of food waste from tourism enterprises in an unplanned way may lead to both environmental and economic damage.

In order to prevent this problem, it is very important to create a waste management plan and enforce it as soon as possible. The use of raw materials that will form waste at a minimum level before being converted into a product will reduce the amount of waste to be generated. In addition, it is possible to reduce the amount of waste to be generated by calculating the multiple usage life of certain substances. Another point is that it is economically beneficial to recycle the waste that is created as part of the tourism industry. Finally, it is essential to ensure that the wastes generated be handled by taking the environmental effects into consideration and obeying the necessary rules and bases. Therefore, the state organs and environmental organizations can play pivotal roles in preventing the tourism regions and the environment from being polluted with the responsibility of their operations (Garrod & Fyall, 1998).

Another important environmental issue is energy consumption. To realize sustainable development, energy resources that do not harm the environment need to be used (Mercan, 2010). Especially, the amount of energy used in the hotel and leisure industry, which serves all day, reaches important levels both environmentally and economically. From this point of view, every work that the utilities will do regarding energy saving will return to enterprises in many positive ways. If considered in the realm of the energy sector, it is an extremely costly and labor-intensive production area. Considering the cost of non-renewable energy sources, in particular, it has an important place in the expenses of the enterprises (Kahraman & Türkay, 2006). In addition to the economical damage of this problem, there is a risk that humanity may be faced with difficulty in time due to overuse of environmentally irreproducible energy resources.

Another problem is that the use of energy sources that pollute the environment is continued, and the use of other living things is restricted by overuse. Nowadays, with the development of technology, the use of energy in the environment is planned, and the losses can be reduced to minimum levels. State control and sanctions can be used to switch to renewable energy sources, or the use of energy resources that can pollute the environment can be reduced, and the necessary measures may be taken to reduce the damage to both the environment and the destination zone (Grabara & Bajdor, 2013).

In recent years, the spread of information through social media accounts within the framework of the development of the internet network has increased the environmental sensitivity. Therefore, it seems that environmental awareness has developed greatly among people. In addition, with the rapid spread of information, it seems that environmental sensitivity has gained a significant role in the ability of businesses to gain competitiveness. With the emergence of this awareness, it is expected that the emergence of environmentalist enterprises will both effect economic gain and increase sustainability (Kade, 2012).

1.3.2 THE SOCIAL DIMENSION OF SUSTAINABLE TOURISM

The social dimension of sustainable tourism deals with the adequacy of education and health services to meet the objective of meeting cultural and spiritual needs of people in terms of language, religion, race equality (Demir, 2014). As for tourism, employers, occupants, consumers and local people should benefit from equal opportunities and the interests of the region should be considered.

The most important partner of the artificial or natural destinations is the local people who lead their lives in that region. The architecture of the intimate relations with tourists, which meet the many requirements of destinations and show the self-sacrifice needed to solve problems, are local people. In particular, the common economic interests of the local people and enterprises are an important contributor to the development and welfare of the region (Albuquerque & McElroy, 1995). Within the framework of the bilateral relations established, tourism enterprises constitute one of the most important social aspects of the sustainability of the local people in terms of human resources (Demir, 2014). In this respect, local people and enterprises start to relate to each other and steps are taken for positive progress of this relationship.

Another aspect of the social dimension of sustainable tourism is the proliferation of social opportunities for the use of tourists in destination areas, where local people and employees benefit from these opportunities. With the increase of these opportunities, it is ensured that positive developments in the sense of preference for new tourists are achieved with the increase of regional development level. With the establishment of such superstructures, it becomes easier to contribute to the sustainability of this center by transforming the destination area into a living center (Carr et al., 2016).

The leading regions preferred by the tourism enterprises are the destination areas that contain cultural influences left by the ancient civilizations. One of the main reasons for this preference is that these regions are curious for domestic and foreign tourists. In other words, tourism plays an important role in the social and cultural heritage of the region, especially in areas of high cultural or artistic value, where local traditions and values play an important role (Grabara & Bajdor, 2013). In particular, it is extremely necessary to transfer the intangible cultural heritage from generation to generation and to keep it from moving forward without deteriorating the heritage by preserving the region from the external influences (Mercan & Özkök, 2013). Having a driving force to pass this heritage onto generations has a vital proposition. Although, it is a tradition of enclosed and conservative societies to pass their heritage on, in recent times, with the development of technology and the increase in the quality of life of people, many people have moved away from their local regions. In order to move this heritage forward in particular, local people should be attracted to the economy to increase the tourism activities and keep attractiveness of the destinations (Katerina & Rakicevik, 2012). In this respect, utmost importance should be attached to the efficient and balanced use of natural and cultural resources (Uğuz, 2011). In particular, the protection and advancement of cultural heritage are of great importance in terms of sustainability for these destination regions.

Another important element in the field of culture is the provision of cultural interaction. Factors such as the perception of tourism by local people, whether they are ready for tourism development, their communication levels with the tourists and the expectations from tourism are very important in terms of tourism development and tourism sustainability in a region (Avcıkurt, 2015). Thus, the mutual love, respect and moral behavior of the local people and tourists can be improved, and the tourism activities in the region can be made permanent as a result of this situation (Carr et al., 2016). As a result of reciprocal communication, cultural exchange is expected to be advantageous for establishing better relationships with local people and tourists.

When we look at the gains obtained from the social dimension of sustainability in terms of businesses:

- businesses attract local tourists as a result of communication with local tourists (Somuncu & Turgut, 2009).
- in terms of human resources, they integrate with the local people and decrease the rate of the work speed.
- in any negative situation in terms of human resources, it is possible to provide personnel from the primary level as a consequence of social relations (Yavuz, 2010).
- they can fulfill their responsibilities by supporting the preservation of cultural heritage (Er & Bardakoğlu, 2016).
- they can benefit the staff, tourists and local people by improving the superstructure possibilities (Seyhan & Yılmaz (2010).

Along with all these achievements, they play an important role in important developments for the local people, such as the prosperity of the local people, cultural awareness, and the increase of cultural interaction.

1.3.3 THE ECONOMIC DIMENSION OF SUSTAINABLE TOURISM

Tourism is one of the largest and fastest growing industries in the world (Frederico, 2003). It is an important economic activity that provides income to the countries. In many countries, it is an increasingly important source of employment and wealth. Maintaining the quality of the environment, raising the living standards, increasing the economic and environmental contributions of tourism will be possible with the concept of sustainable tourism (Avcıkurt & Demirpulat, 2016). The development of sustainable tourism provides and supports the economic, social and cultural well-being of the community where tourism is realized. Therefore, the first measure of the sustainability of tourism in a region is to what extent it contributes to the welfare of the social strata in that area (Somuncu & Turgut, 2009). Besides economic developments in favor of tourism, there are also some changes and transformations within the industry itself. On the one hand, while the preferences, motivation, expectations, and needs of the tourists differ, a change, on the other hand, has occurred in the direction of the preferred tourism regions (Ayaş, 2007).

The economic dimension of sustainable tourism starts with a well-planned operation. It is of utmost importance that the regional preferences of long-term investments are made correctly and cultural heritage and development of that region are taken into account (Carr et al., 2016). Another factor in the construction of long-term investments is that structures to be built are constructed in a way that prevents losing value in time. It is crucial for sustainable tourism to provide economic gains through the discovery of investments that can avoid additional costs later on. While establishing the tourism infrastructure, it is necessary to be supported by many actors because of the reasons such as the highest economic transformation from tourism and the success of the decisions to be taken (Avcıkurt & Demirpulat, 2016).

In general terms, the economic dimension of sustainability is defined as ensuring self-sufficiency of the cost of business without exceeding the amount of income of operation costs (Demir, 2014). In this sense, processes that are not managed with sustainability principle will increase the cost of business in the coming years and will cause financial difficulties for enterprises. It is also necessary to use sustainable resources as a part of considerations regarding the expenditures that will be made later by the enterprises. In addition to the efficient use of resources, the resources used must be estimated in a way that does not cause environmental and social problems, considering other dimensions (Katerina & Rakicevik, 2012).

The main goal of all enterprises, including tourism enterprises, is to make a profit. In order to achieve this, the costs must be reduced to a minimum. By calculating all the details in the establishment phase, this cost has to be calculated so that it will not increase in the following years (Carr et al., 2016). On the other hand, the economic dimension is shaped by other dimensions. Other dimensions, especially the environmental dimension, are economically important. In recent years, it has been observed that environmental conditions are strictly followed and sanctions are quite annoying. It is inevitable that any material that is ignored may cause major changes for a business.

In addition to the measures that can be taken in order to ensure the continuity of economic activities, it will ensure the sustainability of the management with management level that is experienced in the sector (Carr et al., 2016). It is economically important that the management level is not only from the tourism sector, but also knowledgeable about the field of sustainable development and the expectations of the industry. In short, all activities to be carried out in the construction phase and after construction are to be considered as long-term trends in terms of sustainable tourism.

1.4 DEVELOPMENT OF SUSTAINABLE TOURISM

Tourism activities have been developing and becoming a continuously growing sector. Tourism enterprises in our country and in the world have always tried to create new centers by searching new regions. Establishing economies by institutionalizing non-tourism potentials and establishing different attraction features have risks in terms of sustainability of the growth created as a result. During this growth, the process continues without considering many important factors. For many years, many attempts have been made to reduce or eliminate the negative effects of the tourism sector; these initiatives are thought to be able to be overcome by sustainable tourism, which is not harmful to the environment (Grabara & Bajdor, 2013). While trying to prevent irregular tourism development, sustainable tourism development needs to be planned and successfully implemented in order to efficiently overcome daily changes in the turbulent environment around the world (Katerina & Rakicevik, 2012). In this regard, first of all, it is necessary to establish qualified discussions made from an academic point of view by prominent names in the field.

Sustainable tourism debates often remain in an abstract structure, separated from each other, in which the conclusions of assumptions and debates are flawed (Zhenhua, 2003). The fact that an interdisciplinary approach to sustainable tourism research, identification, and development is indispensable avoids the concretization of these debates. In general, researchers spend most of their energy to put forward problems. Very few researchers seem to be interested in solving problems or the origins of problems, and these problems and solution seem to be partial.

As a result of the studies carried out, tourism activities have been spread out to a certain number of months with alternative tourism varieties (Kılıç & Kurnaz, 2010). In this respect, initiatives are being carried out with the hope that they can meet the pleasures, styles, desires, and expectations of all people on earth. The fact that tourism activities cover such a great time and place brings up certain issues in terms of sustainability. The most important aspect of this subject is stability (Angelkova et al., 2012). The existence of a highly variable and fragile structure both in terms of tourism activities and in terms of sustainability strengthens the fact that this issue should be emphasized. The most important aspect of this requirement is the understanding and dissemination of the main theme of sustainable tourism (Kade, 2012).

Four main themes are examined as a starting point for sustainable tourism development. Focal points on the development of sustainable tourism, the stakeholders, in other words, are the role that local people, local governments, tourists and the state play actively. Sustainable tourism development is under the responsibility of all stakeholders (Kade, 2012). All relevant stakeholders should exert extraordinary efforts to prevent the destruction of (social, economic and environmental) all issues related to tourism development that cause adverse impact.

Sue and Ladkin (1997), who conducted research on the local people on this issue, stated that despite the desire of stakeholders to engage in sustainable activities, they lack the intellectual interest in the issue. However, another important factor besides the efforts of people is the government policies. The fact that state policies, one of the largest stakeholders, are far from environmental factors lead to major problems in the development of sustainable tourism. Another factor is the uncertainty of principles which restrict communication with other interested stakeholders. The issue of implementing the principles is a highly anticipated issue for the development of sustainable tourism. The most important obstacles in this regard are the poor implementation of government policies by local governments. The implementation of regional policy, without regard to social and economic indicators, is considered to be the biggest obstacle to the development of sustainable tourism. Another obstacle to be considered is the fact that local people are left out of developmental processes or problems related to communication between the administrations and the local people. While there are many communication elements in today's world where educational opportunities along with social opportunities are improving, the adoption of a one-sided management approach is seen as one of the most important factors behind the inability to ensure the success of sustainable tourism, ignoring all these developments for solving this problem (Ayaş, 2007).

Another important element is tourism enterprises. Problems arising from the fact that business owners and managers have not taken into account the environmental and social benefits of steps taken by economic reasons will cause major problems to be encountered in the future. It is inevitable that bankrupt companies or failed enterprises will emerge as a result of unplanned behavior, especially in the establishment stages of enterprises (Alagöz, 2015). It is anticipated that this problem will be overcome as businesses become aware that sustainable tourism will provide a significant profit for them. However, the increase in costs of the enterprises due to sanctions starting from the establishment phase and from the first year results in the ruling out of this subject (Salah, 2000).

In order for a work to be sustainable, it is at the forefront of the most important issues that stakeholders are under a common roof, making decisions together and controlling decisions to be made. The most important factor here is seen as civil society organizations as well as the state and local governments (Tanja, 2016).

It is necessary to encourage the persons who are in the sectoral structure with incentives and supports for the introduction of sustainable tourism activities with the necessary initiative by government organs. Therefore, the region located in the tourist area of development activities can be presented in a more planned way to the liking of tourists. It is necessary to make efforts to make the regions outside the tourism areas become the centers of attraction with the same determination and support. On the other hand, state institutions play an important role in ensuring people's awareness by giving seminars and courses related to sustainable tourism by creating different educational opportunities for each person and age (Demir, 2014).

Another important element in terms of state institutions is the return and economic gain of the saving policy that sustainable tourism brings. It is expected that the total income of energy use, waste recycling, employment, and other economic components will reach highly important levels in terms of the country's economy. In this sense, both fiscal and social and environmental policies will benefit from government policies (Sezgin & Karaman, 2008).

Local governments are extremely important in terms of sustainable tourism (Uğuz, 2011). Thanks to the implementation of state policies by local governments, it is expected that significant gains will be realized both for the local population and for the tourists. The most important role of local governments for sustainable tourism is to make people conscious. In other words, the creation of a conscious society by providing tourism information for the community and tourism stakeholders (Kade, 2012), along with the explanation of sustainable tourism development, is also a vital for the future.

The creation of a conscious society as a result of the work to be done for tourists and local people, the increase of human interaction with cultural interaction, and the development of human relations are important for the sustainable tourism to move forward and improve (Kavaliauskė & Kočytė, 2014). Among these stakeholders, the most important factor is seen as the human factor. Because it is related to both the service provided and the environment created. In addition, the importance of the attitudes to be attained today on behalf of the human factor, which takes place both in providing services and in receiving services (Carr et al., 2016). From the sustainability of tourism's point of view, the establishment of standards and continuous service areas for the people who are tourists has great importance in terms of recognition and preference of the area to be traveled. In terms of local people, the increase of social welfare and the increase of the purchasing power with the employment of people lead to the end of immigration (Frederico, 2003).

The increase of stability with the continuance of services and service activities for many years and the increase of tourism activities in the coming years are very important from a social and cultural point of view.

On the cultural side, the preservation of existing artifacts built in the region in terms of protection and continuity are the most important aspects of sustainable tourism activities. It is expected that the development of tourism activities will benefit from many aspects by reducing the loss of the most (Er & Bardakoğlu, 2016).

1.5 CONCLUSION

Today, the importance of the concept of sustainable tourism emerging as a result of sustainable development becoming a strategy and emerging as a result of the combination of the concept of sustainability spreading across all sectors with the concept of tourism is extremely clear. It is seen that sustainable tourism is especially at the base of alternative tourism varieties which are emerging with increasing cultural and environmental sensitivity due to people's depression from urban life, working stress and their longing for the past (Karapınar & Barakazi, 2017).

The environment is at the forefront of what is considered important in terms of sustainable tourism. It has been seen that environmental issues cannot be limited to only green, nature, blue-flagged beaches. It has been determined that environmental issues have both economic and social consequences. Especially in the economic sense, much harm can be avoided due to the environmental sensitivity to be created from the establishment phase to the service phase. As a result of the sensitivity shown during the establishment phase, financial difficulties that may arise later will be prevented, and economically difficult situations will be prevented. After the establishment phase, factors such as energy consumption, waste management, and appropriate raw material selection have been found to be important for the enterprise both from the environmental and economic point of view.

It is an important question that the steps taken for sustainable tourism have a great prospect in terms of the social dimension. Particularly in the context of sustainability, the employment of the people of the region and the increase of social and material welfare are very important. This case has been referred to in the literature as to the benefits it brings in economic terms. In addition, education and cultural assets will be upgraded by increasing the sensitivity of the social environment. It may turn out that employers, employees, tourists, and the public attach importance to saving money in the use of non-renewable resources and have increasing levels of sensitivity.

With sustainable tourism, environmental awareness, economic gain, social cohesion issues become important, leading to many stakeholders' gains in many respects. It is especially important to use the facilities of profit-oriented enterprises in a cost-saving manner, to protect the environment in terms of social responsibility, and to increase the sensitivity to cultural assets.

Understanding that these issues, which are vitally important for increasing the lifespan of enterprises and increasing their tourist potential, can lead to a growing awareness of the sustainable tourism.

With the development of sustainable tourism, micro-enterprises, the people of the region and the tourists all get benefits. From macro-perspective, it is seen that the economy of the country and future societies also benefit. It is anticipated that these benefits will increase with the spread of the concept of sustainability throughout all sectors.

KEYWORDS

- **diversification**
- **growth**
- **sustainable**
- **sustainable tourism**

REFERENCES

Alagoz, Gurkan, Gunes, Erkan, & Uslu, Abdullah (2015). Sustainable Tourism Perspective of Hotel Business and Travel Agency Manager: A Field Research, *Gümüshane University Electronic Journal of the Institute of Social Science, 6*(14).

Albuquerque de Klaus & McElroy, L. James (1995). Alternative Tourism and Sustainability, *Alternative Tourism and Sustainability,* 23–51.

Altanlar, Aslı & Akıncı, Güniz, Kesim (2011). Research on The Attitude and Expectations of Tourists and Citizens of the Area for Sustainable Tourism Planning; Case of Akcakoca, Ankara University, *Journal of Environmental Studies, 3*(2), 1–20.

Angelkova, Tanja, et al. (2012). Sustainability and Competitiveness of Tourism, *Procedia-Social and Behavioral Sciences 44*, 221–227.

Avcı, Nilgün (2007). The Importance of Carrying Capacity in Tourism, *Ege Academic Review* 7(2), 485–500.

Avcıkurt, Cevdet (2003). Sociology of Tourism – Tourist Local People Interaction, (1st Edition), Ankara, Detay Publishing.

Avcıkurt, Cevdet (2015). Sociology of Tourism – Tourist Local People Interaction, (4st Edition), Ankara, Detay Publishing.

Avcıkurt, Cevdet and Gudu, Demirbulat, Özge (2016). Perspectives of Tourist Guides to Sustainable Tourism, *Journal of Tourism Theory and Research, 2*(2), 173–191.

Ayaş, Necla (2007). Environmental Sustainable Tourism Development, *Journal of the Faculty of Economics and Administrative Sciences, 9*(1), 59–69.

Carr, Anna, Ruhanen, Lisa & Whitford, Michelle (2016). Indigenous Peoples and Tourism: The Challenges and Opportunities for Sustainable Tourism, *Journal of Sustainable Tourism 24*(8–9), 1067–1079.

Demir, Mahmut (2014). Sustainable Tourism and Hotel Business, Sustainable Tourism Concepts and Applications, ed. Metin Kozak, Ankara, Detay Publishing.

Er, Ozlem ve Bardakoglu, Ovunc (2016). The Assessment of Cultural Heritage as a Product of Sustainable Tourism: The Case of Edirne, *Journal of Business Faculty, 17*(2), 95–111.

Frederico, Neto (2003). A New Approach to Sustainable Tourism Development: Moving Beyond Environmental Protection, *Natural Resources Forum. 27*(3). Blackwell Publishing Ltd, pp. 212–222.

Garrod, Brian & Alan Fyall (1998). Beyond the Rhetoric of Sustainable Tourism?, *Tourism Management 19*(3), 199–212.

Grabara, K. Janusz & Bajdor, Paula (2013). Towards to Sustainable Tourism–Framework, Activities and Dimensions, *Procedia Economics and Finance, 6*, 523–529.

Halis, Muhsin (2009). Basic Business Concepts and Tourism, Tourism Enterprises, ed. Demirkol ve Zengin, 1st Edition İstanbul: Degisim Publishing.

Hasna, M. Abdallah (2007). Dimensions of Sustainability, *Journal of Engineering for Sustainable.*

Hunter, C. (1997). Sustainable Tourism as an Adaptive Paradigm. *Annals of Tourism Research, 24*(4), 850–867.

Holmberg, Johan (1992). Making Development Sustainable: Redefining Institutions Policy and Economics. Island Publishing.

Kade, Sutawa, Gusti (2012). Issues on Bali Tourism Development and Community Empowerment To Support Sustainable Tourism Development, *Procedia Economics and Finance 4*, 413–422.

Kahraman, Nuzhet & Turkay, Oguz (2006). Tourism and Environment, Ankara, Detay Publishing.

Karapınar, Ebru & Barakazi, Mahmut (2017). Evaluation of Cultural Heritage Tourism as a Sustainable Tourism: Gobeklitepe Ruins, *Journal of Current Tourism Research, 1*(1), 5–18.

Katerina, Angelevska-Najdeska & Rakicevik Gabriela (2012). Planning of Sustainable Tourism Development, *Procedia-Social and Behavioral Sciences, 44,* 210–220.

Kavaliauskė, Monika & Kočytė, Rasa (2014). Sustainable Tourism Development in Neringa Region, *Procedia-Social and Behavioral Sciences, 156,* 208–212.

Kaypak, Şafak (2010). Sustainability of Ecological Tourism, *International Alanya Business School Journal, 2*(2), 93–114.

Krajnc, Damjan & Peter Glavič (2005). How to compare companies on relevant dimensions of sustainability, *Ecological Economics 55*(4), 551–563.

Kılıç, Burhan & Kurnaz, Alper (2010). Ecological Farms in Establishing Alternative Tourism and Product Diversity: The Case of Pastoral Valley, *Journal of Business Studies, 2*(4), 39–56.

Kormondy, J. Edward (1996). Concept of Ecology, Prentice Hall. 4th edn, Englewood Cliffs.

Lozano, Rodrigo & Huisingh, Don (2011). Inter-Linking Issues and Dimensions in Sustainability Reporting, *Journal of Cleaner Production 19*(2), 99–107.

Mercan, Ş. Okan (2010). Regional Planning and Touristic Product Formation within the Scope of Sustainable Tourism: A Research on the Altınoluk Case, Dokuz Eylül University, PhD Thesis.

Mercan, Şefik, Okan & Özkök, Ferah (2013). Importance of Local Governments in Sustainable Tourism; Bozcaada Example, *Journal of Social Sciences and Humanities, 5*(1), 409–418.

Priskin, Julianna (2003). Issues and Opportunities in Planning and Managing Nature-Based Tourism in the Central Coast Region.

CHAPTER 2

STRATEGIES OF POST-DISASTER MARKETING: SUSTAINABLE DEVELOPMENT, EXPERIENCE, AND MARKETING

M. E. KORSTANJE[1] and HUGUES SERAPHIN[2]

[1]*University of Palermo, Argentina, Fellow at CERS University of Leeds, UK*

[2]*University of Winchester, UK*

ABSTRACT

The turn of the century has brought news on natural disasters such as quakes, floods, tsunamis, some of them caused by the climate change, but also others as lethal virus outbreaks as Ebola, SARS and Swine Flu by the lack of control of health system, without mentioning terrorism which places tourism industry in jeopardy. In this respect, policymakers and experts accept the idea that new techniques of marketing and programs are needed to protect the organic image of tourist destinations. In this vein, post-disaster marketing situates as a more than interesting sub-discipline that sheds the ink on the acceleration of recovery process timeframe through the articulation of tourism and leisure consumption. This chapter discusses the main strengths and weakness of post-disaster marketing critically to offer an all-encompassing model which can be used by marketing-led scholars in these difficult times. In some respect, post-disaster marketing centers on (dark) tourism to foster a positive discourse revolving around the tragedy, which is conducive to buttress attractiveness in the effaced site.

2.1 INTRODUCTION

The turn of the century brought some unexpected events, many of them not only shocking for tourism industry but also for public opinion (Allan, Adam

& Carter, 2000; Littlefield & Quenette, 2007; Korstanje, 2010). Anthropologists know this tendency as "*milenarism*" which consists of the individual and collective dispositions or reaction in every change of the millennium. These types of states are characterized by atmospheres of extreme fears and hopes which are accompanied by emergent moral forces that lead towards radical religious reformations. In this context, the attacks to *World Trade Center* on September 11 of 2001 is only the prelude for the successive concatenation of natural and made-man disasters such as Katrina Hurricane that hit New Orleans, and the strong quakes in Haiti, Sri Lanka and Chile, without mentioning the outbreak of apocalyptic viruses as SARS, Swine flu and the frightening Ebola. Some voices called the attention on the role of mass media and journalism producing a "spectacle of disaster," which not only served as cultural entertainment for global publics and audiences but also appealed to terror as a mechanism of political indoctrination (Kellner, 2005; Klein, 2007; Virilio, 2010; Innerarity & Solana, 2013; Korstanje, 2016). What seems to be clear is that some of these disasters were caused by climate change effects. Over the recent decades, the successive failures of nations to coordinate efforts in the mitigation of greenhouse gases was directly proportional to the historical center-periphery dependency which was forged in the colonial rule (Chaturvedi & Doyle, 2015). While some nations struggled to reverse the effects of climate change innovating in new energy sources, others are prone to discuss further levels of contamination to protect their domestic industries. In the mid of this mayhem, many specialists and policymakers envisaged a grim future for the tourism industry. The rise of diverse risks (as already noted) that may place tourism in jeopardy, adjoined to the criticism exerted on tourism as a fresh green industry that can ameliorate the effects climate change in the economies, landscapes and cultures, paves the ways for a hotly debate that today remains open (Smith, 1990; Hall & Higham, 2005; Weaver, 2011). As Korstanje and George (2014) put it, policymakers should arrive to a coherent diagnosis of the problem. In so doing, it is vital not only to read the advances of specialized literature but distinguishing the invention of a culture of disaster as a form of entertainment from a real global threat.

As this backdrop, the concept of post-disaster marketing alludes to the uses of tourism and planning to help effaced communities in their recovery process. At a closer look, tourism not only generates further attraction for tourists but also for investors who are in quest of further profits. Of course, this suggests that post-disaster marketing works and dark tourism are inextricably intertwined. In this chapter, we toy with the idea that post-disaster marketing literature emerged as the limitations in the theory of sustainability and risk perception to anticipate disasters. These limitations operated in the horizons of what sociologists dubbed as "the culture of fear" which in the recent years become in a cultural attraction. Thousands of visitors fly to zones of disaster or war only to get an authentic experience. Besides, the epistemological borders between Dark or Thana-Tourism and post-disaster marketing seem to be blurred into a confusing theoretical corpus which is hereby placed under the critical lens of scrutiny.

2.2 DISASTERS AND CLIMATE CHANGE

Specialists and pundits of all stripes are not in agreement of the reasons of climate change, but they agree that this phenomenon is behind the latest natural disasters that whipped mankind at the start of the century. Chaturvedi & Doyle (2015) discussed to what extent the discourse of climate change instills fear in the population instead of laying the foundations to articulate long-lasting sustainable policies in the environmental protection. Though they are not deniers who confront with the theory of climate change is triggered by greenhouse gas emission, they explore the formation of a cultural discourse which is based on the needs of using terror as a disciplinary mechanism of control. In this vein, they coin the term terror climate to denote the orchestration of differential geographical-located programs to foster the submission of the global South. Equally important, climate change is responsible for the millions of dollars in material losses and lost lives causing serious disruptions in the economic organization of nations. The droughts and floods affect not only the houses and the daily life lay-peoples but also harvests which prompt a sudden process of inflation in an economy that aggravates the previous state of employment West faces since 2008 (Ranjan, 2011; Wheeler & Von Braun, 2013). In this respect, Scott, Hall, and Gossling (2012) who are authoritative voices in the study of tourism and climate change said that it is important to discard some myths or biased information around the global warming. The concept of sustainability as it was formulated a couple of decades ago should at least revisited according to new paradigms that help explaining the complexity of this world. Authors emphasize on climate change as a global threat that harms not only the functionality of tourism but the global economy, geography, and the planet as a whole. The question whether climate change happens does not resolve why its effects have been accelerated in the short-run. One thing is the quest for explaining the reasons of climate change while others consist in struggling to reverse its aftermaths. Hence it is important not to lose the sight that sustainable tourism should be differentiated from sustainable tourism development.

While the former signals to those performed leisure activities in the territory oriented to protect the environment the latter refers to the policies tended to strengthen the community's well-being. At the time sustainability ensures a necessary balance to protect not only locals but also the interests of all stakeholders, climate changes defies the already-established sense of sustainability simply because the courses of actions as well as the rational decision-making process are made in the context of extreme uncertainness and volatility. Here a more than pungent question surfaces. To what extent tourism is useful to mitigate the effect of climate change?

Cantered on the accuracy of western technology to measure risks, policymakers should find alternative forms of sustainable tourism that develop more harmonized economies worldwide. However, as widely recognized by them, the tourism industry often falls in a paradox which is very hard to resolve. The process of industrialization from where tourism historically evolved, contributes to the acceleration of greenhouse gases emissions which latter affects the performance of tourism and the world of travels. Albeit the energy sources to be changed, no less true is that serious cultural transformations are needed. This discussion was assertively resolved by Jost Krippendorff (1982, 1986, 1987a, 1987b) in his respective approaches. Since he retains serious concerns on ecological issues, part of his bibliographical legacy was used to discuss the intersection of tourism and ecology. Krippendorff contends that tourism works as an instrument of escapement to revitalize the daily frustrations in workers' minds. Likewise, tourism is neither good nor bad, simply because it corresponds with the cultural values of each society. In this discussion, Krippendorff offers an innovative point of entry. The individual behaviors are culturally formatted by some values which are enrooted in the productive system. The problems of ecology should efficiently be treated in consideration with a substantial change of involving cultural values that caused the problem. The responsibilities of states should not be given by the action or omissions of states, but they should be shared by all citizens. In this vein, Krippendorff adheres to the use of social marketing to change those cultural values which are disposing of modern cultures to contaminate the planet. As stated, Rutty & Scott (2010) interrogate on the rise of temperature worldwide and the potential risks for tourism. Their investigation pursues as main goal the needs of reassessing the claims of much hot for tourist destinations in the Mediterranean region. Per their outcomes, the perception of temperature varies on times and culture adjusting to the perception of holidaymakers. These findings are vital for a coherent assessment or the potential implications for future impacts of climate change. Other seminal articles reveal that often the levels of awareness or familiarity of publics are slim. Basically, interviewees manifest they were familiar with the negative consequences of climate change though they remain indifferent or unfamiliar respecting to what should be done to reverse the situation (Dodds & Graci, 2009). It is important to delve in the critical approach of Peter Burns and Lyn Bibbings (2009) who envisage the end of tourism. There are serious contradictions between the patterns of consumption, which are emulated in tourism, and the needs of reacting against global warming. These complexities, of course, are given to the priority of instrumentalization as guiding cultural value over ethics and other values. While this society prioritizes consumption and mobilities as the best of feasible worlds, this perfect freedom has its costs. A rapid deceleration of the reasons behind global warming would entail the end of mass tourism as we know. Although the literature abounds in this field, offering reports that measure the key indicator at the brink of an ecological crisis no less true is that less attention is given to the real reasons behind the problem. The cult of profits and pleasure-maximization correspond with specific founding values that dissociated mankind from the environment. Without changing these values, any steps would be backfired. In fact, the abundance of reports and descriptive study-cases contributed in the configuration of a great consternation of spectatorship for disasters, but nothing has been done to reverse climate change. Henceforth an increasing number of scholars and policymakers focused their attention on the role of postmarketing literature to overcome the obstacles and epistemological limitations the paradigm of sustainability failed to resolve. To put this in slightly other terms, if we failed to prevent disasters, attacking their underlying causes, it is vital to work in marketing that makes from disaster and suffering a criterion of attraction. This is the rise of post-disaster marketing as a fertile option in the years to come.

2.3 POST-DISASTER MARKETING AND DARK TOURISM

One of the limitations of post-disaster studies remains in the lack of interest in researchers to dissociate the concept of post-disaster consumption from dark tourism. This suggests two important things. On the one hand, we come across with a failure of the sustainability paradigm to make more competitive destinations that balances the green-friendly atmosphere with the laws of the capitalist system. On another, the substantial change of the sense of beautiness in the classic products to adopt a more macabre version of consumption. In fact, the concept of beautiness which can be observed at a landscape or the sojourn in a luxury resort sets the pace to more morbid experiences where visitors are in contact with sites of mass death and suffering which oscillates from concentrations camps (Miles, 2002; Pollock, 2003; Thurnell-Read, 2009, abandoned prisons (Wilson, 2008; McAlister, 2013), or bewitched mansions (McEvoy, 2010) to towns fully destroyed by disasters (Gotham, 2007; Chako & Marcell, 2008).

As the previous argument is given, dark tourism should be understood as the tendency to visit spaces whose main attraction is mass death or traumatic events that captivated the attention of society. As Stone observed, behind dark tourism habits there remain universal forces that usher visitors into their own philosophical finitude. What visitors gave seems to be something else than mere corpses, nor they are moved by morbid reasons. Rather, consumers of dark sites are trying to understand death through the lens of what others have experienced. In other terms, dark tourism consists of disciplining death through others' death (Stone, 2012, 2013; Stone & Sharpley, 2008). Erik Cohen explains that one of the ends of dark tourism consists in giving a lesson to survivors and the community as a whole taking into consideration the accessibility to the territory where the real facts occurred is the key factor that determines these types of phenomena (Cohen, 2011). The discussion revolving around dark tourism is far from being finalized. While some wave claim that dark tourism follows an alienatory nature serving as a mechanism of discipline (Bowman & Pezzulo, 2009; Tzanelli, 2016; Korstanje, 2016), others toy with the belief that dark tourism mediates between the self and the future from where the notion of death operates. In this token, the curiosity for death, for these scholars, is not new, it can be found in medieval times as well as many other non-western cultures. Hence dark tourism should be equated to heritage consumption (Biran, Poria & Oren, 2011; Podoshen, 2013; Hooper & Lennon, 2016). In a recent seminal book, Hooper, and Lennon attempt to solve the dilemma between a radical turn whose position cannot be reconciled to the premises of tourism-management. Editors start from a philosophical contradiction, the terms darkness and tourism are ideologically opposed in the popular parlance. While the former alludes to the pleasure-maximization and the sense of beautiness, the latter alludes to a tragedy. Hooper and Lennon devote their time and efforts in unpuzzling what is the nature and motivation of dark tourism. Dark tourism should be defined as a recently-born tendency where tourists visit spaces of mass-death, suffering or mourning, in order to be in contact with human vulnerability. The existential fragility of others reminds how the proper life should be reinterpreted, and of course, this is one of the main aspects that molds the epistemological borders of dark tourism. This created a diversification of sites, which spans from the classic Auschwitz to new emergent zones as Ground Zero in New York.

In this respect, researchers have faced serious problems to coordinate their results towards unified models that help obtaining a better understanding of this slippery matter. While the attention of the media encouraged the study of dark tourism in almost all western universities, no less true is that the proliferation of these works are not correctly sorted or unified in a coherent conceptual platform. The knowledge dispersion adjoined to the exacerbation of an economic-centered view on dark tourism obscured more than they clarify. The needs for enhancing profits should be morally reconciled by the needs of memorizing death or at least to what extent the spectacles offered by these sites represent the real facts, or the background earlier than the disaster hits the community. Though these points are not being clearly answered throughout the text, it is evident that both, practices which signal to how dark tourism is managed or performed and interpretation, which reveals the motivations of tourists, should be ingrained into an all-encompassing model. We do not share Stone's position (emulated by Hooper and Lennon's assertions) that the curiosity for death we found today can be comparable to the visit of Saint's Thombs in Medieval Days. In this point, Korstanje (2016) has developed a credible and robust examination of the problems professional fieldworkers often fall at the time of investigating these types of topics. First and most important, there is no archeological evidence that proof dark tourism was a timeless drive, which was practiced by other cultures. Though the curiosity for death is enrooted in humankind, dark tourism surfaces in the postmodernity as a result of substantial changes in how the otherness is constructed (Korstanje, 2016). The needs of being special or getting authentic experiences would be of some importance at the time of explaining the rise and rapid evolution of dark tourism as trending behavior. For others scholars, post-disaster marketing plays a vital role strengthening the power of resiliency within society to overcome obstacles and difficult situations (Shondell-Miller, 2008; Burnside, Shondell-Miller & Rivera, 2007; Park & Reisinger, 2010; Shondell-Millar, Gonzalez & Hutter, 2017). Sather Wagsfaff (2016) fleshed out a model to understand dark tourism as an opportunity to foster social cohesion among survivor. Following the example of the WTC attack in New York, she argues convincingly that people make more egalitarian before death and suffering because emotions erode the basis of political hierarchies and classes. However, in some instance officials use these types of space to produce an ideological allegory, she coins as "heritage." While dark tourism wakes up a much profound emotionality within human beings, heritage represents a biased image of facts (Sather Wagstaff, 2016). Following this, it is particularly risky to abuse from these allegories by rememorizing historical facts. This begs the following questions, what would know visitors from WWII or Auschwitz by visiting a museum? Or Americans on the origins of AL Qaeda only visually gazing at the Ground Zero? Unless the real reasons for disasters to be commemorated as they happened the possibilities for disasters to be repeated are high.

At this point, it is also very important to highlight the fact that the development of dark tourism is not a systematic outcome following a disaster. Context, culture, and history are barriers to the systematic development of this form of tourism (Séraphin, 2017).

2.4 RECOVERY MARKETING STRATEGY

2.4.1 OVERVIEW

Following disasters, some destinations have adopted strategies to mitigate their problematic status (Alvarez & Campo, 2014), such as targeting the diaspora as potential tourists (Séraphin & Paul, 2015). Haiti, for instance, changed its Destination Marketing Organisation (DMO) logo in 2012, in order to match its change of image in its marketing strategy (Séraphin, Ambaye, Gowreesunkar & Bonnardel, 2016b). Apart from a logo, there may be other marketing material that can capture the essence of the destination, which can lead to brand identification, impact on purchase intention, and convey positive emotions, trust, brand attachment, and commitment. In this chapter, we are claiming that information and communication technologies (ICT) can help to achieve this objective. Taking the example of Haiti, Séraphin, Butcher, and Korstanje (2016) have provided evidence that ICT and more specifically Virtual Online Learning Material (VOLM) can contribute to educate visitors at the pre-visit stage and more importantly change their negative perception of the destination. Explicit and aggressive marketing strategies (e.g., planned communication activities, an organization of international sports and cultural events, media strategies, exhibitions, changing the target audience, reducing hotel and flight prices) have proven to be ineffective for destinations suffering from a legacy of political instability and disaster (Alvarez & Campo). For this reason, this book chapter has opted for a more subtle strategy, namely through 'education' using online materials. With recent advances in information and communication technologies, there is an interest among marketing researchers and managers to try to understand the challenges and opportunities for emerging markets (Sharma, 2011). This book chapter suggests that education of tourists (Ballengee-Morris, 2002) at the pre-visit stage, using new technology as an opportunity for these markets, may contribute to improving a destinations' image. Existing research on tourists' education focuses on the visit itself, but generally neglects the pre-visit stage (Séraphin, Gowreesunkar & Ambaye, 2016a). Also, the role of technology in tourist education, and their knowledge of the destination are currently little understood (Neuhofer & Buhalis, 2012). However, it has been shown that the Internet provides many opportunities to disseminate knowledge (Liburd & Hjalager, 2010). Compared to more traditional forms of educational delivery, e-learning can improve student understanding of instructions, which in turn can enhance their motivation to learn and subsequently increase their emotional intelligence (Behnke, 2012). Behnke and Greenan (2011, cited in Behnke, 2012: 13) defined this as: "A measure of one's ability to use acquired knowledge, abstract thinking, and problem-solving to interpret and guide personal responses to significant internal and external situations." As for Computer-Assisted Instruction in hospitality management education, it can contribute to stimulating student participation and interest (Law, 2013).

2.4.2 EDUCATION OF CONSUMERS AND IMPACTS ON DECISIONS

Consumer behavior can be defined as "behavior that consumers demonstrate whilst searching for, buying, using, evaluating and disposing of services and products that are expected to satisfy their needs" (Schiffman, Kanuk & Wisenblit, 2010). More specifically, and based on a tourism context, Kirillova et al. (2014) explain that elements that tourists consider as aesthetically pleasing or beautiful (environment, product or services, tourism experience), can contribute to their decision to visit a destination, to return to the same destination and, in the long term, to show loyalty to that destination. The aesthetic case for Haiti rarely reaches potential tourists due to media focus on risks arising from disasters, conflict and the nation's post-colonial status (Séraphin & Nolan, 2014a). Importantly for this study though, Carlson's natural environment model (1979) and the arousal model developed by Caroll (1995) suggest that knowledge and objectivity about the environment can influence appreciation. Their analysis implies that objective knowledge and exposure to more balanced narratives, influence the way potential tourists approach destination choice not just on a cognitive/rational basis, but also on an affective one. Hence the education of tourists can be considered as a way to facilitate a more favorable affective disposition towards post-disaster (post-colonial and post-conflict) destinations. Also pertinent here is Connell's view (cited in Stepchenkova, 2013) that non-promotional communications can be more powerful than more explicitly promotional ones. For specific destinations, specific measures must be taken (Huang, Tseng & Petrick, 2008, cited in Walters & Mair, 2012).

2.4.3 CONTENT ANALYSIS AND STRATEGY

Hsieh & Shannon, (2005) suggest that to analyze content, the research method should "focus on the characteristics of the language as communication with attention to the content or contextual meaning of the text." (...) "The goal of the content analysis is to provide knowledge and understanding of the phenomenon under study." This can then enable the researcher to "interpret meaning from the content of text data' to 'discover and describe the focus of individual, group, institutional, or social attention." According to Weber (1990, cited in Stemler, 2001), this type of research can also "examine trends and patterns in documents" and provide evidence for interpretation (Hammond & Wellington, 2013). Researchers such as Kim, Lee & Jung (2013) used a similar method to analyze advertising for internet gambling in TV programmes. This chapter proposes a tool (or model) called the DRA (Dale, Robinson, and Anderson) model. This model is a combination of two existing and complementary frameworks. On the one hand, the Dale and Robinson (2001) framework, and on the other hand, the framework developed by Anderson (2001, cited in Walters & Mair, 2012). Dale and Robinson (2001) forwarded a three-domain model of tourism education and an action plan for its implementation that would meet the needs of all stakeholders (learners, training providers, and employers). As for Anderson (2001 – cited in Walters & Mair, 2012), he put together a framework to test the effectiveness of a disaster recovery message drafted by a DMO. Dale and Robinson's model (2001) entitled "Three Domains Model of Tourism Education," argues that in order for a tourism educational programme to meet the needs of the industry and stakeholders involved (namely learners, education institutions and employers), it needs to: (a) develop interdisciplinary skills for a broad understanding of the industry; (b) provide expertise in a specific area in terms of skills; and (c) explore niche markets. As for Anderson (2001, cited in Walters & Mair, 2012), the art of convincing a target audience relies on three elements: (a) an "ethos" element, based on the credibility and trustworthiness of the source of the message; (b) a "pathos" element, linked with the emotional appeal of the message and finally, (c) a "logos" element, referring to the capacity of the message to inform. "Ethos, logos, and pathos are interconnected, influence each other, and all affect the effectiveness of persuasion" (Walters & Mair, 2012: 95). We are here, Therefore, suggesting that a good online (Social media and others) recovery message should follow the DRA model. From a methodological point of view marketers need to do: (1) some content analysis using software like TextStat (2) classify the keywords following the DRA model.

2.5 CONCLUSION

The main aims of post-disaster recovery campaign marketing are to restore a positive destination image, correct negative images and in some cases, create a new image and restore confidence in the minds of potential visitors. This book chapter provides conceptual evidence that online materials, if well managed by marketers, can play a similar role to disaster recovery messages and contribute to educating potential visitors with the objective of changing their perception and encourage them to visit these destinations. This chapter adds to this the fact that content (using the DRA model) can also influence users' decision-making process. Undoubtedly the rise of dark tourism and post-disaster market exhibits the problems of tourism-management not only to prevent global risks but also the conceptual limitations of the precautionary principle to anticipate disasters. This chapter is aimed at giving new fresh reflections which would be useful for policymakers, practitioners and marketing experts in the years to come. The concept of post-disaster management and marketing offers a fertile ground to be continued in the next approaches.

KEYWORDS

- **consumption**
- **dark tourism**
- **death**
- **disasters**
- **post-disaster marketing**
- **suffering**

REFERENCES

Adam, B., Allan, S., & Carter, C. (Eds.). (2013). *Environmental Risks and the Media.* Abingdon, Routledge.

Alvarez, M. D. & Campo, S. (2014). The influence of political conflicts on country image and intention to visit: A study of Israel's image, *Tourism Management. 40*, 70–78.

Ballengee-Morris, C. (2002). Cultures for sale: Perspectives on colonialism and self-determination and the relationship to authenticity and tourism, *Studies in Art Education. 43*, 232–245.

Baudrillard Kellner, D., (2005). Globalization and Terrorism: Some Comments on Recent Adventures of the Image and Spectacle on the Occasion of Baudrillard's 75th Birthday. *Baudrillard Studies. 2*(1), 1–7.

Behnke, C. (2012). Examining the relationship between emotional intelligence and hospitality student attitudes toward E-learning, *Journal of Hospitality & Tourism Management. 24*(2), 12–20.

Biran, A., Poria, Y., & Oren, G. (2011). Sought experiences at (dark) heritage sites. Annals of tourism research. *38*(3), 820–841.

Bowman, M. S., & Pezzullo, P. C. (2009). What's so 'dark' about 'dark tourism'? Death, tours, and performance. Tourist Studies. *9*(3), 187–202.

Burns, P., & Bibbings, L. (2009). *The End of Tourism? Climate Change and Societal Challenges.* 21st Century Society. *4*(1), 31–51.

Burnside, R., Shondell-Miller, S., & Rivera, J. D. (2007). The impact of information and risk perception on the hurricane evacuation decision-making of greater New Orleans residents. *Sociological Spectrum. 27*(6), 727–740.

Carroll, N. (1995). On being moved by nature: Between religion and natural history. In: S. Kemal and I. Gaskell (eds.). Landscape, Natural Beauty, and the Arts. Cambridge, UK: Cambridge University Press, 244–266.

Chacko, H. E., & Marcell, M. H. (2008). Repositioning a tourism destination: The case of New Orleans after Hurricane Katrina. *Journal of Travel & Tourism Marketing. 23*(2–4), 223–235.

Chaturvedi, S., & Doyle, T. (2015). An Introduction: A Critical Geopolitics of 'Climate Fear/ Terror': Roots, Routes and Rhetoric. In *Climate Terror.* Basingstoke: Palgrave Macmillan UK, 1–20.

Cohen, E. H. (2011). Educational dark tourism at an in popular site: The Holocaust Museum in Jerusalem. *Annals of Tourism Research. 38*(1), 193–209.

Dale, C. & Robinson, N. (2001). The theming of tourism education: A three-domain approach, *International Journal of Contemporary Hospitality Management. 13* (1), 30–34.

Dodds, R. & Graci, S. (2009). "Canada's tourism industry-mitigating the effects of climate change: a lot of concern but little action." *Tourism and Hospitality Planning & Development. 6*(1), 39–51.

Gotham, K. F. (2007). (Re)branding the big easy tourism rebuilding in post-Katrina New Orleans. *Urban Affairs Review. 42*(6), 823–850.

Hall, C. M., & Higham, J. E. (Eds.). (2005). *Tourism, Recreation, and Climate Change. Vol. 22.* Bristol: Channel View Publications.

Hammond, M. & Wellington, J. (2013). *Research Methods. The key Concepts*, Abingdon: Routledge.

Hooper, G., & Lennon, J. J. (Eds.). (2016). *Dark Tourism: Practice and Interpretation.* Abingdon: Routledge.

Innerarity, D., & Solana, J. (Eds.). (2013). *Humanity at Risk: The Need for Global Governance.* New York: Bloomsbury Publishing USA.

Kirillova, K., Fu, X., Lehto, X. & Cai, L. (2014). What makes a destination beautiful? Dimensions of tourist aesthetic judgment, *Tourism Management, 42*, 282–293.

Klein, N. (2007). *The Shock Doctrine: The Rise of Disaster Capitalism*. New York: Macmillan.

Korstanje, M. (2010). "Commentaries on our new ways of perceiving disasters." *International Journal of Disaster Resilience in the Built Environment*, *1*(2), 241–248.

Korstanje, M. (2016). *The Rise of Thana Capitalism and Tourism*. Abingdon: Routledge.

Korstanje, M. E., & George, B. (2012). Global warming and tourism: chronicles of apocalypse?. *Worldwide Hospitality and Tourism Themes*. *4*(4), 332–355.

Krippendorf, J. (1982). Towards new tourism policies: The importance of environmental and sociocultural factors. *Tourism Management, 3*, 135–148.

Krippendorf, J. (1986). The new tourist – Turning point for leisure and travel. *Tourism Management, 7*, 131–135.

Krippendorf, J. (1987a). Ecological approach to tourism marketing. *Tourism Management*. *8*, 174– 176.

Krippendorf, J. (1987b). *The Holiday-Makers: Understanding the Impact of Travel and Tourism*. Oxford: Butterworth-Heineman.

Liburd, J. & Hjalager, A. M. (2010). Changing approaches towards open education, innovation, and research in tourism, *Journal of Hospitality and Tourism Management*, *17*, 12–20.

Littlefield, R. S., & Quenette, A. M. (2007). Crisis leadership and Hurricane Katrina: The portrayal of authority by the media in natural disasters. *Journal of Applied Communication Research*. *35*(1), 26–47.

McAlister, J. F. (2013). Collecting the Gaze: Memory, Agency, and Kinship in the Women's Jail Museum, Johannesburg. *Women's Studies in Communication*. *36*(1), 1–27.

McEvoy, E. (2010). "West End Ghosts and Southwark Horrors": London's Gothic Tourism, London Gothic: Place, Space, and the Gothic Imagination, Vol. 140.

Miles, W. F. (2002). Auschwitz: Museum interpretation and darker tourism. *Annals of Tourism Research*. *29*(4), 1175–1178.

Neuhofer, B. & Buhalis, D. (2012). *Understanding and Managing Technology-Enabled Enhanced Tourist Experiences*, 2nd Advances in Hospitality and Tourism Marketing and Management Conference.

Park, K., & Reisinger, Y. (2010). Differences in the perceived influence of natural disasters and travel risk on international travel. *Tourism Geographies*. *12*(1), 1–24.

Podoshen, J. S. (2013). Dark tourism motivations: Simulation, emotional contagion, and topographic comparison. Tourism Management. *35*, 263–271.

Pollock, G. (2003). Holocaust tourism: Being there, looking back and the ethics of spatial memory. *Visual Culture and Tourism, Berg, Oxford*, 175–189.

Rajan, R. G. (2011). *Fault Lines: How Hidden Fractures Still Threaten the World Economy*. Princeton: Princeton University Press.

Rutty, M., & Scott, D. (2010). Will the Mediterranean become too hot for tourism? A reassessment. *Tourism Planning & Developing, 7* (3), 267–281.

Sather-Wagstaff, J., (2016). *Heritage that Hurts: Tourists in the Memoryscapes of September 11*. Abingdon: Routledge.

Schiffman, L. G., Kanuk, L. L. & Wisenblit, J. (2010). *Consumer Behavior* (10th Ed.), London: Pearson Prentice Hall.

Scott, D., Hall, M., & Gossling, S. (2012). *Tourism and Climate change*. Abingdon: Routledge.

Séraphin, H. (2013). *The Contribution of Tour Guides to Destination Understanding and Image. The Case of Haiti via an Analysis of 'Bonjour Blanc: A Journey Through Haiti,'* International Research Forum on Guided Tours, Breda University of Applied Sciences (Netherland).

Séraphin, H. (2017). Tourism, and terrorism in France: The limits of Dark Tourism, *Worldwide Hospitality and Tourism Themes*, doi: 10. 1108/WHATT-09-2016-0044.

Séraphin, H., & Nolan, E. (2014). Voodoo in Haiti: A Religious Ceremony at the Service of the 'Houngan' Called 'Tourism,' in Frost, W. & Laing, J. (Eds.), *Rituals and Traditional Events in the Modern World*, New York: Routledge.

Séraphin, H., Ambaye, M. & Gowreesunkar, V. & Bonnardel, V. (2016b). A Marketing Research Tool for Destination Management Organisations' Logo Design, *Journal of Business Research,* doi: 10.1016/j. jbusres.2016.04.074.2016b.

Séraphin, H., Gowreesunkar, V. & Ambaye, M. (2016). The Blakeley Model applied to improving a tourist destination: The case of Haiti, *Journal of Destination Marketing Management*, doi: 10.1016/j.jdmm.2016.07.004.2016a.

Sharma, P. (2011). Country-of-origin effects in Developed vs. Emerging Markets: Exploring the contrasting roles of materialism and value-consciousness. *Journal of International Business Studies. 42*(2), 285–306.

Shondell Miller, D. (2008). Disaster tourism and disaster landscape attractions after Hurricane Katrina: An auto-ethnographic journey. *International Journal of Culture, Tourism and Hospitality Research. 2*(2), 115–131.

Shondell-Miller, S., Gonzalez, C., & Hutter, M. (2017). Phoenix tourism within dark tourism: rebirth, rebuilding, and rebranding of tourist destinations following disasters. *Worldwide Hospitality and Tourism Themes. 9*(2), 1–20.

Smith, K. (1990). Tourism and climate change. *Land Use Policy. 7*(2), 176–180.

Stemler, S. (2001). An overview of content analysis. *Practical Assessment, Research & Evaluation. 7*(17), 1–10.

Stepchenkova, S. & Zhan, F. (2013). Visual destination images of Peru: Comparative content analysis of DMO and user-generated photography, *Tourism Management, 36,* 590–601.

Stone, P. (2013). Dark tourism scholarship: a critical review. *International Journal of Culture, Tourism and Hospitality Research. 7*(3), 307–318.

Stone, P. R. (2012). Dark tourism and significant other death: Towards a model of mortality mediation. *Annals of Tourism Research. 39*(3), 1565–1587.

Stone, P., & Sharpley, R. (2008). Consuming dark tourism: A thanatological perspective. *Annals of Tourism Research. 35*(2), 574–595.

Thurnell-Read, T. P. (2009). Engaging Auschwitz: an analysis of young travelers' experiences of Holocaust Tourism. *Journal of Tourism Consumption and Practice, 1(1),* 26–52.

Tzanelli, R. (2016). Thanatourism and Cinematic Representations of Risk: Screening the End of Tourism. Abingdon: Routledge.

Virilio, P. (2010). University of Disaster. Cambridge: Polity Press, p. 17.

Walters, G. & Mair, J. (2012). The effectiveness of post-disaster recovery marketing messages – The case of the 2009 Australian bushfires, *Journal of Travel & Tourism Marketing. 29* (1), 87–10.

Wheeler, T., & Von Braun, J. (2013). Climate change impacts on global food security. *Science. 341*(6145), 508–513.

Wilson, J. Z. (2008). *Prison: Cultural Memory and Dark Tourism*. London: Peter Lang, p. 49.

CHAPTER 3

RESPONSIBLE TRAVEL AND TOURISM ADVENTURE: EVIDENCE FROM MALAWI AS A TOURIST DESTINATION

JAMES MALITONI CHILEMBWE[1] and VICTOR RONALD MWEIWA[2]

[1]*Mzuzu University, Private Bag 201, Luwinga, Mzuzu, Malawi*

[2]*Malawi Institute of Tourism, P.O. Box 2673, Blantyre, Malawi,*
E-mail: chilembwe1975@gmail.com

ABSTRACT

Adventure tourism has become one of the growing sectors in the travel and tourism industries in the developing worlds. However, Malawi's tourism is at the blossoming stage with lots of adventure tourism products still untouched, to be explored, and experienced. The unspoiled nature of the adventure tourism products appeals to adventure travelers to travel responsibly to a less traveled destination offering unique experiences as a must visit tourism destination. Adventure travelers can benefit more from the activities through sustainable management approaches to ensure maximum protection and conservation of the environment and exiting products. The responsibility of managing products is for both residents and travelers for sustainability. There is a need to take control of all threats including poaching, product management, soil erosion and travelers' acceptance to be grouped and timetabled in recommended carrying capacities to avoid destroying the environment and product marketing. Responsible travel and tourism, moreover, brings in great tourism opportunity offerings to the locals around the tourism destination ranging from employment, sponsored community projects and infrastructure development in the destination area. It also helps to preserve culture and traditions as some adventure tourists are much more interested to interact with local people and learn of their way of living. However, using two case studies approach, in the Liwonde Wildlife

Reserve (LWR) and Nyika Wildlife Reserves (NWR) the results reveals that both reserves are marketed as tourism destinations rather than adventure tourism destinations. The study suggests that there several adventure tourism activities unexplored requiring intensive marketing to adventure travelers. It further suggests that the adoption of modern marketing strategies like online and digital marketing can enhance putting Malawi as an adventure tourism destination on the world map.

3.1 INTRODUCTION

Responsible travel is one of the closely related terms in tourism and is ethically based (http://www.responsibletravel.org). Terms as ethical tourism, geo-tourism, ecotourism, responsible tourism, pro-poor tourism, and sustainable tourism may all point to responsible tourism due to its identically. Responsible tourism is not a product but rather an approach (Spenceley, 2014). Studies suggest that an "integrated approach" should aim to incorporate all stakeholders in order to enhance collective responsibility towards tourism (Mason, 2003, 2008; Mille & Twining-Ward, 2006; Bello, Lovelock, & Carr, 2016). Moreover, stakeholders such as tour operators, accommodation managers, holidaymakers; attraction managers, travelers, planning authorities at national, regional and local level are important to achieve responsible travel. On the other hand, adventure tourism is one of the largest and fast-growing sectors in the global tourism industry and as an outdoor recreation (Buckley, 2006; Cater, 2013; Varley & Taylor, 2013); adventure tourism is a hot trend in the tourism industry (http://www.responsibletravel.org). Other authors have pointed out that the sector is rapidly growing (Khan, 2010; Khadka, 2012). The popularity of the sector is a result of tourists seeking unusual holidays, different from the typical beach vacation. Furthermore, other researchers suggest that ordinary people are no longer contended to spend their holidays away from their usual places of residents to resting on a sunny beach but choosing a destination they will experience some kind of adventure (Annon, 2012; Honey & Gilpin, 2009; The George Washington University and Adventure Tourism Trade Association, 2013). Apart from being involved in adventure experience a team; however, others have opted for individual exploration due to more experienced to highly endangered adventure activities (Varley & Taylor, 2013).

In recent years, moreover, global adventure tourism has grown exponentially worldwide (UNWTO, 2014; Giddy & Webb, 2016) with tourists visiting destinations previously undiscovered. As a result, new destinations have been marketing themselves as unique, appealing to those travelers looking for rare, unusual and incomparable experiences. The adventure travel usually includes two of the following components: a physical activity, a cultural exchange or interaction, and the engagement with nature (UNWTO, 2014; Rantala, Rokenes, & Volkonen, 2016). Therefore, the features of adventure tourism products should include geographic setting, and the activities rely on the natural environmental features such as mountains, rivers, forests, parks and games reserves (Buckley, 2006). According to Giddy & Webb (2017) adventure tourists seek experiences that highlight the natural and cultural assets that make the place more distinct from any other. Although some travelers prefer lesser endangered adventure tourism activity, the emphasis should be placed on the provision of more healthy-physical activities (Bauer, 2017; Dreams, 2017; Siegel, 2017; SNV Netherlands Development Organization, 2017), allowing a deeper cultural exchange with different people and the appreciation for a guilty free holiday (Honey & Gilpin, 2009; Heyniger, 2011). Responsible tourism comes with its own benefits as such adventure tourism is one among favored sectors in tourism (UNWTO, 2014; Andrews, 2015). There are health benefits in additional to economic and social-cultural interactions. Moreover, as cited by others, for example; attracting high-value customers, supporting local economies, and creating a positive growth for tourism and destination economies as well as encouraging sustainable practices (UNWTO, 2014).

Based on the reflections made above, the concept of responsible travel is inseparable to tourism which encompasses adventure travel. However, adventure tourism has received a relatively a little attention. Although, countries like South Africa, Kenya, Uganda, Tanzania have had rigorous promotion and marketing of adventure tourism activities; sadly, they are yet to take issues of responsible travel seriously. Furthermore, there more adventure product on the queue waiting to be discovered. Malawi too is not exceptional. Moreover, more studies of adventure focused and reported heavily on the accidents or injuries as a result of adventure real activities; hence little attention has been paid to the marketing of adventure or responsible travel (Buckley, 2006; Flaherty & Choi, 2016). It is also argued that although adventure tourism is a growing sector in the world tourism industry for mass tourists, the sector is yet to market more on individualized adventure activities which remains little known to consumers (Buckley, 2004). At present, commercial adventure tourism is derived from the outdoor recreation literature but, the sector deserves research attention in its own right. It may, Therefore, not wrong to suggest that adventure theory has not readily been explored, hence showing indications that there also more responsible activities and products, not only to be uncovered by researchers but also to be marketed for commercial tourism (Buckley, 2004; Mullins, 2014).

In the study at hand, and putting reflections on the background above, it is in the interest of the authors to analyze two cases of *Liwonde* Wildlife Reserve (LWR) and *Nyika* Wildlife Reserve (NWR) in Malawi. The study has explored adventure tourism products and also appreciated the nature, quality and evaluates the tourists experience on the products. This current study explores the viability of responsible travel and adventure tourism in LWR and NWR for the potential of marketing Malawi as an adventure tourism destination.

3.2 A REVIEW OF ADVENTURE TOURISM AND RESPONSIBLE TRAVEL

The commercial tourism firms have considered adventure tourism products, and the environment (nature) in which the adventure products exist as a cowfish segment (Muller & Cleaver, 2000). And tourism being an umbrella concept encompassing travel in which responsible tourism is under discussion. Tourism covers all sectors which include "adventure tourism, ecotourism, responsible travel, and nature tourism," among others. Tourism history dates back to 1811 or before (Buckley, 2006) and continues to change in its definition. It is interpreted as the activities of persons traveling to and staying in places outside their usual environment for not more than one consecutive year for leisure, business and other purposes (Page, 2007). Tourism encompasses travelers away from home, the business and the people who serve them by expending or otherwise making the travel easier or more entertaining but be mindful about being responsible travelers (Lundberg. Khrishnamurthy, & Mink, 1995). Adventure tourism has been defined by a number of authors and organizations as a type of niche tourism involving exploration or traveling to remote or wilderness destination with perceived and possible risk, requiring specialized skills and tends to be associated with high-level involvement and activity by the participants, most of it was outdoor (http://www.responsibletravel.org; Lundberg, Khrishnamurthy, & Mink, 1995; Cater, 2013; Varley & Taylor, 2013). Considering levels of excitement and risks involved; adventure tourism becomes an excitement, risk, and discovery to mental and physical experiments in nature by stepping out the comfort zones, requiring a significant physical or mental effort that may involve some kind of risk but always a calculated risk (Lundberg, Khrishnamurthy, & Mink, 1995; Khadka, 2012). These definitions entail that the activities involve high risks, require skill and high level of participation both mental and physical.

Over the year's adventure tourism has gained and continues to gain favors from the tourists as it involves healthy-physical activities which allow for a deeper cultural exchange with different people, and an appreciation for a guilty free holiday (Heyniger, 2011). To strength, the case why the segment is a cloud puller, good nature travel blog (Andrews, 2015), outlines ten reasons why adventure tourism is gaining favors among the sectors of tourism which are more health benefits apart from economic and further said that there are no age limits for adventure activities (Andrews, 2015). Furthermore, adventure tourism is described as an alternative to mass tourism as in philosophy and attitude (Sanyal, 2009). This is in terms of values and approaches. Adventure tourist tends to be more concerned with the preservation of the natural environment, wildlife, authenticity of local traditions and cultures. As a result, there are several core characteristics of adventure such as "uncertain outcome; danger and risk; challenge; anticipated rewards; novelty; stimulation and excitement; escapism and separation; exploration and discovery; absorption and focus; and contrasting emotions" (Swarbrooke et al., 2003; Annon, 2012). Adventure tourist is one who simply takes part in adventure activities, one who takes an adventure as principal excitement (Buckley, 2010). An adventure tourist as a traveler who participates in strenuous, outdoor vacation travel, typically to remote places renowned for their natural beauty and physical attributes, involving hazardous activities (http://www.responsibletravel.org; Cater, 2013; Varley & Taylor, 2013). According to Swarbrooke (2002), adventure tourism is for the intrepid traveler and usually involves people traveling off the beaten track into fragile environments and/or communities that are socially vulnerable. Something important to note here is that activities and setting are generally linked to outdoors, wilderness, remote, unusual or exotic locations (Swarbrooke, 2002; Bello, Lovelock, & Carr, 2016). Moreover, Swarbrooke et al. (2003) and Wilks & Page (2003) stressed that adventure tourism activities often include a small group of participants, visiting an exotic and remote location, which are often under somewhat primitive conditions, and utilizing the services of a guide, leader, and outfitter. On the same note adventure tourism is characterized by its ability to provide the tourist with relatively high levels of sensory stimulation, usually achieved by including physically challenging experiential components with the (typically short) tourist experience (Muller & Cleaver, 2000; Bauer, 2010; The George Washington University and Adventure Tourism Trade Association, 2013). Adventure tourists expect a certain level of risk although somehow a calculated risk.

3.2.1 ADVENTURE TOURISM PRODUCTS IN THE CONTEXT OF RESPONSIBLE TRAVEL

A product in its generic sense as a thing, a place, a person, an event, or an organization which satisfies the needs of a person which is offered has an intrinsic value for the customer (Sanyal, 2009). Some important points to note from the definition; the product must be offered, satisfy the needs of the customers and must be offered at a value (Gardiner & Kwek, 2017). Moreover, adventure tourism products or activities are extreme like sports because of the nature of the products although the levels of involvement are different (Yeoman, 2008). Adventure tourism marketing study shows that adventure tourism products are classified according to the type of activity adventure tourists are engaged in and not the degree of difficulty or risk (Bauer, 2010; Khadka, 2012; The George Washington University and Adventure Tourism Trade Association, 2013). However, it adds the element of risks involved, excitement and the experience (Khadka, 2012). The products have been categorized as adventure tourism products as hard and soft (Bauer, 2010; Khadka, 2012). Hard adventure involves high-risk activities that may require full-time participation (McCarthy, 2006; Cater, 2013). Hard adventure involves risks and challenges and requires persons to be physically and mentally healthy (Khadka, 2012). These activities or products includes; long distance backcountry trekking, downhill mountain biking, paragliding, mountaineering and climbing or hiking, canoeing and kayaking and heli-skiing holidays (Khan, 2010; Cater, 2013). On the other hand, soft adventure products include activities people engage in and do not want any danger and want a lower risk factor (McCarthy, 2006; Khadka, 2012). The people are not restricted by any physical or mental health requirements. Activities include bird watching, hiking on the gradually changing terrain, kayaking, bicycle touring, camping, cross country, skiing, riding, surfing, all provide some examples of soft adventure tourism products which included; wilderness jeep safaris supervised and escorted trekking, cycling holidays, sailing holidays (motorized water sports), and windsurfing (Font & Carey, 2005; Andrews, 2015).

3.2.2 MANAGING RESPONSIBLE TRAVEL AND ADVENTURE TOURISM DESTINATIONS

Long-term competitiveness in tourism calls for meaningful and appropriate management approaches where a multi-stakeholder partnership is an important element of promoting tourism (Buckley, 2006; UNWTO, 2014). A multi-stakeholder partnership management will ensure sustainable management of the resources and this will lead to the management of resources in such a way that economic, social, and aesthetic needs can be fulfilled while maintaining cultural integrity, essential ecological process, and biological diversity and life support system (Cater, 2013). The concept of sustainability has been applied in nature tourism at a variety of scales and in both the public and private sectors (Towards Earth Summit, 2002). Again, Globe 90 Conference held in Canada produced strategies for sustainability, of which, five goals of sustainable tourism can also be applied in the adventure tourism and included: (a) to develop greater awareness and understanding of the significant contribution tourism can make to the environment and the economy of the country; (b) to promote equity and development; (c) to improve the quality of life of the host community; (d) to provide high quality of experience for the visitor; (e) to maintain the quality of the environment. Management of information between accommodation facilities and adventure tourists is another key element in the managing of an adventure tourism destination. There is also an outline of important information which is conveyed to visitors upon arrival as a way of briefing (Swarbrooke, 2002; Buckley, 2010). The briefings are commonly reserved for the first day and the information typically include; safety and security; eating areas; tents and camping practices if the clients are camping; and cleaning, washing, litter control, movements; and other environmental management issues. Going on an activity, the tour guide always makes a briefing, and the instructions are repeated on several times. Sustainable management approach helps in establishing the right numbers of people visiting the location at a time. Carrying capacity is a notion which recognizes that both natural and man-made attractions have upper limits in their capacity to absorb visitors (Goodwin, 2005; Sanyal, 2009; Kotler & Armstrong, 2012; Bello, Lovelock, & Carr, 2016; SNV Netherlands Development Organization, 2017). Others, for example; bring in another important element of monitoring (Mille & Twining-Ward, 2006). Monitoring involves regularly assessing the state of an issue or phenomenon; thus in our case the adventure tourism products and the environment in which these products exists, relate to particular goals, expectations, and objectives.

3.2.3 MARKETING RESPONSIBLE TRAVEL AND ADVENTURE TOURISM PRODUCTS

Marketing is the process by which companies' value for the customer and build a strong customer relationship in order to capture value from customers in return (Kotler & Armstrong, 2012). Furthermore, marketing adventure tourism products require a representational and description of products and creating an experience that exceeds expectations (Mason, 2008). Also, as stated that it is good to consider when the tourists come into contact with the product, experiencing it and expectations of a destination as an opportunity for marketing (Khadka, 2012). Understanding the nature of the adventure products will help marketers to develop marketing strategies that will promote the products and not assuming that because the products are authentic and exciting, tourists will find them (Font & Carey, 2005). An increased growth in the adventure tourism market over the years has been facilitated by the growing demand for adventure tourism (ACSEDU, 2017; Khadka, 2012). Destinations may be marketed by regional or local tourism promotion agencies or associations, public, but they depend on private commercial operators to provide the retail products at the destination and generally also to run adventure events (Font & Carey, 2005; Buckley, 2010; Kotler & Armstrong, 2012; Briggs, 2016; Lake, 2017). Marketing of adventure tourism products should focus on the satisfaction of the customers' needs and wants, and also focusing on the future customer who needs to be identified and anticipated (Carter, 2006; Khadka, 2012). A concept of integrated marketing communication tools can be used on the development of strategies to ensure yielding successful results. Integrated marketing communication is a management concept, designed to make the five components of the promotional mix such as advertising, sales promotion, online marketing, public relations and direct marketing which can effectively work together in a unified force to promote adventure tourism products (Carter, 2006; Kotler & Armstrong, 2012). Moreover, the literature shows that there is plenty of research on tourism marketing, but not much has been published in the area of adventure tourism (Buckley, 2010). Adventure tourism product marketing is basically from general tourism destination marketing but with a focus on the adventure products, the quality of the products and the quality of marketing (Carter, 2006; Page, 2007; Bauer, 2010; Buckley, 2010). As tourism is now more than ever recognized as a major economic contributor in many destinations worldwide, it entails that there is a need to do more on adventure tourism marketing. Responsible travel and adventure tourism also add value for foreign exchange and supports export industries, environmental, social, cultural, historic resources, and protection (Antigua and Barbuda Tourism Development, 2003). Tourism as a worldwide phenomenon affects a vast number of people and places and has the potential to touch many more (Boniface & Cooper, 2005). Adventure activities have a lot of benefits to the personal health of the person doing the activities. For example, another author outlined ten benefits that one can get from the adventure sports which include burning of calories and keeps the blood flowing and stretches the body little by little that's helping us becoming stronger (Andrews, 2015). Other reasons include a great way of having fun while providing with new experiences and opportunities to live life to the fullest, escaping from the daily routine. The thrill to life and the experiences become great memories to cherish.

3.3 METHODOLOGY

3.3.1 CASE STUDY AREA A: LIWONDE WILDLIFE RESERVE (LWR)

The first study was conducted in Liwonde Wildlife Reserve. It is located on the upper Shire River plain, and the main part is on the eastern side of the Shire River and a small part on the western side of the river. The wildlife reserve (inside) has three accommodation facilities inside the park, namely; *Mvuu* Camp, Bushman's Baobab Lodge and *liwonde* Safari Camp (Figure 3.1). Outside there is Ngom's Lodge, located a few meters from the main southern entrance gate to the reserve, Hippo view lodge located close to the Liwonde Bridge at Liwonde Town, and the Njobvu Village Cultural Lodge in the Lingwalanya Village about 4 km from the western entrance gate. The 584 km^2 (about 220 square miles) park is about 160 km from Blantyre, about 250 km from Lilongwe, 105 km from Mangochi and just 6 km from Liwonde Township (Department of Tourism, 2015–16; The Eye Magazine Malawi, 2016).

3.3.2 CASE STUDY AREA B – NYIKA WILDLIFE RESERVE (NWR)

The second study was conducted at within the Nyika Wildlife Reserves of which the big part of it, is in Rumphi District. The reserve is the largest in Malawi. It occupies a track of mountainous plateau and associated with hills and escarpments in northern Malawi. Its area is about 3200 km^2 and is centered upon 10° 33″S 33° 50″E. The park lies astride the *Chitipa*, *Karonga* and *Rumphi* districts of the northern region. Part of its western boundary coincides with the Malawi-Zambia border, and the section is contiguous with the Zambian Nyika Wildlife Reserve (Department of Tourism, 2015–16; The Eye Magazine Malawi, 2016).

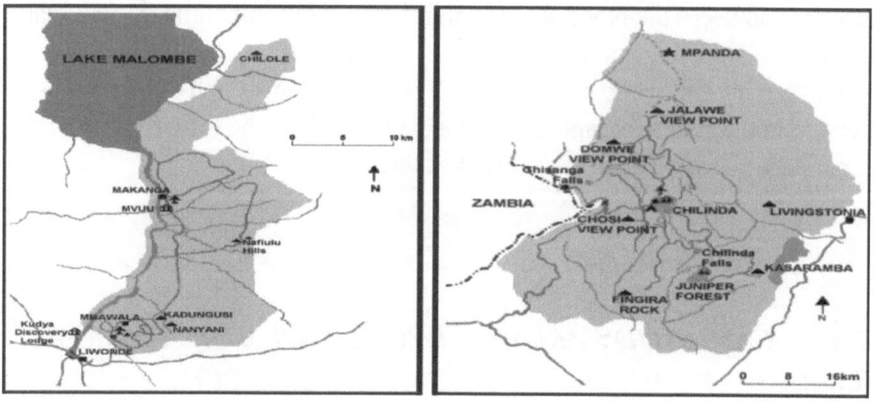

FIGURE 3.1 **(See color insert.)** Showing Location of the Study Area. Map A: *Liwonde* Wildlife Reserve. Map B: *Nyika* Wildlife Reserve. Source: Safari Guide Africa (2017).

3.4 RESULTS AND DISCUSSION

The study had 14 respondents; 7 from each wildlife reserve. These respondents were from the camps and lodges inside and around the two parks. The gender distribution in frequency and percentages of the respondents to the study has shown that gender was balanced due to the nature of the work in the wildlife reserves. There were more men representing 11 (78.6%) as compared to women (21.4) three women, representing 21.4% and 11 men representing a 78.6%. The participants were free to express their views and opinions on the subject under study. It was discovered from the study that adventure tourists prefer accommodation facilities when they go to *Nyika* and *Liwonde* Wildlife Reserves because the two parks are large and offer a wide range of adventure experiences. Results have shown that about 50% of adventure tourists like full accommodation, about 28.57% go for all types of accommodation in the park, about 14.29% go for basic accommodation (more like backpacking) while 7.14% prefer camping. In terms of product classification, 7(50) respondents indicated that the wildlife reserves have both hard and soft adventure tourism products. The other 7(50) also indicated the wildlife reserves have only the soft type of adventure activities. The researchers noted that the respondents failed to distinguish between hard and soft adventure as most of them could manage every park activity. Tourists too have been selecting activities that seem to be manageable and interesting to them. But from the researchers' observation, the wildlife reserves have both hard and soft adventure tourism products. Table 3.1 summarizes the products or activities and also provides levels of risk in the Nyika and

Liwonde National Park basing on their own understanding and experience of the products.

TABLE 3.1 Product Classification and Levels of Risk (researchers' data, 2017)

Product name	**Results in percentage (100%)**	**Product classification**			**Risks**	**Results in Percentage (100)**
	Hard	*Soft*	*Hard*	*Soft*	Level	Vote
Boat Safari	14	86	No	Yes	Low	7.1
Bushwalk (Walking Safari)	50	50	Yes	Yes	High	42.9
Game drive	0	100	No	Yes	Low	6.2
Bird watching	14	86	No	Yes	Low	22.4
Village walks (interaction)	21	79	No	Yes	Low	7.1
Rhino tracking	64	36	Yes	No	High	14.3

As indicated in Table 3.1, adventure tourism products like boat safari, bush walk or walking safaris, game drives, bird watching, and village walks and interactions are categorized as soft whereas rhino tracking and bush walk (walking safaris) are in the category of hard although the bush walks looks balanced 50% hard and 50% soft. However, the discrepancy comes in due to one's familiarity and experience towards safari walking. Over 78.6% of inexperienced tourists in safari walking express too as hard. In terms of the level of risk, the results revealed that bush walk is the most challenging adventure activity (42.9%) followed by bird watching (28.6), and Rhino tracking (14.3%). Moreover, boat safari, village walks, and interaction are in the least challenging activities during adventure tourism.

3.4.1 EXPERIENCES ON THE ADVENTURE PRODUCTS (ACTIVITY)

3.4.1.1 BOAT SAFARI

This offers a great opportunity for the tourists to see aquatic animals such as crocodiles and the hippos at a very close distance than in any other activity or product (Figure 3.2). It offers a great experience seeing and learning their behaviors in their own territories. Land animals are one of the best things to see while on a boat safari. Depending on lucky, they may see a crocodile catching land animals and what happens when a crocodile got a catch. The tourists may also have a chance to see elephants bathing, but the most

FIGURE 3.2 Boat Safari. Source: Central Wilderness Safari (2017).

interesting is when they find a head of elephants crossing from one side to the other. The land animals to see on the boat safari include; the Kudu, Zebras, Antelopes, Elephants, Bush backs, Warthogs and many more.

3.4.1.2 BUSH WALK

During this activity, tourists are always accompanied by an experienced safaris guide and a game ranger on this activity. Tourists are not allowed to do the walking safari on their own or with a guide without a game ranger. When they are set to start the activity, the safari guide gives a briefing, explaining all safety measures and what the tourists are supposed to do in a given circumstance. It is the most adventurous and dangerous activity especially in the big game territories because they are exposed to different wildlife including elephants, snakes, buffaloes and some dangerous animals (Figure 3.3).

3.4.1.3 GAME DRIVES

Just like the other activities, before the game drive starts a safari guide who is also a driver brief the tourists on the safety measures, how they must behave, what they must do and not do while on a safari. On a game drive, the tourists are accorded a great opportunity to view the wildlife at a very close point, while listening to the information the safari guide relays to them regarding the particular animals on site (Figure 3.4).

FIGURE 3.3 Walking Safari. Source: Central Wilderness Safari (2017).

FIGURE 3.4 On a game drive. Source: Central Wilderness Safari (2017).

3.4.1.4 BIRD WATCHING

About 400 of the 700 species of bird found in Malawi making it a paradise for bird watchers. Bird watching just like the other activities starts with briefing the tourists. It is done on foot and if vehicles are used is only to transfer clients of the tourists from the camp or lodge to a point where the journey for the bird watching experience will start from on foot. The activity is offered in combination with other activities whether game drive, walking safari or boat safari (Figure 3.5).

3.4.1.5 VILLAGE WALK AND INTERACTIONS

Villages around the wildlife reserves offer the adventure tourists an opportunity to experience Malawi's culture and traditions; the way people live in a real village set up. The tourists interested to experience a village walk are taken outside the park into a village for a village walk, traditions dance, and meals. One of the villages with a distinctive and deep cultural and traditions where most of the tourists are taken to is the *Lingwangwa* Village.

3.4.1.6 RHINO TRACKING

Like other products, when set for the activity, led by an experienced guide with a tracker and a game ranger, the tourists are briefed on the safety and how the tourists are supposed to go and signs they must follow while on

FIGURE 3.5 Lilian – Lovebird perched. Source: Central Wilderness Safari (2017).

the Rhino Tracking, and as well as the behaviors of the animals and the other animals they will meet and see in the sanctuary. Rhinos are dangerous animals with poor eyesight but with very good smell.

3.4.1.7 CANOEING

Offered at Bushman's Baobab Lodge, the tourists are briefed and led by an experienced safari guide on a canoe and dugout on fiber glasses into the stream and sail through into the Shire River. Canoe safari is dangerous experience comparing with a boat safari as it relies on personal fitness to peddle against water flow and the tourists are highly exposed to dangerous crocodiles and hippos which might easily attack and kill them (Figure 3.6).

3.4.2 CONTRIBUTIONS TOWARDS SUSTAINABILITY OF ADVENTURE TOURISM

In a way of assessing the contributions that the adventure tourism providers are making towards the sustainability of adventure tourism products in the wildlife reserves, the research study found out that there are no extra monies paid for the promotion of adventure tourism products in the wildlife reserves apart from one percent tourism levy that is collected by the Department of Tourism for marketing and promotion of tourism in Malawi. Despite no collection of extra monies for promotion of adventure tourism, on positive note, the study discovered that the facilities together with the African Parks are engaged in different community projects which include: sponsoring environmental conservation projects, conserving and

FIGURE 3.6 Canoe Safari on the Shire River. Source: Baobab (2017).

promoting culture and traditions, sponsoring children in schools, and as well as constructing school blocks in the communities around the park in the process educating the local people the importance of caring about the environment and the wildlife. Through this project, they train the people in the surrounding village to protect the wildlife and the environment in the national park. This gesture concurs with other researchers who comment about community participation and benefits (Buckley, Shakeela, & Guitart, 2014; Bello, Lovelock, & Carr, 2016; Pomfret & Bramwell, 2016; Moorhouse, D'Cruze, & Macdonald, 2017). Empowering the communities through businesses so that they should not poach from wildlife reserves. This kind of approach is meant to make the development of tourism ecologically supportable in the long term and the importance that lies in its motive to conserve the resources and increase the value of local culture and tradition (Sanyal, 2009).

3.4.3 BENEFITS OF PROMOTING ADVENTURE TOURISM PRODUCTS

Tourism is now more than ever recognized as a major economic contributor in many destinations worldwide, which adds value for foreign exchange and supports export industries, environmental, social, cultural, historic resources, and protection (Antigua and Barbuda Tourism Development, 2003). It is also indicated that tourism as a worldwide phenomenon affects a vast number of people and places and has the potential to touch many more (Boniface & Cooper, 2005). Adventure tourism activities offer a lot of benefits which include the personal health of a person doing the activities. As outlined earlier, there several benefits of adventure tourism activities to an individual which include burning of calories and keeps the blood flowing and stretches the body little by little that's helping us becoming stronger (Andrews, 2015). Other reasons include a great way of having fun while providing us with new experiences and opportunities to live life to the fullest, escaping from the daily routine. The thrill to life and the experiences become great memories to cherish. The study also found out that it will also benefit the accommodation facilities as they increase the number of people sleeping in their facilities and in the process, increases in the revenues through accommodation and activities. This will also bring about employment to the communities and jobs to qualified Malawians, economic benefits and as well conservation of the environment.

3.4.4 MARKETING ADVENTURE TOURISM PRODUCTS IN MALAWI

While marketing tourism products that are not a sole activity but involve several actors, there are also a number of strategies to use. As shown by the study, wildlife reserves are being marketed as a tourist destination but not an adventure tourism destination. The findings indicate that the players (the providers) of these products use the internet through their websites to market their facilities and the activities (products) they offer to their clients. It is on their websites that they display pictures of their accommodation facilities and pictures of the activities during the consumption of the activities. Brochures and leaflets are distributed in travel and tour operator's offices so that the customers visiting these offices may have an interest. Magazines, for example, the eye magazine and other print and television media (The Eye Magazine Malawi, 2016). The adventure tourism providers also rely much on tour operators to market their accommodation facilities and also the Department of Tourism website in Malawi. Therefore, it is pointed out that marketing materials such as websites and brochures are often being the first point of contact that a customer has with a product or business, and as such can be a key determinant in the initial levels of interest towards a product or activity (Price & Driscoll, 2012). The use of image and wording within marketing materials should strategically be used to focus on the satisfying the customer needs and wants as this will increase or decrease the interest or desire of the customers (Maattanen-Bourke, 2008; Price & Driscoll, 2012). Furthermore, there was a general consensus that marketing adventure tourism products require a representational and description of products and creating an experience that exceeds expectations (Maattanen-Bourke, 2008; Mason, 2008; Price & Driscoll, 2012). In addition, the different marketing strategies that can be used, it is good to consider when the tourists come into contact with the product, experiencing it and expectations of a destination as an opportunity for marketing (Khadka, 2012). Internet was considered to be the most effective tool that can be used to market adventure tourism products in NWR as it will display images and wording of products and adventure products are typically marketed by intermediaries, who provide access to the market in an effective way (UNWTO, 2014).

3.4.5 IDENTIFIED WAYS OF MARKETING ADVENTURE TOURISM IN NWR AND LWR

Marketing is the process by which companies' value for the customer and build a strong customer relationship in order to capture value from customers in return (Kotler & Armstrong, 2012). The study suggests that the best way of marketing adventure tourism and products in the wildlife reserves is through a number of actions as summarized as: (a) Advertising through different media, presentation of ideas, images and video clips of tourists experiencing the products in the wildlife reserves and promotion of these products on television and internet. (b) Sales promotion, a short-term strategy to encourage people to visit the wildlife reserves to experience the products. For example, by reducing park entrance fees for the locals and maintaining the participating fees in different activities. Just to encourage people to visit the park. The providers to organize familiarization tours which will include tour operators, media personnel and foreign tour operators to sample the products. (c) Public relations and publicity which involves all players in the wildlife reserves to be designing programs that will promote, sustain and protect the image of the products and destination. For example: sponsoring community environmental projects, sponsoring children in school and enhancing community participation in the conservation and protection of the environment and wildlife. (d) Personal selling may involve face-to-face interactions with prospective customers, for example, tour operators and organizations to visit and experience the products in the wildlife reserves. (e) Direct marketing relies on communicating directly to clients, typically through direct mail, online marketing, websites, social media, and attendance at consumer shows. This point is also in agreement with the author (UNWTO, 2014), who suggests that a mixture of web-content adventure tourism marketing can pull interested group towards your direction. (f) Branding wildlife reserves as an adventure tourism destination. Branding is a pull strategy that can be used to pull customer towards the products.

The potential growth of responsible and adventure tourism in Malawi depend so much on the modern marketing strategies (Khadka, 2012; Kotler & Armstrong, 2012; Lake, 2017). Malawi as a tourism destination with an adventure tourism product offers both hard and soft adventure products. A specific example of a hard adventure product is canoeing (Maattanen-Bourke, 2008). Canoeing is done on the Shire River and is offered at Bushman's Baobab Lodge. The other products offered in the park are in the category of soft adventure which includes bird watching, game drives, walking safaris, boat safaris, camping, and others. As part of responsible travel, it requires in the marketing of adventure activities to include messages of responsibilities and sustainable management approach to the resources for the environment and wildlife protection and conservation in the wildlife reserves (Maattanen-Bourke, 2008; Khadka, 2012). This enhances the potentiality to the growth of adventure tourism in Malawi as a country which has many adventure hot spots across the country which needs to be identified and promoted. As added by another author (Khadka, 2012), sustainability of the resources will mean continuously experiencing the nature, culture, and lifestyles of the local people who promote the environmental and cultural understanding. Moreover, marketing outdoor adventure tourism is about putting the right contents or pictures to the right people and at the right time (Weber, 2001; Canwell & Sutherland, 2003).

3.5 CONCLUSION

Adventure tourism business has pointed out why the sector is growing in the global tourism industry. The rapid growth and popularity of the adventure sector have been facilitated due to changes in market demand as tourists are now preferring and seeking unusual holidays, different from the typical beach vacation (Khadka, 2012). The growth on consumer demands also requires the incorporation of responsible travel by those who participate in wildlife reserves activities. Adventure as an activity is viewed as organization type of niche tourism involving exploration or traveling to remote or wilderness destination with perceived and possible risk, requiring specialized skills and tends to be associated with high-level involvement, and mostly outdoor activity (Weber, 2001; Font & Carey, 2005; Khan, 2010; Khadka, 2012). The activity may be soft or hard. The sector relies on the environment; the adventure tourism products usually include two of the following components: a physical activity, a cultural exchange or interaction, and the engagement with nature (UNWTO, 2014). Therefore, wildlife reserves in Malawi have adventure tourism products ranging from birding, boat safari, game drives, bush walk, rhino tracking, canoe safari and as well as village interactions. These products are managed in a way that they do not bring challenges to the environment that would be detrimental to the destinations. Sustainable management is employed to ensure continued processes and enjoyment of the benefits that are realized from the products through the management of information that every client upon arrival is briefed on how to behave and making sure they conserving the environment by following the rules and regulations outlined in the briefing (Buckley, 2010). The adventure tourism products in Malawi are largely marketed by tour operators, and of course, the facilities have their own internet where they market their facilities as well as the tourism products. Finally, the study suggests that, with a vigorous application of modern marketing strategies, the adventure tourism in Malawi stands a better chance to position itself on a global map. Moreover, this requires marketing joint efforts by both public and private tourism firms. Tourists also have a big role to play through destination marketing dissemination using technology and word of mouth recommendation (Paul & Geoffrey, 2009). There are a lot of adventure tourism destinations in Malawi, and more are yet to identified, marketed, and promoted to make Malawi as an adventure tourism destination.

KEYWORDS

- **adventure tourism**
- **adventure travelers**
- **Liwonde Wildlife Reserve**
- **Malawi's tourism**
- **Nyika Wildlife Reserves**

REFERENCES

ACSEDU. (2017). *Marketing adventure tourism. Ecotourism and adventure tourism.* Available at: http://www.acsedu.com/courses/ecotourism-and-adventure-activities-courses.aspx [accessed Aug 19, 2017].

Andrews, C. G. (2015). Ten Reasons why adventure travel is good for you, 2015; *Good Nature Travel.* Available at: http://goodnature.nathab.com/ten-reasons-why-adventure-travel-is-good-for-you/ [accessed Jul 4, 2017].

Annon. (2012). *Trends in Alternative Tourism and Tailor-Made Products – Ecotourism, Adventure Tourism, Cultural, Historical Tourism Cross Tour Fair. A Danubian Al-Tour-Nature Dream, 2012*; Ethan Maxwell. Rousse Bulgaria.

Antigua and Barbuda Tourism Development. (2003). *Tourism's Economic Impacts Increasing the Contribution to Prosperity*. KPMG Consulting LP: Antigua.

Bauer, A. (2010). *Marketing Adventure Tourism: What Works and What Doesn't.* S.N. 81234. Munich, Germany.

Bauer, I. (2017). The plight of the cute little lambs: Travel medicine's role in animal welfare. *Journal of Ecotourism*; *16*(2), 95–111. Available at: http://dx.doi.org/10.1080/14724049.2016.1206551 [accessed 25 Jul, 2017].

Bello, F. G., Lovelock, B., & Carr, N. (2016). Enhancing community participation in tourism planning associated with protected areas in developing countries: Lessons from Malawi.; *Journal of Tourism and Hospitality Research. 3*(5), 407–423.

Boniface, B. G., & Cooper, C. (2005). *Worldwide Destinations: The Geography of Travel and Tourism*. Butterworth-Heinemann: Oxford.

Briggs, P. (2016). *Bradt Malawi Edition 7. Guilford, Connecticut 06437-0480.* The Globe Pequot Press Inc: USA.

Buckley, R. (2006). *Adventure tourism research: A guide to the literature*. Griffith University: Cold Cost, Australia.

Buckley, R. (2006). *Adventure Tourism*; CABI International: Wallingford, Oxfordshire.

Buckley, R. (2004). *Skilled Commercial Adventure: The Edge of Tourism*. In New Horizons in Tourism; Singh, T.V., Ed. CABI International: Oxford, 37–48.

Buckley, R., Shakeela, A., S., & Guitart, D. (2014). Adventure tourism and local livelihoods. *Annals of Tourism Research, 48*, 266–291.

Buckley, R. C. (2010). *Adventure Tourism Management*. Butterworth-Heinemann: Oxford.

Bushmans Baobab. (2017). Canoeing Safari on Shire River. Available at: www.bushmansbaobab.com [accessed 7 Jul, 2017].

Canwell, D., & Sutherland, J. (2003). *Leisure and Tourism*. Nelson Thorms Ltd: Cheltenham, UK.

Carter, C. (2006). World adventure capital. In; *Adventure Tourism*. CABI: Wallington, 429–442.

Cater, C. (2013). Meaning of Adventure. *In Adventure tourism: Meaning, Experience and Learning*. Taylor, S., Varley, P., & Johnston, T., (ed.). *Contemporary Geographies of Leisure, Tourism and Mobility*, Routledge: London, 7–18.

Central wilderness Safari. (2017). Boat and walking safari. Available at: www.caws.com [accessed 8 July 2017].

Centre for Responsible Travel. (2017). Trends and statistics 2016 CREST. Available at: http://www.responsibletravel.org [accessed 2 May, 2017].

Department of Tourism, (2015–16). Ministry of Information, Department of Culture. *Destination Malawi 2015–16*. The Official Directory Malawi, p. 7.

Dreams, H. (2017). Life of a Travel Agent–Challenges and Privileges, https://www.himalayan-dreams.com/life-of-a-travel-agent-the-challenges-and-privileges [accessed 11 July 2017].

Flaherty, G. T., & Choi, J. (2016). The 'Selfie' Phenomenon: Reducing the risk of harm while using smartphones during international travel. *Journal of Travel Medicine, 23*(2), 1–3.

Font, X. & Carey, B. (2005). *Marketing Sustainable Tourism Products*. Leeds Metropolitan University: Leeds, UK, 2005.

Gardiner, S., Kwek, A. (2017). Chinese participation in adventure tourism: A study of generation Y international students' perception. *Journal of Travel Research, 56*(4), 495–504.

Giddy, J. K, & Webb, N. L. (2017). Environmental attitudes and adventure tourism motivation. *Geo Journal*; 2017, Springer Science. Available at: https://link-springer-com.gcu.idm.oclc.org/content/pdf/10.1007%2Fs10708–017–9768–9.pdf [accessed 25 May 2017].

Giddy, J. K., & Webb, N. L. (2016). The influence of the Adventure tourism: From motivation to experience. Current Issues in Tourism; 2016. Available at: http://dx.doi.org/10.1080/13683500.2016.1245715 [accessed 25 Jul, 2017].

Goodwin, H. (2017). Responsible Tourism and the Market, Occasional Paper No. 4, 2005. *International Centre for Responsible Tourism*. University of Greenwich: United Kingdom. https://www.responsibletravel.com/ImagesClient/certification.pdf [accessed 11 July 2017].

Heyniger, C. (2011). The rising prominence of adventure and its staying power. *Adventure Trade Magazine*.

Honey, M. & Gilpin, R. (2017). *Special Report–Tourism in the Developing World: Promoting Peace and Reducing Poverty,* 2009, USIP. https://www.usip.org/sites/default/files/tourism_developing_world_sr233_0.pdf [accessed 11 July 2017].

Jenkins, H. P. (2017). *Introduction to adventure tourism*. Available at: http://www.slideshare.net/hillarypjenkins [accessed 16 May, 2017].

Khadka, S. (2012). *Marketing of adventure and nature tourism: a case study of Shambala Trekking Agency in Nepal, Ostrobothnia*. Central Ostrobothnia University of Applied Sciences, 2012.

Khan, M. A. (2010). Adventure tourism in India. *International Journal of Multidisciplinary Research and Development*, p. 374.

Kotler, P., & Armstrong, G. (2012). *Principles of Marketing*, 14th ed., Pearson Education Inc.: New Jersey, 2012.

Lake, L. (2015). *Marketing 101: What is branding*. 2015. Available at: http://www.marketing.abou.com/cs/brandmktg/a/whatisbranding.htm [accessed 13 Jun, 2017].

Lundberg. D., Khrishnamurthy, M., & Mink, H. (1995). *Tourism Economies Perspectives*. John Willey and Sons Inc: Canada.

Maattanen-Bourke, O. (2008). *The story of Malawi's Liwonde National Park: On the Wild Side*. Central African Limited, Blantyre.

Mason, P. (2003/2008). *Tourism Impacts, Planning and Management*. Butterworth-Heinemann: Oxford.

McCarthy, E. *Adventure tourism*. Teagasc Rural Development Centre Athenry Co: Galway, 2006.

Miller, G., & Twining-Ward, L. (2006). Monitoring an approach to sustainable tourism. In: *Tourism management Dynamics, Trends, Management and Tools*. Butterworth-Heinemann: Burlington, 51–55.

Moorhouse, T., D'Cruze, N. C., & Macdonald, D. W. (2017). Unethical use of wildlife in tourism: What's the problem, who is responsible, what can be done? *Journal of Sustainable Tourism*; *25*(4), 505–516. Available at: http://dx.doi.org/10.1080/09669582.2016.1223087 [accessed 25 Jul, 2017].

Muller, T., & Cleaver, M. (2000). Targeting the Canzus baby boomer explorer and adventurer segments. *Journal of Vacation Marketing*, *6*(2), 154–149.

Mullins, T. (2014). A socio-economic case for skills in outdoor adventure. *Journal of Experimental Education*, *37*(2), 129–143.

Page, S. L. (2007). *Tourism Management: Managing for Change*. Elsevier: Oxford.

Paul, W., & Geoffrey, S. N. (2009). Value, Satisfaction and behavior intentions in an adventure tourism context. *Annals of Tourism Research*, *36*(3).

Pomfret, G, & Bramwell, B. (2016). The characteristics and motivational decisions of outdoor adventure tourists: A review and analysis. *Current Issues in Tourism*, *19*(6), 1447–1478. Available at: http://dx.doi.org/10.1080/13683500.2014.925430 [accessed 25 Jul, 2017].

Price, S., & Driscoll, C. O. (2012). *Perceptions of Risk in Irelands Emerging Adventure Tourism Industry*. Institute of Technology: Tralee.

Rantala, O., Rokenes, A., & Volkonen, J. (2016). Is adventure tourism a coherent concept? A review of research approaches on adventure tourism. *Annals of Leisure Research*, *13*, 1–15. Available at: http://dx.doi.org/10.1080/11745398.2016.1250647 [accessed 25 Jul, 2017].

Safari Guide Africa. (2017). *Destination Malawi*. Available at: http://www.safariguideafrica.com/destination/malawi/ [accessed 8 July 2017].

Sanyal, A. (2009). *Tourism management: Step by step by Government to promote tourism in India*. 2009. Available at http://anandasanyal.blogspot.co.uk/2009/06/steps-taken-by-govt-to-promote-tourism.html [accessed 2 Jul, 2017].

Siegel, L. (2017). The eco-guide to responsible travel. The Observer, 2017, Jan 29. Online.

SNV Netherlands Development Organization. (2017). The Market for Responsible Tourism Products: With Special focus on Latin America and Nepal, 2017. http://www.

responsibletravel.org/docs/The%20Market%20for%20Responsible%20Tourism%20 Products.pdf [accessed 11 Jul, 2017].

Spenceley, A. (2014). *Introduction: Responsible Tourism in the Southern Africa.* Taylor and Francis, London. Available from: ProQuest E-book Central, [accessed 5 July 2017].

Swarbrooke, J. (2002). *The Development and Management of Visitor Attractions*, 2nd ed. Butterworth-Heinemann: Oxford.

Swarbrooke, J., Beard, C., Leckie, S., & Pomfret, G. (2003). *Adventure Tourism: The New Frontier.* Butterworth-Heinemann: Oxford.

The Eye Magazine Malawi. (2016). The Insider's Guide to Malawi. *The Eye*. March–May, 2016, p 73.

The George Washington University and Adventure Tourism Trade Association; *Adventure Tourism Market Study*. The George Washington University: Washington, 2013.

Towards Earth Summit. (2002). *Sustainable Tourism: Turning the Tide, Economic Briefing No. 4*, 2002.

UNWTO. (2014). Global report on adventure tourism. AM Reports, UNWTO: Madrid, 2014.

Varley, P., & Taylor, S. Introduction. (2013). *In Adventure Tourism: Meaning, Experience and Learning.* Taylor, S., Varley, P., Johnston, T., ed. *Contemporary Geographies of Leisure, Tourism and Mobility*, Routledge: London, 1–3.

Weber, K. (2001). Outdoor adventure tourism: A review of research approaches. *Annals of Tourism Research. 28*(2), 360–377.

Wilks, J., & Page, S. J. (2003). *Managing Tourist Health and Safety in the New Millennium.* Elsevier: Oxford, 2003.

Yeoman, I. (2008). Tomorrow's tourist: Scenarios and trends. In: *The Sports Tourist: The Urban Rat Race*. Butterworth-Heinemann: Oxford.

CHAPTER 4

WHAT IS THE IMPORTANCE OF SOCIAL MEDIA SHARINGS FOR TOURISTS? THE ROLE OF ARGUMENT QUALITY AND SOURCE CREDIBILITY

BEKIR BORA DEDEOĞLU

Tourism Faculty, Nevsehir Hacı Bektaş Veli University, Nevsehir, 50300, Turkey, E-mail: b.bora.dedeoglu@nevsehir.edu.tr

ABSTRACT

To show interest in a destination and have information about the destination is of critical importance in terms of its preferability. Despite of this critical importance, the factors affecting the destination brand awareness and destination involvement were mainly examined within the scope of traditional mediums. Being one of the most important natural advertisement and promotion means, the effect of consumer-generated content in social media on these factors has not yet been emphasized and examined sufficiently. Because of the increase in the time spent in social media mediums, it becomes highly possible that the contents generated in these mediums influence the individuals, as well. Therefore, the contents generated by individuals in social media mediums should be examined in more detail. In this chapter, it will be examined how the importance attached to different types of sharing by consumers affects the destination brand awareness and destination involvement levels of tourists. Moreover, it will be investigated whether these effects would differ in accordance with tourists' perceptions of argument quality and source credibility. Thus, significant clues on social media management and marketing will be provided to destination management and marketing organizations.

4.1 INTRODUCTION

Along with the increasing variety of products and services, the variety of resources from which individuals can obtain information does increase, as well. Among information resources, the role of the Internet and social media mediums has become more important for consumers. Consumers who want to obtain information about a product and service are likely to utilize the Internet and social media mediums at an optimal level. Especially the contents generated in social media mediums by individuals have become crucial for purchasing decisions, preferences and behaviors of consumers (Cox et al., 2009; Litvin et al., 2008; Ye et al., 2011). The fact that the contents generated by others in social media mediums shape consumers' decisions and preferences at a significant scale has created the necessity for an organization to focus on these points.

At this point, organizations (e.g., Sheraton, Starwoods) that care about the contents generated by others in social media platforms have started to take particular steps through creating their own websites to enable the communication of consumers with the facility and other consumers (Sigala, 2009). Such kind of steps that ensures the integration of consumers into the process not only allows individuals to express their opinions about product and services freely but also make it possible for them to obtain information from and be influenced by these opinions. Therefore, the contents generated by others could help people reach a specific awareness level about the brands (product and services).

Nevertheless, social media mediums are the platforms through which individuals not only obtain information about product and services but also satisfy their motives such as curiosity about others, expressing their opinions, feeling socialized, gaining and improving prestige/status/popularity (Le and Ma, 2012; Leung, 2009; Li, 2005; Smock et al., 2011; OECD, 2007; Stoeckl et al., 2007; Urista et al., 2009; Park et al., 2009). The contents generated by others for self-realization and increasing their own popularity could prompt other consumers' feeling to experience the same product or service and so, could increase their interest in and awareness of the content. The generation process of the contents derives from individuals' own motives, and organizations are not involved in this process. Accordingly, it can be indicated that content-generation processes in which organizations are not involved are also important. Due to all these reasons, it can be expected that the generation of contents in social media by others is an important factor for increasing consumers' awarenesses and involvement levels with regard to a specific brand.

Another important aspect of the relationship of social media contents with the involvement and the brand awareness is the argument quality and source credibility. As a matter of fact, it is indicated by such researchers as Petty et al. (1983), Shan (2016), Sussman and Siegal (2003), and Wathen and Burkell (2002) that argument quality and source credibility are quite important in terms of the impact of the message conveyed. At this point, even though the sharing by other users increase the brand awareness through prompting the involvement of consumers, it is an important question whether this determinative effect differs in accordance with the source credibility and argument quality. More clearly, although it is expected that positive perceptions of consumers regarding the contents of the sharings would increase their involvement level as to the product, the directive effect of consumers' perspectives related to the quality of the social media sharings and to source credibility should be investigated.

Despite the fact that the brands have started to pay more attention to the activities in social media that prompts the feelings of the consumers, the attempts in this regard are not sufficient. Especially for the tourism sector, it can be indicated that management and marketing organizations are too passive in this field (Pike and Page, 2014). It can be stated that it is an important shortcoming in the literature that the importance of social media sharings which is a quite important variable in terms of both the brand awareness—the first step of the creation of the destination brand—and the destination involvement has not been examined in detail. In addition, the moderating role of the argument quality and source credibility on the effect of the importance attached to social media sharings on destination involvement and destination brand awareness not been examined in detail. Thus, this point is another shortcoming in literature. In this regard, the primary purpose of this chapter is to investigate the effect of the importance attached to social media sharings on the brand awareness and the destination involvement. Besides, it is also aimed at examining the moderator role of the argument quality and source credibility on this structural relationship. This chapter supports the importance of social media mediums for the brand, and informations and suggestions on the role the argument quality and the source credibility are provided to the destination management and marketing organizations, as well.

4.2 CONCEPTUAL FRAMEWORK

4.2.1 SOCIAL MEDIA SHARINGS (SMSS) AND DESTINATION BRAND AWARENESS

Social media sharing of others have become more important for organizations recently. However, despite of its increasing importance, it is observed that the impact of social media sharings on various components has not yet been examined. The examination of SMSs as a general structure within the scope of customer-generated content might result in the ignorance of particular components which are quite important for organizations (Gretzel and Yoo, 2008). Therefore, it could become necessary to examine the customer-generated content in more detail. At this point, Shao (2009) categorized the customer-generated contents as consuming, participating and producing behaviors. Consuming behavior covers the actions of observing, reading and recognizing, and individuals under no circumstances contribute to the content in this behavior. On the other hand, participating sharing refers to the interaction between users and the content, e.g., the assessment of the contents and posting comments. Lastly, producing behavior is the set of actions which cover the sharing of such components as text, images, audio, and video as well as the generation of the content (Shao, 2009). Even though this categorization by Shao is quite important in terms of examining and understanding the customer-generated content, it is necessary to examine CGC from different perspectives.

As a matter of fact, by way of active utilization of social networks as well as their own official websites, organizations can enable consumers to generate content in social media mediums. In the event that the term of customer participation is taken into consideration as it is referred in consumer behavior literature, information, assessments, etc. provided by others in social media mediums of the organization can be regarded as the participant sharing (Dedeoğlu, 2016). These kinds of behaviors make significant contributions to customer participation activities of the organizations such as the improvement of the existing products and the development of new products (Bonner, 2010). Whereas they indicate that consumers show participant sharing, they can also demonstrate SMSs shown by consumers in a non-participant way. Non-participant sharings indicate the sharing behaviors to which the organizations are not included in and which individuals show directly in their own accounts, profiles or their friends' profiles (Dedeoğlu, 2016). As the participant sharings are a kind of means where the organizations are involved in and through which customers can directly have information or impression, they are so important so that the organizations would be effective on their customers' decisions. Nevertheless, it would be necessary for them to take steps, in other words, to make even low-cost investments to encourage customers for showing participant sharing. On the other hand, non-participant sharings of the individuals are more motived by their desires and instincts compared to the participant behavior (Kim and

Jogaratnam, 2003; Küçükergin and Dedeoğlu, 2014) and it could be stated that this does not create any direct cost on the business.

In addition, because the non-participant sharing is not shown with the purpose of directly obtaining information about the brand and the (producer) brand itself is not included in the relevant process (Dedeoğlu, 2016), it could be perceived by the consumers sincerer and could become a more effective factor (Brown and Hayes, 2008). Although various studies have been carried out on social media or user-generated content so far, it is observed that this important point has not been examined. However, it can be expected that the non-participant sharing would be as effective as the participant sharing on desires, interests, motivations, and behaviors. Thus, figuring out which one of the sharing is more effective for the brands would contribute to the social-media oriented marketing strategies.

Especially for the brands, social media can be an important factor in terms of both raising the awareness and reaching a big audience only through a simple effort (Safko, 2012; Weinberg, 2009). Social media marketing can contribute to the brands in two ways: the firm-generated content and the customer-generated content. Sometimes it would be necessary for organizations to make high-cost investments to raise the customer awareness on their brands. Besides radio and TV channels, billboards in the neighborhoods of consumers and other various methods such as the printed media materials can be used for raising the awareness (Gartner, 1994). However, firm-generated contents in social media are more cost-effective than these mediums. Besides the cost-effectiveness, they allow organizations to simultaneously convey their messages or the familiarity they would like to create to all continents only in a few seconds. In addition to its above-mentioned features, customer-generated content is considered as more sincere and reliable by the potential consumers (Brown and Hayes, 2008; Tan, 2011).

The touristic products, especially, have a reverse supply process which means consumers purchase the products without seeing and trying them. Therefore, it can be expected that customer-generated content which covers such characteristics as sincerity and reliability would be more important in terms of increasing the brand awareness. However, as stated before, customer-generated contents can be divided into two groups such as those generated directly or indirectly with the encouragement of businesses and those generated without the involvement of the businesses. In the present research, these content generation behaviors are examined as participant and non-participant sharing (Dedeoğlu, 2016). In the light of those mentioned, it is expected that the importance attached to participant and non-participant sharings would affect the destination brand awareness positively. Accordingly, the hypotheses are developed as follows:

- H1: The importance attached to the participant sharing in social media affects destination brand awareness positively and significantly.
- H2: The importance attached to the non-participant sharing in social media affects destination brand awareness positively and significantly.

4.2.2 SMSS AND INVOLVEMENT

Involvement which is an important term in terms of consumers' purchasing processes (Rochschild, 1984, Hoyer and MacInnis, 2008) is defined from the marketing perspective as any consumer's level of involvement in any object (Huang et al., 2014; Boisvert and Ashill, 2011). Howard and Sheth (1969) examine the involvement as a structure that a customer classifies and interrelates whereas Myers (1968) defines the involvement as the psychological, social or economic interest of a consumer resulting from his/her relationship with the product category during his/her brand-choice process. On the other hand, Day (1970) defines the involvement as the general involvement level regarding a specific object or the object's centrality for ego structure of an individual whereas Mitchell (1979) states that the involvement is an individual dimension and internal state variable indicating the scale of stimulation, interest or motives stimulated by a specific stimulant or a case. From a different perspective, Antil (1984) indicates that involvement in the case of significance or involvement stimulated by a specific stimulant or a situation whereas Zaichkowsky (1985) defines the involvement as an individual's interest in any object in the light of his/her basic needs, values, and benefits. In brief, as stated by Engel et al. (1993), consumers' involvement in product and services is perceived important when their needs, aims and values come together, and then the consumers have an involvement.

What needs to be taken into consideration here is the factors effective on the involvement levels of consumers. More clearly, basic needs aims and value judgments which increase their involvement level to a specific object should be taken into notice. In literature, there exist many studies on the involvement levels of consumers. Vakratsas and Ambler (1999) researched the impact of promotion inputs on the involvement level, and Vaughn (1980) developed the FCB matrix to explain the relationship between the advertising and the involvement. On the other hand, Rossiter and Percy (1987), instead of focusing on the brand attitude-based approach, assumed that brand awareness is a necessary communication target. It can be indicated that it is mainly focused on the promotion activities in the mentioned studies. Rossiter et al. (1991) stated that brand attitude, in the absence of brand awareness, could result in the insufficient advertisement, and assumed that brand awareness is a necessary communication objective for advertising, and Therefore, examined it as the antecedent of the brand attitude. Nevertheless, the possibility that involvement levels of individuals could emerge before gaining brand awareness should not be ignored.

Changing communication tools, especially social media mediums, could increase individuals' involvement levels through ensuring that they care about the brand (object) even if they do not have much information about it (Arndt, 1967; Bansal and Voyer, 2000; Goossens, 2000). For instance, even if any sharing posted in any of social media mediums (e.g., Facebook, Instagram) is related to a brand about which an individual knows nothing or few, it still could prompt other individuals' desire to experience the same brand because of such reasons as being recognized and loved by the society. Under such circumstances, an increase in their involvement levels could be expected. Therefore, it is appropriate to say that customer-generated contents are among the prominent advertising tools today. Besides being perceived by individuals (consumers) who follow them (Akehurst, 2009, Gretzel et al., 2008; Park et al., 2007), customer-generated contents can be generated and shared with the purposes of being recognized by or gaining status within the society (Lampel and Bhalla, 2007; Lakhani and Hippel, 2003; Waskoand Faraj, 2005). This sharing could be the stimulant which creates or motives the involvement for a consumer who comes across CGC by chance (e.g., while scrolling down on Facebook) (Mitchel, 1979) or who sees/reads/watches CGC with the aim of satisfying his/her specific needs. As a matter of fact, because these sharings can be posted by the individual in order to gain prestige and status or to be recognized by the society, they can prompt the feeling of having the same experiences for those who have seen them. In this regard, it can be expected that the importance attached to non-participant and participant sharings (contents) would affect the consumer involvement in a positive and significant way. Accordingly, the hypotheses are developed as follows:

- H3: The importance attached to the participant sharing in social media affects consumer involvement in a positive and significant way.
- H4: The importance attached to the non-participant sharing in social media affects consumer involvement in a positive and significant way.

4.2.3 ARGUMENT QUALITY AND SOURCE CREDIBILITY

Although it is expected that the contents generated in social media mediums would increase the involvement and awareness levels of consumers, two of the points to be taken into consideration are the argument quality and source credibility. Besides the accuracy and adequateness of the messages, the fact that the source through which the messages are conveyed is perceived reliable is quite important so that the contents of the messages are influential and persuasive (Chaiken, 1980; Petty et al., 1983; Shan, 2016; Sternthal et al., 1978; Sussman and Siegal, 2003; Wathen and Burkell, 2002). As a matter of fact, these two points and their impacts on consumer attitudes were examined in literature within the scope of the argument quality and the source credibility. Argument quality is defined by Eagly and Chaiken (1993) as the credibility nature or the power of the persuasive proof and by Bhattacherjee and Sanford (2006) as the persuasive power of an argument included in an information-based message. On the other hand, source credibility is defined by Ohanian (1990, p. 41) as "a term commonly used to imply a communicator's positive characteristics that affect the receiver's acceptance of a message" and by Hovland et al. (1953) as the perceived expertise and trustworthiness of the source.

In the studies carried out within the scope of ELM, the terms of argument quality and source credibility were examined within the framework of central and peripheral routes (Bansal et al., 2008; Lee, 2012; Shih et al., 2015; Petty and Cacioppo, 1986a; Shu and Scott, 2014; Sussman and Siegal, 2003). ELM model refers to an individual's action of further thinking so as to determine the accuracy of arguments on a subject rather than taking all possibilities into consideration by him/herself (Cacioppo and Petty, 1984). In other words, ELM can be described as the comprehensibility of the information process during which an individual pays attention to the message and evaluates the quality of the message (Qing et al., 2007). ELM model is composed of central and peripheral routes (Cacioppo and Petty, 1984; Cacioppo et al., 1985). The central route is the one that an individual mostly prefers under the circumstances where his/her ability and motivation to precisely examine the arguments on the subject are relatively high whereas the peripheral route is the one that an individual prefers under the circumstances where his/her ability and motivation are low, and the attitudes are determined by the positive or negative clues which are directly related to the message or allow the presence of a very simple moderator for the credibility of the message (Petty and Cacioppo, 1986a). More clearly, when an individual accepts a message through central route, s/he examines the message quite carefully, thinks over it and tries to link it to his/her existing knowledge (Cacioppo et al., 1985); on the other hand, when s/he prefers the peripheral route, s/he can have a reasonable point of view on the accuracy of the argument without thinking over it in deep (Cacioppo et al., 1985). At this point, taking into consideration the ELM model, it can be expected that the message-based argument quality and source credibility perceptions of a consumer could have a differentiating effect on the objectives of raising brand awareness or brand involvement. Accordingly, in terms of the argument quality, consumers would examine the arguments in more detail and get more information if they found the quality of arguments in social media sharings low, or, would not examine them in detail if they found the quality high, and in this regard, it can be expected that the perceptions on argument quality would have a moderating role on the impact of SMS on involvement and brand awareness.

However, in terms of source credibility, Aristotle's Rhetoric can be taken into notice. Aristotle's rhetoric contains such three engaged attractiveness components as ethos, pathos, and logos. According to Ethos, one of those components, the source has a character-based specific attractiveness, and this attractiveness increases the efficiency of the message from the source (Leith, 2011). Such factors as the formation of this attractiveness, the age, the experience level on the topic, perceived honesty, and outspokenness of the source can play an active role in the formation of this attractiveness (Percy and Elliott, 2009). On the other hand, an individual's reputation or sociality could also be a component of the source credibility (Bansal et al., 2008; Burgoon et al., 2000; Tseng and Fogg, 1999). At this point, the source credibility examined as a peripheral route in the ELM model can be examined within the scope of the ethos appeal. In this regard, considering that source credibility is a function of the source attractiveness (Patzer, 1993), it can be stated that source credibility has a moderator role in the relationship between SMSs and the destination and between SMSs and the destination brand awareness. Accordingly, the fact that individuals' high source credibilities in social media might cause that their sharings are perceived attractive. In this case, it can be expected that social media sharings have a stronger impact on individuals' involvement levels and brand awarenesses for tourist's groups where the source credibility is perceived high. Accordingly, the hypotheses and the model are developed as follows:

- H5a: The source credibility has a moderating effect in the relationship between the importance attached to the participant sharing and the destination brand awareness.
- H5b: The source credibility has a moderating effect in the relationship between the importance of non-participant sharing and the destination brand awareness.
- H5c: The source credibility has a moderating effect on the relationship between the importance of the participant sharing and destination involvement.
- H5d: The source credibility has a moderating effect on the relationship between the importance of non-participant sharing and destination involvement.
- H6a: The argument quality has a moderating effect in the relationship between the importance of the participant sharing and the destination brand awareness.
- H6b: The argument quality has a moderating effect on the relationship between the importance of non-participant sharing and the destination brand awareness.
- H6c: The argument quality has a moderating effect on the relationship between the importance of participant sharing and destination involvement.

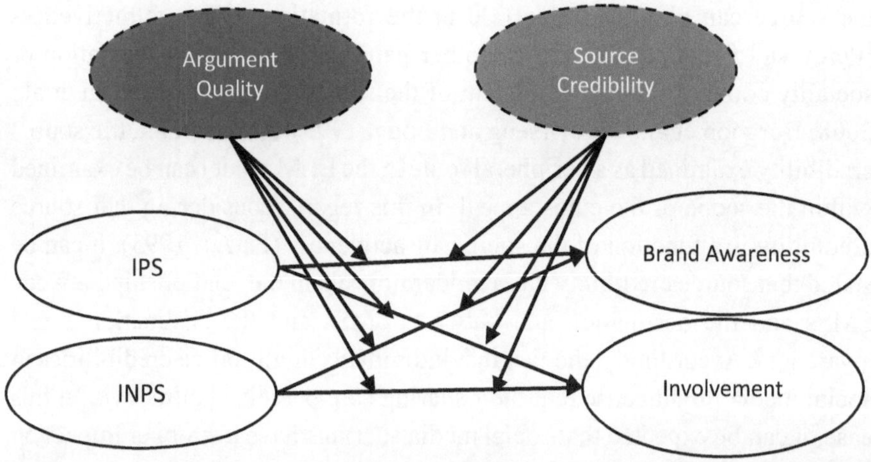

FIGURE 4.1 Research model (Notes: IPS: Importance attached to participant sharing; INPS: Importance attached to non-participant sharing).

- H6d: The argument quality has a moderating effect on the relationship between the importance of non-participant sharing and destination involvement (Figure 4.1).

4.3 METHODOLOGY

4.3.1 INSTRUMENT

For the research questionnaire it was benefitted from the studies of Bailey and Pearson (1983) and Sussman and Siegal (2003) with regard to the statements used for the argument quality measurement; from the studies of Wu and Schaffer (1987), Sussman and Siegal (2003) and Bhattacherjee and Sandford (2006) with regard to the statements used for the source credibility measurement; from the study of Ferns and Walls (2012) with regard to the statements used for the brand awareness measurement; and finally, it was benefitted from the studies of Mittal (1995), Gursoy and Gavcar (2003) and Prayag and Ryan (2012) regarding the involvement. Considering the study of Dedeoğlu (2016), ISMS scale was measured via 11 items under two dimensions as importance attached to participant sharing and importance attached to non-participant sharing.

As the questionnaires were planned to be conducted on tourists from different nations, they were prepared in German, Russian, English, and Turkish taking into consideration the nationalities of the tourists visiting Alanya. The back-to-translation method was used for the translation of the questionnaires (Brislin, 1970, 1976). In this regard, English was chosen as the source language for the questionnaires. Since the original versions of the phrases selected from the literature were in English, it was aimed at keeping them in original. At the first stage of the translation, the questionnaire the source language of which is English was sent to an expert for proofreading. Then, the questionnaire (in English) was sent to the translators knowledgeable both in the source (English) and target (German, Russian, and Turkish) languages. After that, questionnaires in German, Russian, and Turkish translated by the translators were re-translated into English by different translators with the same professional skills (in source and target languages), and the translations of the questionnaires in different languages were compared. As a result, it was observed that all questionnaires were translated in a way that all items have the same meaning. Therefore, it was decided to use the questionnaires translated at the first stage. Nevertheless, a pre-test was carried out on 10 persons from each questionnaire group so as to prevent possible mistakes. As it was observed as a result of the pre-test that the items in questionnaires are clear and understandable, the questionnaires were used as they were prepared at the first stage.

4.3.2 SAMPLING AND DATA COLLECTION

The research population is composed of domestic and foreign tourists in Alanya. As most of the tourists visiting Alanya (Turkey) are foreign tourists, foreigners were selected as a part of the population. On the other hand, it would be necessary to attract local tourists to compensate for the gap by considering the possible decrease in the number of international tourists. At this point, it is necessary to examine the assessments of tourists so that national and local authorities could encourage local tourists further. Therefore, domestic tourists were selected as the other part of the population.

The questionnaires were conducted by the interviewers through the methods of personal interviews and the drop-and-collect method. Since 172 out of 650 questionnaires in total were not filled in an appropriate way, the remaining 478 questionnaires were used for the analysis.

4.3.3 DATA ANALYSIS

Structural equation modeling (SEM) was utilized for testing the developed model. However, at first, missing values were determined and provided, then the extreme values were determined and eliminated, and at last, normal distribution assumption was checked. In order to provide the missing values, the imputation method was preferred, and it was benefitted from the mean substitution method within the scope of this method. On the other hand, Mahalanobis distance was examined for the determination of outliers, and no extreme value was noticed (Mahalanobis (28) 73.252, $p < 0.001$). Lastly, as the maximum likelihood method was utilized for the estimation of SEM, skewness, and kurtosis values were checked, and the normal distribution assumption was examined. Because the lowest and highest values for skewness coefficient are –0.545 and 0.130, respectively and those for kurtosis coefficient are –1.231 and 0.281, respectively, it could be indicated that the normal distribution assumption is verified (Kline, 2011). As the two-step approach recommended by Anderson and Gerbing (1988) was preferred for the implementation of SEM, the measurement model was checked at first, and then the structural model was examined.

So as to examine the moderating effect of the argument quality and the source credibility, a multiple group analysis was conducted (Byrne, 2010). Therefore, the respondents were divided into two groups based on their responses to the items discerning the argument quality and the source credibility through the K-Means cluster analysis. The groups were called as a high-argument quality group and low-argument quality group; high-source credibility group and low-source credibility group.

4.4 FINDINGS

4.4.1 *DEMOGRAPHIC FINDINGS*

About 51.5% of the respondents (f = 246) are female, and 47.2% of them are tourists at and over 42 ages. The married respondents comprise 48.3% of all respondents. Most of the respondents are composed of those who graduated from the undergraduate program (27.2%), the high school (25.9%) and the associate degree program (24.7%). 27.6% and 19.2% of the respondents are from Germany and Russia, respectively, whereas 31.6% of them use German and 21.1% uses Russian as the official language. 32% of the respondents have visited Germany for the first time, and 22.6% of them do not have any children. Most of the respondents (35.4%) stayed in a 4-star hotel, and 23.6% of them stayed in a 5-star hotel. Lastly, having examined the incomes of the groups, it was realized that the number of participants was almost the same in different income groups.

4.4.2 *MEASUREMENT MODEL*

Following the two-step approach by Anderson and Gerbing (1988), a measurement model was tested via the confirmatory factor analysis prior to the structural model. As it was observed after CFA that INV4 ($\beta = 0.49$) and DBA2 ($\beta = 0.46$) items were below the standardized loading value 0.50, they were removed from the model. According to the goodness-of-fit indices, the proposed measurement model, which was finalized after re-performing the CFA, was found to fit the data well ($\chi^2 = 385.928$; df = 215; $\chi^2/df = 1.795$; CFI = 0.97; TLI = 0.96; RMSEA = 0.041).

Table 4.1 shows the standardized loading sand t-values of each indicator. All indicators had the significant standardized loadings at $p < 0.05$ and t-values of the individual indicators ranged from 12.09 to 20.16 for all data sets. In addition, the average variance extracted (AVE) value was greater than

0.50 which indicates that the variance captured by the respective constructs is greater than the variance due to measurement errors (Dillon & Goldstein, 1984). The variance extracted scores of constructs ranged from 0.58 to 0.88, which suggested adequate convergent validity (Fornell and Larcker, 1981; Hair et al., 2009). The composite reliabilities indicate internal consistency, meaning that the measures all consistently represent the same latent construct (Fornell and Larcker, 1981; Hair et al., 2009). The composite construct reliability for each construct ranged from 0.80 to 0.90. Table 4.1 also indicates that the average variance extracted for each construct was greater than the squared correlation coefficients for the corresponding inter-constructs. This, in turn, confirmed the discriminant validity (Fornell and Larcker, 1981).

4.4.3 STRUCTURAL MODEL

SEM was conducted to test the relationships among the constructs. According to the goodness-of-fit indices ($\chi^2 = 276.450$; df = 99; $\chi^2/df = 2.792$; CFI = 0.96; TLI = 0.95; RMSEA = 0.061), the structural model was parsimonious. The structural results of the proposed model are shown in Figure 4.2.

The hypothesized relationship between IPS and the destination brand awareness (Hypothesis 1) was supported by the corresponding estimate of 0.30 ($t = 5.507$; $p < 0.001$), showing that IPS significantly affects the destination brand awareness. The standardized path coefficient of the relationship between INPS and destination brand awareness (Hypothesis 2) was 0.18 ($t = 3.394$; $p < 0.001$), which indicates that INPS was a significant predictor of the destination brand awareness. Hypothesis 3 was also

TABLE 4.1 Result of Measurement Model

			Low	0.39	4.332*				
H_{6a}	Argument Quality	IPS →DBA	High	0.37	5.694*	425.902 (210)	428.952 (211)	3.05 (1) p = 0.081	Yes
			Low	0.19	1.969***				
H_{6b}		INPS →DBA	High	0.14	2.252***	425.902 (210)	426.758 (211)	.86 (1) p = 0.355	No
			Low	0.25	2.601**				
H_{6c}		IPS →Inv	High	0.32	5.126*	425.902 (210)	431.812 (211)	5.91 (1) p = 0.015	Yes
			Low	0.03	$.291^{ns}$				
H_{6d}		INPS →Inv	High	0.34	5.298*	425.902 (210)	428.527 (211)	2.63 (1) p = 0.105	No
			Low	0.45	4.851*				

Notes: *$p < 0.001$; **$p < 0.01$; ***$p < 0.05$; NS: Not significant; IPS: Importance attached to participant sharing; INPS: Importance attached to non-participant sharing; DBA: Destination brand awareness; Inv: Involvement.

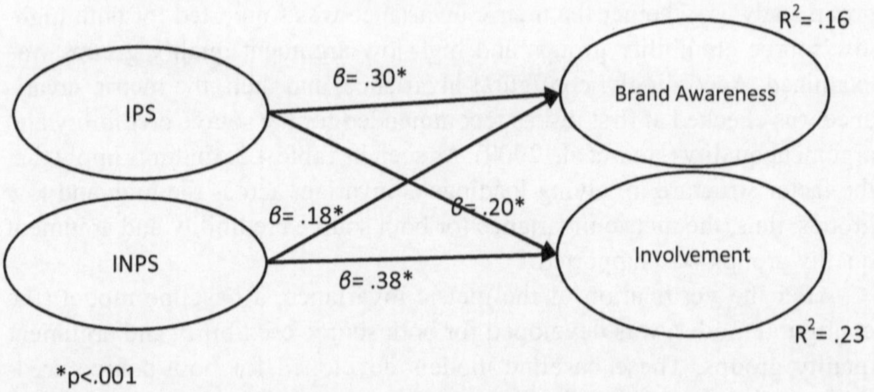

FIGURE 4.2 Result of structural model (Notes: IPS: Importance attached to participant sharing; INPS: Importance attached to non-participant sharing).

supported by an estimate of 0.20 (t = 3.968; p < 0.001), suggesting that IPS was a significant antecedent of involvement. Also, the hypothesized relationship between INPS and involvement (Hypothesis 4) was supported ($\beta = 0.38$; $t = 6.866$; $p < 0.001$).

4.4.4 THE MODERATING EFFECTS OF SOURCE CREDIBILITY AND ARGUMENT QUALITY

In order to be able to examine the moderating effect of source credibility and argument quality, metric invariance related to high and low groups of these variables has to be supported at first. Therefore, before the multiple

TABLE 4.2 Results of Metric Invariance Test for Source Credibility and Argument Quality Groups

Moderator	Models	χ^2	df	RMSEA	CFI	$\Delta\chi^2$(df)	Result
Source Credibility	C.I.	371.278	196	0.043	0.96	ΔX^2(12)–12.11 $p = 0.437$	Supported
	M.I.	383.390	208	0.042	0.96		Supported
Argument Quality	C.I.	371.015	196	0.043	0.96	ΔX^2(12)–11.37 $p = 0.498$	Supported
	M.I.	382.380	208	0.042	0.96		Supported

Notes. C.I: Configural invariance; M.I: Metric invariance; df: degree of freedom.

group analysis, whether the metric invariance was supported for both high-low source credibility groups and high-low argument quality groups was examined. Accordingly, configural invariance, and then, the metric invariance was checked at first just as recommended for the source credibility and argument quality (Hair et al., 2009). As seen in Table 4.2, findings imply that the factor structure involving loadings is invariant across the high and low groups, thus, the metric invariance for both source credibility and argument quality groups was supported.

After the verification of the metric invariance, a baseline model (the configural model) was developed for both source credibility and argument quality groups. These baseline models developed for both source credibility and argument quality groups were compared with a set of nested models, in which a particular parameter of interest is constrained to be equal across.

As shown in Table 4.3, the chi-square test revealed that the effects of IPS on the destination brand awareness and the involvement were significantly different between the high and low argument quality groups ($\Delta\chi^2$(df) = 3.05 (1), p = 0.081; $\Delta\chi^2$(df) = 5.91 (1), p = 0.015). More clearly, the effect of IPS on the destination brand awareness (β = 0.37; t= 5.694; p < 0.001) and the involvement (β = 0.32; t = 5.126; p < 0.001) was found to be more determinative on the argument quality for the individuals with high levels of perception. On the other hand, it was observed that the source credibility had a moderating effect on the relationship between importance attached to non-participant sharing and the destination brand awareness ($\Delta\chi^2$(df) = 7.19 (1), p = 0.007) and between importance attached to participant sharing and the involvement ($\Delta\chi^2$(df) = 7.20 (1), p = 0.007). More clearly, for the tourists with high level perception of source credibility, the impact of importance attached to non-participant sharing on the brand awareness (β = 0.28; t = 4.191; p < 0.001), and the impact of importance attached to participant sharing on the involvement (β = 0.29; t = 4.551; p < 0.001) were found to be more determinative. Accordingly, H_{5b}, H_{5c}, H_{6a}, and H_{6c} were supported.

4.5 DISCUSSION AND IMPLICATIONS

Even though it is emphasized that the contents generated in social media by other consumers are important for brand choices, uni-directional perspective regarding the contents generated in social media by consumers hinders revealing the importance of this component. Therefore, social media sharings were examined two-dimensionally in the chapter as participant and

TABLE 4.3 Result of Structural Invariance for Source Credibility and Argument Quality Groups

Hyp.	Moderator	Relation	Group	SFL	t	Baseline	Nested	$\Delta\chi^2$(df)	Supported
H_{5a}	Source Credibility	IPSàDBA	High	.30	4.549*	428.863 (210)	429.131 (211)	.27 (1) p = .605	No
			Low	.29	3.180*				
H_{5b}		INPS àDBA	High	.28	4.191*	428.863 (210)	436.056 (211)	7.19 (1) p = .007	Yes
			Low	-.03	-.328^{ns}				
H_{5c}		IPS àInv	High	.29	4.551*	428.863 (210)	436.059 (211)	7.20 (1) p= .007	Yes
			Low	.05	.560^{ns}				
H_{5d}		INPS àInv	High	.34	5.226*	428.863 (210)	428.981 (211)	.12 (1) p= .731	No
			Low	.39	4.332*				
H_{6a}	Argument Quality	IPS àDBA	High	.37	5.694*	425.902 (210)	428.952 (211)	3.05 (1) p= .081	Yes
			Low	.19	1.969***				
H_{6b}		INPS àDBA	High	.14	2.252***	425.902 (210)	426.758 (211)	.86 (1) p= .355	No
			Low	.25	2.601**				
H_{6c}		IPS àInv	High	.32	5.126*	425.902 (210)	431.812 (211)	5.91 (1) p= .015	Yes
			Low	.03	.291^{ns}				
H_{6d}		INPS àInv	High	.34	5.298*	425.902 (210)	428.527 (211)	2.63 (1) p= .105	No
			Low	.45	4.851*				

Notes. *$p<.001$; **$p<.01$; ***$p<.05$; ^{NS}: *Not significant IPS: Importance attached to participant sharing; INPS: Importance attached to non-participant sharing; DBA: Destination brand awareness; Inv: Involvement.*

non-participant sharing, and their role on the involvement as well as brand awareness was also investigated. In addition, the moderating effect of the argument quality and the source credibility on the impact of the importance of social media sharings on the brand awareness and the involvement levels of consumers was also examined.

4.5.1 THEORETICAL IMPLICATIONS

According to research findings, it was found that INPS and IPS had a positive and significant effect on both the destination brand awareness and involvement. Especially it was observed that INPS had a strong impact on involvement whereas IPS had the same strong impact on destination brand awareness. Therefore, considering the fact that the places where participant sharing is shown are the platforms in which the organization is involved, it can be expected that the individuals paying attention to IPS are more likely to visit these platforms with the intention of obtaining information (Dabholkar, 1990). More clearly, compared to those paying attention to INPS, it is further possible for tourists paying attention to IPS to take action with the purpose of getting more information. As a matter of fact, IPS is shown in social media mediums where organizations are involved in, and it is more likely for tourists visiting these platforms to get impression and information about the product and services of the organization. This case may explain the reason why destination brand awareness is affected more by IPS.

On the other hand, involvement occurs as a result of being exposed to stimulants. Individuals' involvement in a specific object is prompted by a specific stimulant and can occur as a result of individuals' paying attention to that object in accordance with their internal value judgments (Goossens, 2000). At this point, the generation of the content in social media mediums by others in the form of a non-participant could be based on the fact that the content is perceived as a specific stimulant and that the tourists pay attention to the product and service as a result of this stimulant's being shaped by their value judgments. In addition, considering that the involvement levels of consumers rise when they want to experience specific products and services (Celci and Olson, 1988), it can be indicated that INPS is a more powerful stimulant for the possible tourists and affects their involvement levels further. Therefore, it was observed that social media contents generated by others were a kind of stimulants and especially those generated in individuals' own platforms and pages where organizations are not involved in had a more determinative effect on the involvement levels of tourists.

According to the results of the multiple group analysis in the current research, no moderating effect of the argument quality on INPS's relationship with the destination brand awareness and the involvement was found. This could result from the fact that the primary objective for tourists paying attention to the non-participant sharing was not the information to be obtained through that channel. It is more likely for them to care about the brand itself rather than obtaining information about it, or just to have it. As a matter of fact, it can be understood from the findings of the present research that the importance attached to non-participant sharing is more effective on the involvement rather than brand awareness.

On the other hand, the effect of the IPS on the brand awareness and the involvement was found to be more determinative for the individuals with the high-level perception of the argument quality. In other words, it was realized that argument quality had a moderating effect on the importance attached to participant sharing's relationship with brand awareness and involvement. It could be assumed that the primary intention for tourists visiting the platforms where the participant sharing is shown is to obtain information. This assumption could be based on the fact that participant sharings are generally examined considering the impact of the instinct to get information. More clearly, as participant sharing contents are generally generated by the individuals having had experience with that product and service, individuals might find these contents more reliable and in better quality. Therefore, these individuals mostly use such kind of contents to get information. At this point, the fact that individuals perceive the argument quality high increases the effect of the participant sharing behavior on the brand awareness and the involvement level. On the contrary, the fact that individuals perceive the argument quality low results in the lower impact of importance attached to participant sharing on the brand awareness and the involvement levels.

For tourists with the high-level perception of source credibility, the impact of importance attached to non-participant sharing on the brand awareness and the impact of importance attached to participant sharing on the involvement were found to be more determinative. As a matter of fact, it was also observed that when tourists perceived the source credible, the importance attached to non-participant sharing did not affect the brand awareness and importance attached to participant sharing did not affect the involvement in a significant way. Accordingly, it can be indicated that source credibility does play a significant role in terms of the increase of the brand awareness by the non-participant sharing and of the involvement levels by the participant sharings. In brief, tourists' high-level perceptions of source credibility cause brand awareness to be affected by the importance attached to non-participant sharing and the involvement to be affected by the importance attached to participant sharing positively. This could be based on the high popularity and the attractiveness of the source in social media. As emphasized before, the experience level, the age and the prestige of the source increase its credibility. Accordingly, although it is assumed that the primary instinct of tourists paying attention to the participant sharing is to obtain information, it is expected, as already revealed in the present research, that high perception of source credibility and participant sharing of the source would affect the involvement levels of tourists more, considering the fact that source credibility is an important factor and it is very linked to the popularity and the attractiveness. Similarly, tourists could consider that the individuals are reliable (source credibility) because of their attractiveness and popularity, and thus, non-participant sharing could have a stronger impact on the brand awareness.

4.5.2 PRACTICAL IMPLICATIONS

Organizations should encourage the generation of more informative contents in platforms where they are involved in. In this way, it would become possible for consumers to have a specific level of awareness of the organization. Therefore, in terms of the generation of informative contents, organizations could provide promotions or develop prize systems in order to encourage the individuals, who have experience in or opinion about the organization, to generate contents in the platforms where they are involved in. For instance, organizations could provide particular incentives to consumers such as giving bonus points to be used during the holiday in the organization on the condition that they make a comment or evaluation about their experiences in the organization on its own official website or the page of a moderator holiday-purchasing website.

On the other hand, even though the effect of the importance of INPS on DBA is lower than that of IPS, it should not be ignored that it affects DBA in a positive and significant way. The direct cost of INPS is lower than that of IPS. As a matter of fact, in order to ensure the efficiency of IPS, it could be necessary for organizations to employ qualified persons to follow the evaluations in the websites as well as in their own websites and social media accounts; however, the employment of new personnel is an additional cost for the organizations. Nevertheless, as INPS occurs as a result of individuals' own instincts (to be liked, recognized by the society, etc.), its cost for the organizations is quite low. Therefore, organizations should not ignore INPS in order to increase the brand awarenesses of the consumers. Besides, organizations provide such opportunists in the platforms in which they are not involved that would prompt individuals not only to generate contents by making comments or evaluations but also to generate them in a way that they prompt the willingness of others to have the same experience.

At this point, considering that the involvement levels of tourists are positively affected by INPS within the scope of the social learning theory, organizations should provide incentives to encourage tourists to share their experiences and opinions on their own platforms. In this regard, for instance, tourists could be encouraged to put hashtags (#) while generating contents in the social media mediums about their experiences in the organization. Or, as another example, a standard or a discounted vacation could be offered to the individuals after the contents generated with a hashtag by those individuals had reached a specific number of sharing or likes determined by the organization itself. In addition, suitable physical opportunities for taking pictures in the destination could be provided to the tourists. For instance, cameras located at certain points could be used by tourists freely or paying a reasonable price so that they could take the pictures of beautiful landscapes in the destination from different aspects. This opportunity would enable tourists to use more materials for their sharings.

4.5.3 LIMITATIONS AND FUTURE RESEARCH

This research is subject to several limitations that offer opportunities for future researches. First, the results are based on a sample in Alanya (Turkey). Therefore, the research should be carried out on different samples so as to be able to generalize the research model. Second, the statements of the participants with regard to items in the scale were recorded under actual experience situations. However, considering that their holidays had not been finished, the probability that any different experience would change their perceptions and opinions should not be ignored. For this reason, questionnaires in future studies could be conducted in airports and on passengers who leave the destination. Third, the current research did not include a control variable that representing tourist' previous experiences with a hosting city or nation. This situation could affect the destination-related perceptions of tourists. Therefore, future studies should reveal the moderating effect of tourists' past experiences with the destination. Fourth, the term of involvement was used in the present research only as recommended by Mittal (1995). In future studies, the impact of social media contents on other dimensions of the involvement should be investigated.

KEYWORDS

- **argument quality**
- **consumer-generated contents**
- **destination brand awareness**
- **involvement**
- **social media sharings**
- **source credibility**

REFERENCES

Akehurst, G. (2009). User-generated content: the use of blogs for tourism organizations and tourism consumers. *Service Business*, *3*(1), 51–61.

Anderson, J. C., & Gerbing, D. W. (1988). Structural equation modeling in practice: A review and recommended a two-step approach. *Psychological Bulletin*, *103*(3), 411.

Antil, J. H. (1984). Conceptualization and operationalization of involvement. Advances in consumer research, *11*(1), 203–209.

Arndt, J. (1967). Role of product-related conversations in the diffusion of a new product. *Journal of Marketing Research*, *4*(3), 291–295.

Bailey, J. E., & Pearson, S. W. (1983). Development of a tool for measuring and analyzing computer user satisfaction. *Management Science*, *29*(5), 530–545.

Bansal, G., Zahedi, F., & Gefen, D. (2008). The moderating influence of privacy concern on the efficacy of privacy assurance mechanisms for building trust: A multiple-context investigation. *ICIS 2008 Proceedings, 7.*

Bansal, H. S., & Voyer, P. A. (2000). Word-of-mouth processes within a services purchase decision context. *Journal of Service Research*, *3*(2), 166–177.

Bhattacharjee, A., & Sanford, C. (2006). Influence processes for information technology acceptance: An elaboration likelihood model. *MIS Quarterly*, 805–825.

Boisvert, J., & Ashill, N. J. (2011). How brand innovativeness and quality impact attitude toward new service line extensions: the moderating role of consumer involvement. *Journal of Services Marketing*, *25*(7), 517–527.

Bonner, J. M. (2010). Customer interactivity and new product performance: Moderating effects of product newness and product embeddedness. *Industrial Marketing Management*, *39*(3), 485–492.

Brislin, R. W. (1970). Back-translation for cross-cultural research. *Journal of Cross-Cultural Psychology*, *1*(3), 185–216.

Brislin, R. W. (1976). Comparative research methodology: Cross-cultural studies. *International Journal of Psychology, 11*(3), 215–229.

Brown, D., & Hayes, N. *Influencer Marketing: Who Really Influences Your Customers?*; Elsevier, Butterworth-Heinemann: Oxford, 2008.

Burgoon, J. K., Bonito, J. A., Bengtsson, B., Cederberg, C., Lundeberg, M., & Allspach, L. (2000). Interactivity in human-computer interaction: A study of credibility, understanding, and influence. *Computers in Human Behavior*, *16*(6), 553–574.

Byrne, B. M. (2010). *Structural Equation Modeling Using AMOS: Basic Concepts, Applications, and Programming*. Taylor and Francis: New York.

Cacioppo, J. T., Petty, R. E., & Stoltenberg, C. D. (1985). Processes of social influence: The elaboration likelihood model of persuasion. *Advances in Cognitive-Behavioral Research and Therapy*, *4,* 215–274.

Celsi, R. L., & Olson, J. C. (1988). The role of involvement in attention and comprehension processes. *Journal of Consumer Research*, *15*(2), 210–224.

Chaiken, S. (1980). Heuristic versus systematic information processing and the use of source versus message cues in persuasion. *Journal of Personality and Social Psychology*, *39*(5), 752.

Cox, C., Burgess, S., Sellitto, C., & Buultjens, J. (2009). The role of user-generated content in tourists' travel planning behavior. *Journal of Hospitality Marketing & Management*, *18*(8), 743–764.

Dabholkar, P. A. (1990). How to improve perceived service quality by increasing customer participation, In Proceedings of the 1990 Academy of Marketing Science (AMS) Annual Conference (pp. 483–487). Springer International Publishing, 2015.

Day, G. S. (1969). A two-dimensional concept of brand loyalty. *Journal of Advertising Research*, *9*(3), 29–36.

Dedeoğlu, B. B. (2016). The Relationship between Social Media, Involvement and Destination Brand Equity, PhD Dissertation, Institute of Social Sciences, Akdeniz University, Antalya, 2016.

Dillon, W. R., & Goldstein, M. (1984). *Multivariate Analysis Methods and Applications*; Wiley: New York.

Eagly, A. H., & Chaiken, S. (1993). *The Psychology of Attitudes*; Harcourt Brace Jovanovich College Publishers: Fort Worth, Texas.

Engel, J. F., Blackwell, R. D., & Miniard, P. (1993). *Understanding the Consumer, Seventh Ed.,* Dryden: Fort Worth, Texas.

Ferns, B. H., & Walls, A. (2012). Enduring travel involvement, destination brand equity, and travelers' visit intentions: A structural model analysis. *Journal of Destination Marketing & Management*, *1*(1), 27–35.

Fornell, C., & Larcker, D. F. (1981). Evaluating structural equation models with unobservable variables and measurement error. *Journal of Marketing Research*, 39–50.

Gartner, W. C. (1994). Image formation process. *Journal of Travel & Tourism Marketing*, *2*(2–3), 191–216.

Goossens, C. (2000). Tourism information and pleasure motivation. *Annals of Tourism Research*, *27*(2), 301–321.

Gretzel, U., Kang, M., & Lee, W. Differences in consumer-generated media adoption and use: A cross-national perspective. *Journal of Hospitality & Leisure Marketing*, (2008). *17*(1–2), 99–120.

Gretzel, U., & Yoo, K. H. (2008). Use and impact of online travel reviews. *Information and Communication Technologies in Tourism*, 35–46.

Gursoy, D., & Gavcar, E. (2003). International leisure tourists' involvement profile. *Annals of Tourism Research, 30*(4), 906–926.

Hair, J. F., Black, W. C., Babin, B. J., & Anderson, R. E. (2009). *Multivariate Data Analysis: A Global Perspective, Seventh Ed.,* Upper Saddle River: Prentice Hall.

Hovland, C. I., Irving K. J., & Kelley, H. H. (1953). *Communication and Persuasion*; New Haven, CT: Yale University Press.

Howard, J. A., & Sheth, J. N. (1969). *The Theory of Buyer Behavior* (*Vol. 14);* New York: Wiley.

Hoyer, W. D., & MacInnis, D. J. (2008). *Consumer Behavior, 5th edition*, South-Western, USA: Cengage Learning.

Huang, L., Gursoy, D., & Xu, H. (2014). Impact of personality traits and involvement on prior knowledge. *Annals of Tourism Research, 48,* 42–57.

Kim, K. Y., & Jogaratnam, G. (2003). Travel motivations: A comparative study of Asian international and domestic American college students. *Journal of Travel & Tourism Marketing, 13*(4), 61–82.

Kline, R. B. (2011). *Principles and Practice of Structural Equation Modeling, 3rd edition,* Guilford Press: New York.

Küçükergin, K. G., & Dedeoğlu, B. B. (2014). The importance of employee hospitability and perceived price in the hotel industry. *Anatolia, 25*(2), 254–267.

Lakhani, K. R., & Hippel, E. (2003). How open source software works: Free user-to-user assistance. *Research Policy, 32*(6), 923–943.

Lampel, J., & Bhalla, A. (2007). The role of status seeking in online communities: Giving the gift of experience. *Journal of Computer-Mediated Communication, 12*(2), 434–455.

Lee, C. S., & Ma, L. (2012). News sharing in social media: The effect of gratifications and prior experience. *Computers in Human Behavior, 28*(2), 331–339.

Lee, W. K. (2012). An elaboration likelihood model based longitudinal analysis of attitude change during the process of IT acceptance via education program. *Behavior & Information Technology, 31*(12), 1161–1171.

Leith, S. (2011). *You Talking to Me? Rhetoric from Aristotle to Obama*; Profile Books.

Leung, L. (2009). User-generated content on the internet: an examination of gratifications, civic engagement and psychological empowerment. *New Media and Society, 11*(8), 1327–1347.

Litvin, S. W., Goldsmith, R. E., & Pan, B. (2008). Electronic word-of-mouth in hospitality and tourism management. *Tourism Management, 29*(3), 458–468.

Mitchell, A. A. (1979). Involvement: A potentially important mediator of consumer behavior. *Advances in consumer research, 6*(1), 191–196.

Mittal, B. (1995). A comparative analysis of four scales of consumer involvement. *Psychology & Marketing, 12*(7), 663–682.

Myers, J. G. (1968). *Consumer Image and Attitude*; University of Berkeley, CA: Institute of Business and Economic Research.

OECD. (2007). Web, Participative: User-Created Content. Directorate for science, technology and industry, Committee for information, computer and communications policy, Working Party on the Information Economy.

Ohanian, R. (1990). Construction and validation of a scale to measure celebrity endorsers' perceived expertise, trustworthiness, and attractiveness. *Journal of Advertising, 19*(3), 39–52.

Park, D. H., Lee, J., & Han, I. (2007). The effect of online consumer reviews on consumer purchasing intention: The moderating role of involvement. *International Journal of Electronic Commerce, 11*(4), 125–148.

Park, N., Kee, K. F., & Valenzuela, S. (2009). Being immersed in social networking environment: Facebook groups, uses and gratifications, and social outcomes. *Cyber Psychology and Behavior*, *12*(6), 729–733.

Patzer, G. L. (1983). Source credibility as a function of communicator physical attractiveness. *Journal of Business Research, 11*(2), 229–241.

Percy, L., & Elliott, R. (2009). *Strategic Advertising Management, Third Edition*; New York: Oxford University Press.

Petty, R. E., & Cacioppo, J. T. (1984). The effects of involvement on responses to argument quantity and quality: Central and peripheral routes to persuasion. *Journal of Personality and Social Psychology*, *46*(1), 69.

Petty, R. E., Cacioppo, J. T., & Schumann, D. (1983). Central and peripheral routes to advertising effectiveness: The moderating role of involvement. *Journal of Consumer Research, 10*(2), 135–146.

Petty, R., & John T. C. (1986). *Communication and Persuasion: Central and Peripheral Routes to Attitude Change*; Springer Science & Business Media.

Prayag, G., & Ryan, C. (2012). Antecedents of tourists' loyalty to Mauritius The role and influence of destination image, place attachment, personal involvement, and satisfaction. *Journal of Travel Research, 51*(3), 342–356.

Qing, T., Ng, B. Y., & Kankanhalli, A. (2007). Individual's Response to Security Messages: A Decision-Making Perspective. In Decision Support for Global Enterprises, Springer US, pp. 177–191.

Rossiter, J. R., & Percy, L. (1997). *Advertising, Communications and Promotion Management,* Second edition, Singapore: Irwin McGraw-Hill.

Rossiter, J. R., Percy, L., & Donovan, R. J. (1991). A better advertising planning grid. *Journal of Advertising Research,* 11–21.

Rothschild, M. L. (1984). Perspectives on involvement: current problems and future directions. *Advances in Consumer Research, 11*(1), 216–217.

Safko, L. (2012). *The Social Media Bible: Tactics, Tools, and Strategies for Business Success*; John Wiley & Sons, New Jersey.

Shan, Y. (2016). How credible are online product reviews? The effects of self-generated and system-generated cues on source credibility evaluation. *Computers in Human Behavior, 55,* 633–641.

Shao, G. (2009). Understanding the appeal of user-generated media: A uses and gratification perspective. *Internet Research, 19*(1), 7–25.

Shih, H. P., Lai, K. H., & Cheng, T. C. E. (2015). A Dual-Process Model to Assess User Attitudes and the Likelihood of Electronic Word-of-Mouth Adoption. In *Proceedings of the 19th Pacific Asia Conference on Information Systems (PACIS2015).*

Shu, M., & Scott, N. (2014). Influence of social media on Chinese students' choice of an overseas study destination: An information adoption model perspective. *Journal of Travel & Tourism Marketing, 31*(2), 286–302.

Sigala, M. (2009). E-service quality and Web 2.0: expanding quality models to include customer participation and inter-customer support. *The Service Industries Journal*, *29*(10), 1341–1358.

Smock, A. D., Ellison, N. B., Lampe, C., & Wohn, D. Y. (2011). Facebook as a toolkit: A uses and gratification approach to unbundling feature use. *Computers in Human Behavior, 27*(6), 2322–2329.

Sternthal, B., Phillips, L. W., & Dholakia, R. (1978). The persuasive effect of scarce credibility: a situational analysis. *Public Opinion Quarterly, 42*(3), 285–314.

Stoeckl, R., Rohrmeier, P., & Hess, T. Motivations to produce user-generated content: Differences between webloggers and video bloggers. *BLED Proceedings, 30*, 2007.

Sussman, S. W., & Siegal, W. S. (2003). Informational influence in organizations: An integrated approach to knowledge adoption. *Information Systems Research, 14*(1), 47–65.

Tan, Y. (2011). *Chinnovation: How Chinese Innovators Are Changing the World*; John Wiley and Sons.

Tseng, S., & Fogg, B. J. (1999). Credibility and computing technology. *Communications of the ACM, 42*(5), 39–44.

Urista, M. A., Dong, Q., & Day, K. D. (2009). Explaining why young adults use MySpace and Facebook through uses and gratifications theory. *Human Communication, 12*(2), 215–229.

Vakratsas, D., & Ambler, T. (1999). How advertising works: what do we really know? *The Journal of Marketing*, 26–43.

Vaughn, R. (1980). How advertising works: A planning model. *Journal of Advertising Research*, 27–33.

Wasko, M. M., & Faraj, S. (2005). Why should I share? Examining social capital and knowledge contribution in electronic networks of practice. *MIS Quarterly*, 35–57.

Wathen, C. N., & Burkell, J. (2002). Believe it or not: Factors influencing credibility on the Web. *Journal of the American Society for Information Science and Technology, 53*(2), 134–144.

Weinberg, T. (2009). *The New Community Rules: Marketing on the Social Web*; Sebastopol, CA: O'Reilly.

Wu, C., & Shaffer, D. R. (1987). Susceptibility to persuasive appeals as a function of source credibility and prior experience with the attitude object. *Journal of Personality and Social Psychology, 52*(4), 677.

Ye, Q., Law, R., Gu, B., & Chen, W. (2011). The influence of user-generated content on traveler behavior: An empirical investigation on the effects of e-word-of-mouth to hotel online bookings. *Computers in Human Behavior, 27*(2), 634–639.

Zaichkowsky, J. L. (1985). Measuring the involvement construct. *Journal of Consumer Research*, 341–352.

CHAPTER 5

MOUNTAIN TOURISM DESTINATION DEVELOPMENT WITH STRATEGIC SOCIAL MEDIA

ARJUN KUMAR LIMBU[1] and PRITHBI BAHADUR LIMBU[2]

[1]*Sr. Program Officer, Kathmandu Environmental Education Project, Nepal, E-mail: arjunlimbu@outlook.com*

[2]*Central Department of Sociology, Tribhuvan University, Nepal, E-mail: prithbilimbu@outlook.com*

ABSTRACT

Himalayan or Mountain Tourism in Nepal, since its early beginnings in 1951, has impeded socially, geographically, educationally, due to inadequate, poor marketing and promotion methods and styles, despite the apparent potential for tourism. With developments in technology over the last few decades, and the growing number of young educated tourism entrepreneurs being familiar with the social media use, have been promoting their individual markets, trekking companies and tours using these new methods. In general, there is a wide gap in international market approaches between Kathmandu trekking companies and village tourism operators in the use of social media. For the equitable distribution of tourism to have its effect economically, rural or village tourism (Home-stay operators) need proper training and need to understand its use and importance in attracting clients. The social media tools strategy should be developed solid framework scope, which can become the productive focal point for generating a healthy economy. The Social Media Pyramid (SMP) model is proposed as a model for use not only in the Tamang Heritage Trail (THT), Rasuwa district, Nepal, to realize how an area can make use of social media strategies for their destination promotion and development. This SMP model can connect with local to national to international travel associates and give a total promotion of their particular tourism destination to the world. In this model approach, every such unit can connect with each of the individual tourism operators or professionals. Such many units can connect as a neural-network for sustainable tourism destinations enabling peace, security, and communication. Nepali Rural mountain tourism operators can adopt this model for sustainable mountain tourism in this, a dynamic tourism market.

5.1 INTRODUCTION

Tourism has been an important source of employment for Nepal and a significant contributor to the national economy through foreign exchange. Nepal has officially been open for international tourism since 1951. Since then, Nepal has attracted tourists as a preferred destination for its highest mountains as Himalayas, bio-climate, nature, diverse ethnoculture and adventure in world tourism with a variety of tourism products on offer. Globally, Nepal is synonymous to the Himalaya or Mountains. Nepal is more than mountains because of its people, who are inhabitants. For hundreds of years, the government of Nepal has never prioritized the access for rural, marginalized mountain people's development, until tourism influenced them economically. However, still mountain communities need the direct economic benefits from tourism, and equally, they need more openings towards the outside world, and their young generations are only the ones that can change because of education, access of technology.

A majority of the people in the mountain areas of Nepal lives in abject poverty. Mountain Tourism in Nepal has been essentially demand driven, and local people accepted with whatever resources available to tourists for food, accommodation, transport, and other needs of visitors from around the world. Mountain Tourism development is a concept to encompass the characteristics of addressing the values of differences in perspective to the environment and its economy (Banskota & Sharma, 1998).

In Nepal, the concept for a 'Great Himalayan Trail' was outlined in the ADB/SASEC Tourism development plan. The proposed trail starts in the Karnali (West), and continues through Mustang, Manang, Manaslu, Ganesh Himal, Rolwaling, Everest, and Kangchenjunga (SNV/Nepal & ICIMOD, 2006). Nepalese mountaineering is renowned for adventure tourism due to its prominent peaks. The government has taken the policy to make it special destinations by managing mountaineering sector in an environment-friendly manner. The trend shows increasing mountaineers and revenue generation (Nepal Tourism Statistics, 2015).

In promoting sustainable tourism, the essential point is to benefit the local people, with a continuous flow of tourists to their destinations each year. However, the impediment that many tourism destinations face is that not all regions have the spectacular features that tend to attract tourists. The most popular destination: Everest, Annapurna and Langtang Regions, have thus far been the major attractions in Nepal. This is only the 'tip of the iceberg' as far as the potential for the development of tourism in Nepal is concerned. These areas, the local people and their culture attract tourists, traveling in mountainous rural areas.

It can be seen, that with growth and technology accessibility, especially mobile internet services, Nepali educated youth have been attracted toward the use of social media in connecting mostly with friends and family members. In Nepal, particularly in urban areas, entrepreneurs, institutions and organizations have slowly been moving toward social media for their personal marketing and promotion with the use of photo postings, videos, etc.. Keeping widgets is very common as it is effective for business growth. In this new age, social media can, and will play a vital role in bringing much needed sustainable tourism and with it prosperity to this potential scope of Tourism in Nepal.

More and more people are using social media (Agichtein et al., 2008). According to Graham (2005), social media are anything where users can participate, create and share content. Castells (2007) recognized the revolution enforced by social media and pointed out that we are all living in a completely new communication era. According to Nielsen (2009), two third of the people on the internet are already using social media, and it is becoming more and more over time. This also gives companies opportunities, because social media enables companies to talk with their customers (Mangold & Faulds, 2009) and Therefore, reach a big potential group. This group of consumers can experience the companies' products, as well as get to know the people who use those (Mangold & Faulds, 2009).

Social Media is an accessible, easy, direct, instantaneous method to connect with completely new and seemingly limitless possibilities that tourism professionals can tap into. Rural people, being for the most part far less familiar with the emerging new technologies compared to urban dwellers, who have benefited economically more from tourism activity. It is an intent that a larger share of the benefits of tourism will begin to flow to the rural areas, and indeed, where most of the tourism activity takes place. Social media tools and their use in the marketing and promotion of a destination and their individual trekking tours, companies go together simultaneously to enhance the relationship.

5.2 CONTENT MARKETING STRATEGIES AND SOCIAL MEDIA (SM)

Social Media is a digital means of disseminating information for varied intentions. Content making and usage in social media is very crucial in marketing strategy. Content without value is wasted information. The S.M strategy should be the increase in friendly contacts and engage members. Bringing as many as possible into the loop and keeping effective content, i.e., Value creating Content, these are the major considerations for strategists.

The prime characteristic of the sustainable tourism, the foremost necessity, is a unique local product that has the capacity to attract tourists. This can be natural (mountains, rivers, valleys), artificial (temples' statue, view towers, sculptors, man-made rock climbing), and anthropological (cultural). Visual objects have proven to be of great importance in content marketing, promoting strategies worldwide, with the use of photos, videos and shared emotions through various means of social media. All parts of Nepal have these anthropological/cultural/ natural attractions in abundance; all that needed is a proper marketing SM tools, strategy effectively.

With the use of social media, everybody can be famous, a new way of communicating (Castels, 2007). Time magazine voted in 2006 "You" as a person of the year. Despite the old saying of Thomas Carlyle "the history of the world is but the biography of great men." He believed that it is the few, the powerful and the famous who shape our collective destiny as a species (Time, 2006).

However, the content should be the fusion of the anthropological and natural/artificial factors since the human factor should not and cannot be disregarded. The content should be such that the harmonious, amiable relation between the traveler and local people is displayed and highlighted in this new media. Human relations are of the first importance in content marketing strategies. Human beings communicate; recall, reiterate in their communion with each other. Since not all tourism areas can be spectacularly attractive, the strategy that tourism policymakers should apply is a human bonding between the traveler and local people. Our common humanity, which transcends differences; compassion and mutual respect, will always attract clients (tourists) to return repeatedly to Nepal. Therefore, the clear strategy to adopt is to bring about a connection at the human level. This strategy should be applied in the mountainous communities.

This human relation as a primary element of marketing strategy is the 'people to people' relation between tourists and local entrepreneurs. It especially requires local entrepreneurs to take time to build harmonious conversations coupled with honest and good services, rather than accepting them as a client with a mercenary habit. '*A home away from home*' theme aptly applies in this context.

5.3 SOCIAL MEDIA PYRAMID MODEL (SMP)

Application of this SMP model tends to unite local people reinforcing a community spirit building stronger connections in a united mission. The work involves different layers for destinations to use Social Media effectively; it also brings additional jobs, peace, security, and most importantly builds a platform for the future generations by promoting tourism globally.

To conceptualize how SMP can work in destinations like Nepal, an excellent example is the Tamang Heritage Trail (THT), Rasuwa district, Nepal. The THT circuit connects a number of villages making a loop trail starting from Syaphrubesi and passes through Goljung, Gatlang, Chilime, Tatopani, Brimdang, Thuman, Timure and Briddim. Goljung, a traditional Tamang village, offers traditional culture and costumes and the ancient Monastery adds a traditional touch as shown in Figure 5.1.

In each village, there are home-stay accommodations, hotels, and teahouse operators. This becomes the first lower level of a Pyramid model. Each village represents a unit. All tourism stakeholders of this unit need to have their individual social media tools, but common to all. For a unit, there should be one social media tool with the village name where each member or stakeholder connects. With this configuration, tourism entrepreneurs can promote and market their individual products. This represents an individual layer.

The next level consists of each unit's individual social media where all villages connect to each other's unit. For instance, Tatopani Trail, Gatlang Village, Briddhim, Syabrubesi. This is a unit layer.

The top layer is the administrative functional part, which can continue to post contents like photos, updates, and news about the whole Tamang Heritage Trail. The management or administrative working system should employ the 'One body-Many Mind' system. The one body refers to the aggregate of the destination development, and many minds refer to all individuals as members of a destination. It unites, for instance, all the Tamang Heritage Trail's (THT) villages. This is a functional layer of SMP as shown in Figure 5.2.

The SMP model contains both horizontal and vertical management systems. It can be an open and closed system in using social media for the marketing and promotion for their region, and for their individual businesses

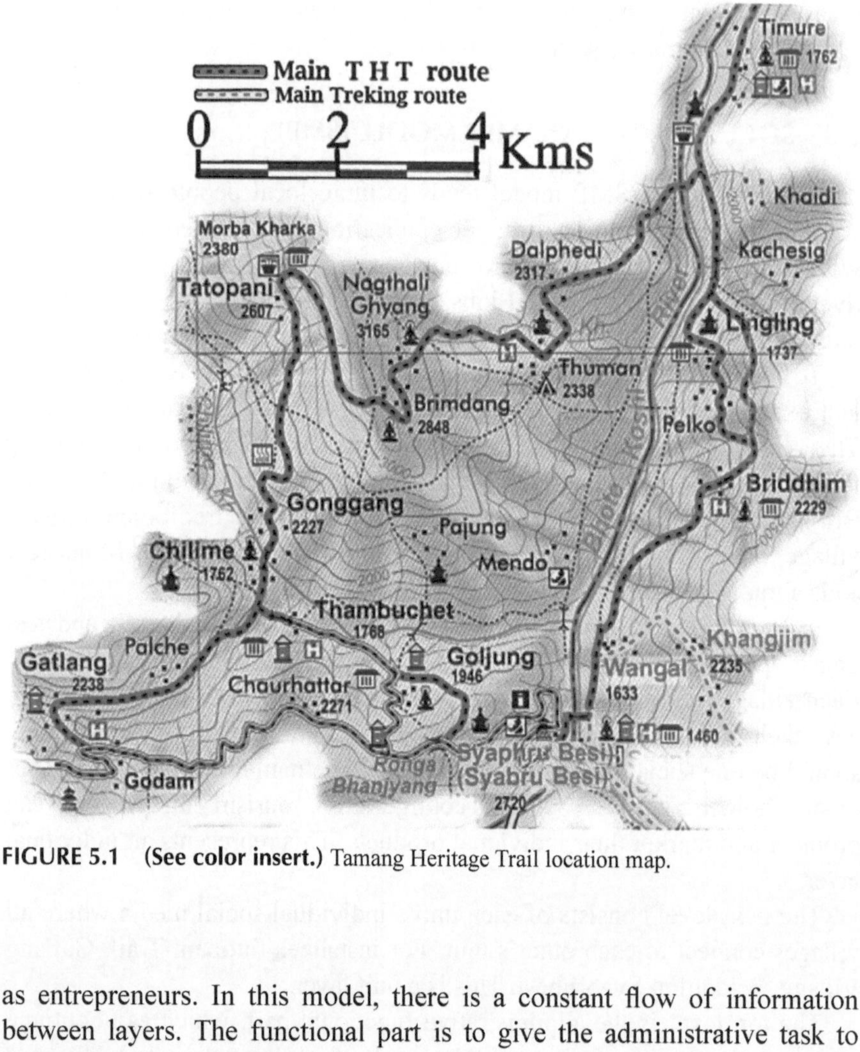

FIGURE 5.1 **(See color insert.)** Tamang Heritage Trail location map.

as entrepreneurs. In this model, there is a constant flow of information between layers. The functional part is to give the administrative task to those who play a pivotal role in the swiftness of a trekking region. The SMP integrates all entrepreneurs in their mission, goal, and objectives by dealing collectively with challenges. It means security, peace, and appreciation for the tourists traveling in their region. The SMP should establish a channel for the new training necessary for each participating tourism stakeholder regarding social media and its use, benefits, and application. The functional unit can connect, communicate with other local, national and international tour, trekking and mountaineering operators in order to promote and market their region as a tourism destination.

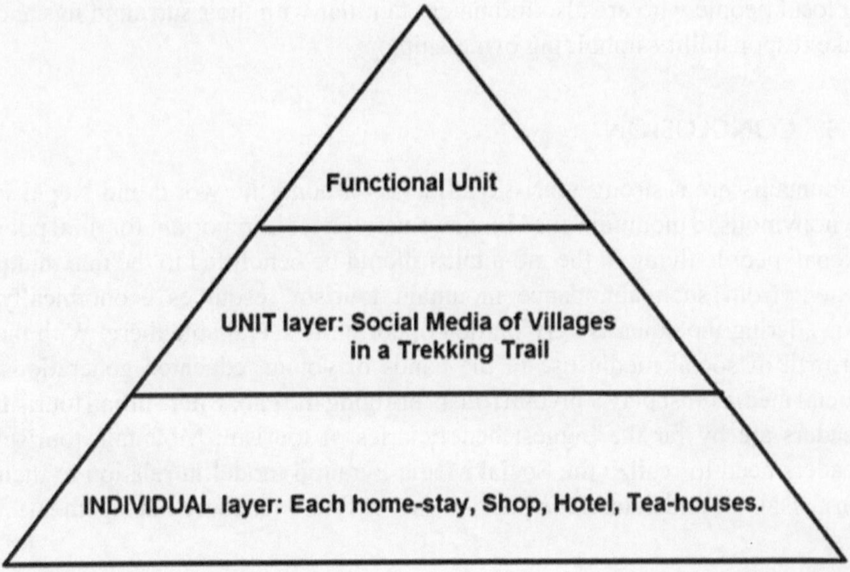

FIGURE 5.2 Social media pyramid.

5.4 SOCIAL MEDIA DATABASE STRATEGY

Besides the change in use, there will also be a change in organizational communication processes, and that is where the structuration of Poole and McPhee (1985) theory comes into play. The process of structuration is about the development maintenance and change of structures. Despite the importance of structures in organizations, Poole's theory states that structures are not as permanent as others might think (Dainton, Zelley, 2005).

The functional unit should keep a record as a record of accomplishment database on a monthly, seasonal or yearly basis for the analysis of the tourism growth, economic and development of their region as a touristic destination. Data variables can be associated with the tourists' number on a regular basis. The theme may be on Like, Share, Posts, Reply, Message, domestic/international tourists on multivariate analysis that provide rich information. The data mining of social media is not something of the distant future as regards the tourism industry in Nepal. With the appropriate provision of training, the marketing and promotion, management, photography, computer technology use and software (basic programming, Photoshop, and web designing), map reading and navigation, statistics, database management, and research methodology. These skills as well provide additional jobs to local people who are also intimately familiar with their surroundings and take responsibility-upholding ownership.

5.5 CONCLUSION

Mountains are a strong touristic attraction around the world and Nepal is synonymous to mountain-the Himalaya natively. It is important for rural poor Nepali people living in the mountains should be benefitted to the maximum extent from such abundance mountain tourism resources economically, considering the other as very limited opportunities available there. With the growth of social media use in the hands of young, educated generations, social media must play a pivotal role in bridging the gap, where urban tourism leaders are by far the biggest beneficiaries of tourism. Mountain tourism leaders need to realize the Social Media Pyramid model in relation to their profession to understand the possibilities of future database management.

KEYWORDS

- **Himalaya (Mountains)**
- **networking**
- **social media (SM)**
- **tourism destination**

REFERENCES

Agichtein, E., Castillo, C., Donato, D., Gionis, A., & Mishne, G. (2008). Finding High-Quality Content in Social Media. *Proceedings of the International Conference on Web Search and Web Data Mining,* 183–194.

Banskota, Kamal & Sharma, Bikash (1998). Mountain Tourism for Local Development, Training Manual for Programme Designers and Implementers, ICIMOD and CREST.

Castells, M. (2007) Communication, Power and Counter-power in the Network Society. *International Journal of Communication 1,* 238–266.

Dainton, M. & Zelley, E. D. (2005). Applying Communication Theory for Professional Life, A Practical Introduction.

Mangold, W., & Faulds, D. (2009). Social Media: The new hybrid element of the promotion mix. *Business Horizons, 52(4),* 357–365.

Ministry of Cultural Tourism, and Civil Aviation. Government of Nepal, http://www.tourism.gov.np.

Nepal Tourism Statistics (2015). Government of Nepal Ministry of Culture, Tourism & Civil Aviation Planning and Evaluation Division Statistics and Research Section Singha Durbar, Kathmandu. http://www.tourism.gov.np.

Nielsen (2009). Global Faces and Networked Places.

Poole, M. S., McPhee, R. D. (1983). A structurational analysis of organizational climate. In L. L. Putnam & M. E. Pacanowsky (eds.). *Communication and Organization: An Interpretive Approach.* Thousand Oaks: Sage Publications.

SNV/Nepal & ICIMOD (2006). Great Himalayan Trail, Preparatory Study.

Time Magazine (2006). You—Yes, You—Are TIME's Person of the http://www.time.com/time/magazine/article/0,9171,1570810,00.html (accessed on 16 March 2019).

CHAPTER 6

PARADIGM SHIFT IN DESTINATION BRANDING PROCESS FROM THE PERSPECTIVE OF USER-CREATED CULTURE-HERITAGE-RELATED CONTENT IN SOCIAL MEDIA

SAMIK RAY

Reginal Level Guide, Govt. of India, Department of Tourism, Eastern Region, Kolkata, India

ABSTRACT

In the last few decades, visit to the social networking sites and social media is growing faster. Users of those sites and media stay connected online with friends and acquaintances; interact over any topic including tourist destinations, attractions, information, and products; share user-created contents like information, ideas, live experiences, comments, photos, videos, reviews, etc. Indeed most of the contents are culture-heritage and contemporary lifestyle experiences which those users gathered while on travel. Culture and heritage are the matter to be experienced; regardless of primary motivation and tangibility status. State of reality and authenticity of cultural experiences on travel or tour are posited and tested by uploading and posting user-created contents like documents, photos, pictures, and videos of experiences on the walls of Social media.

Concurrently it has been found that cultural attractions and events are considered as strong tourism magnets for centuries. Thus it could be the unique selling proposition (USP) or unique branding proposition (UBP) of destinations. The recent trend of posting user-created cultural experience contents about travel in the social media magnifies the tourism market; identifies the culture-heritage centric brand or image of destination; and makes the image building or branding activities to stand in contrast to the entire conventional marketing paradigm by marginalizing the hegemony or the direct and high degree control of tourism service producers or authorities in this regard. Therefore, a paradigm shift in destination branding process becomes evident with the growing pursuit of posting culture-heritage experience contents of travels on holiday or for business on the walls of social media. The present chapter will study this shift along with its impact on destination branding in particular and over tourism as a whole.

6.1 INTRODUCTION

Deindustrialization, a decline of manufacture based economy, extreme work pressure and fast but suffocated life of post-1950s together made the transformation to post-industrial society possible. It was marked by demand escalation for services related to leisure or recreation. Travel and tours then spread dramatically within cross-sections of people as a wider social phenomenon at large. Mobility due to tourism caused an enormous makeover in the society and national economies of many regions by virtue of its multiplier effects (Ray, 2014), and shoved the destinations to be open to encounters and interactions among travelers-tourists, hosts, and tourism entrepreneurs leading to changes, crises, and challenges in the socio-economic sphere of destinations. The most important challenge dealt with was the commoditization of destination as a whole and its space, place, culture, landscape and so on.

Both tourists and marketers perceive a destination as a tourism product. They reveal the destination's attribute in terms of cognitive and affective basis (Baloglu, and McCleary, 1999; Boo & Baloglu, 2009). In the post-1950s, several destinations strive to make a way into the tourism market. Competition rises among those destinations. Destination product is not like other consumable products; rather it is somewhat different and complex. Thus, the necessity of effective marketing, branding and image building of destination for tourism is understood.

6.2 TOURISM DESTINATION

6.2.1 WHAT DOES IT CONNOTE?

Words like the place, location, and destination are frequently used to refer to a geographical area. Expressions are apparently synonymous by usage but bear distinct contextual semantic value. Neither all places – locations are destination nor are all destinations tourism destination. While a destination is defined as a place or location where one is going or directed; a tourism destination is described as a geographical location or place that has the ability to meet tourists' or visitors' wants, needs, and requirements. Spatially it is a geographical area (Silva, Mendes, Guerreiro, 2001) or place; could be a country, a rural or urban location, a region (Davidson, & Maitland, 1997); where travelers and tourists visit and stay for recreation and leisure or non-leisure activities, consume different services, experience and gaze tangible and intangible attractions like inherent or exhibited natural and cultural novelties of the location, and use the amenities.

6.2.2 PRODUCT SENSE

Tourism destination is not a regular product since it is commonly defined by coherent structure of tourism services and products' offer (Silva, Mendes, Guerreiro, 2001) that include both man-made and pre-existing or given sets of tangible and intangible attractions, services, and assets related to natural and cultural heritage, space, goods, concepts, accessibility, activity, and amenity. It is described as an amalgam of tourism products, services, and experiences (Leiper, 1995; Buhalis, 2000) and revealed as the sum total of complex, sophisticated service and product offerings presented to the place consumers (Rainisto, 2003) like entrepreneurs, investors, tourists and travelers for buying, using, experiencing, and consumption. As a product tourism destination consists of both tangible and intangible attributes (Hailin, & Hyunjung 2011) and components that formed a comprehensive, composite, and multifaceted provision of a product, services for a highly integrated experiential consumption. (Silva, Mendes, Guerreiro, 2001; Cazes, 1994; Thibaut, & La ville 1994) Sometimes tourism destinations are termed as place product when a group of different and varied stakeholders; either independently or collectively; involve in shaping a normal location or destination into a commercially viable product of tourism. Broadly speaking major stakeholders are government, travel-tour organizations and enterprises, local residents, and entrepreneurs from accommodation as well as transport sectors (García, Gómez, & Molina, 2012) who participate; either in isolation or together; quite effectively in preparation, production, marketing, and supply of varied products and services for tourism consumption. They venture on possible offers for the visitors; putting effort for visitors' satisfaction; (Ashworth, & Voogd, 1995; Ray, 2005) attempt to develop varied tourism business using assets and resources of the location and so on.

6.2.3 CHALLENGES IN MAKING A DESTINATION VIABLE FOR TOURISM

In the highly competitive scenario, the scale of success in making a destination viable in terms of tourism largely depends on the quality, quantity, and diversity attributes of offers and image. Among offers destinations' given set of assets and attractions; which consist of both tangible and intangible (Hailin, & Hyunjung 2011) geographical, natural, cultural, historical, economic, and socio-demographic elements; are key to image or brand building. Attributes of destinations' assets and attractions thereby contribute largely to shaping a normal destination workable for tourism, but organizations and individuals involved in shaping have no control over those attributes and its management. Over the time attributes may change due to environmental reasons and cross-cultural encounters and interactions between strangers and hosts. Destination organizers have little control over those changes. Both uncontrolled and little controlled issues put a barrier in turning a normal destination into a tourism location.

Destination seems to be the most complicated entity when it is treated as the tourism product because of the variety and complexity in stakeholders' fabric. Stakeholders involved in production, development, marketing, and management of tourism destination with their specific needs, interests, and goals. Thereby getting those together under one umbrella to make a place or location developed and marketed as tourism destination would certainly be a challenge.

Most of the destinations are also highly sensitive to certain natural, geographical, political and socio-economic contexts or happenings. Climate is a constant chronic context that determines and controls tourist and travelers flow to a destination. Natural calamities like flood, earthquake, famine, epidemic; socio-political happenings like rebellion, refugee migration, terrorism, military oppression, political sanction; socio-economic crises like an economic recession or depression, liquidity crisis, demonetization, drop in GDP are non-chronic, unforeseen, temporary context. Like chronic contexts non-chronic too have bearings on tourist flow (Ray, 2005). All those contexts develop either constant or temporary seasonality syndrome. This syndrome restricts the act of converting a place to a tourism location and then it's marketing too as organizations and individuals engaged in image building and marketing have no control over it.

Attractions, space, idea, concept, goods, services, activities, and amenities together make a bouquet of interrelated offers. A destination put forward those offers as its product in the world tourism market. Offers are largely experiential and vital in destinations' image building for tourism. Among offers services are the most important. Service production and its consumption experience are extremely heterogenic as psychological settings, and socio-economic background of all consumers is never being alike; productions with the same mindset constantly cannot be possible; physical settings of production, supply, and consumption too cannot be similar always. Consumption experience of the same service thus varies consumer to consumer. So heterogeneity as well holds back destinations' brand and image building activity for tourism to an extent.

There are other constraints besides those being mentioned here. While some constraints are regional or local others are global. Some are inherent and intrinsic. Others are extrinsic, accidental, incidental, and inadvertent. So the magnitude of constraints in shaping a destination as tourism location cannot be overlooked and becomes more crucial when it coupled with high competition. Both together affect image building, branding, and marketing of a tourism destination immensely. Shaping and commoditizing a destination for tourism in the perspective of little or no control condition over most of the constraints and reality of competitiveness are possible only by putting emphasis or focus on the destination and its consumer.

Visitors, tourists, travelers, and excursionists are considered as consumers when a destination as a whole is a commodity. Shaping, image building, branding, and marketing strategy have to understand the destination and what it can offer. Alongside knowledge about changing perspective and varieties of tastes, choices, demands of consumers are of paramount importance too as in the 21st-century consumers are increasingly value the individual gaze, visits undertaken for experiences (Ray, 2017), natural and sociocultural sustainability of the destination (Middleton, & Hawkins, 1998), and time and cost advantage of consumption.

Multidimensional character enables tourism destinations to produce varied products in accordance to the demands, expectations, and needs of various market segments. It may improve the position of destination in a competitive market but cannot create the destination identity as product offers are almost similar to some other destinations. Positioning around nearly alike product offers to make consumers' choice decision difficult and restrict to price offers only. The most important element of pricing a destination product offer is its unique quality or attribute. Uniqueness in product offer differentiates a destination from the rests; facilitates the price difference; makes destination's position stronger in the highly competitive situation too. Concurrently one destination cannot match all the demands, needs, and wants of all the target markets as tourism services and activities of a destination are restricted mainly by given sets of assets, resources, and attractions. So understanding and focusing unique difference in product offer are important for better positioning and unique point of identity that could be the USP or UBP of destination. Hence differentiation based on uniqueness in product offer facilitates a destination to develop its image or brand as a competitive advantage in the market.

6.2.4 IMAGE BUILDING OR BRANDING

6.2.4.1 BACKDROP

Branding is practiced over thousands of years but the term and concept by which we relate it today were not known. The earliest evidence of it could be traced in the act of marking livestock and herbs, but towards differentiation watermarks, hallmarks, blind stamps were used since the 13th century. Use of a mark at a wider scale as a brand was felt with the growing mass market. By the late 19th and early 20th century, it got connected to the sense of trademark widely. Since 1940s, manufacturers began to feel that consumers develop a social, psychological and anthropological association with the brands; they were using; though researches and studies on tourism destination branding became apparent only in the late 20th century.

6.2.4.2 APPROACHES

As tourism destination offers an amalgam or combinations of varied products and services; branding and image building become different and difficult from any singular product branding. Branding a tourism destination involves creation of a name and tagline, designing a symbol or logo and formation of any sign or combination of those to establish emotional and commercial relation between destination and tourists (Kavaratzis, 2007) to pledge memorable or unforgettable experience at destination as expected by the tourists, represent a promise of value, facilitate differentiation in the market, differentiate a destination from its competitors, build image, and create identity of the destination (Blain, Levy, & Ritchie, 2005; Hailin, & Hyunjung 2011; Morgan, Pritchard, & Piggott, 2002; Kotler, & Gertner, 2002; Kotler, 1991; Keller, 1993). Thus, the brand is not merely a visual symbol; rather it is the core value, reputation

(Gilmore, 2002), and image of destination rooted in reality and truths that facilitate destinations' positioning or repositioning in the market with an aim to convey a clear idea or sense of the destination. In practice a handful of initiatives able to understand the social, economic, emotional, and ethical value of brand beyond mere using of graphic symbols like the logo, catchphrases, and slogans.

In the conventional system tourism destination branding is developed usually within the direct control of producers, product developers, marketers, and anyone or any group else responsible and accountable for marketing, branding, and promotion of destination. Brands are frequently linked to attributes of given or existing and manmade sets of destinations' attractions, services, and assets.

Brunei is called as 'The Abode of Peace' but to pull the tourists, Brunei uses an expression, 'Kingdom of unexpected treasure,' a brand catchphrase or slogan. This syntactic construction attempts to create a specific promise imagery of the hustle and bustle free escape amid unspoiled lush jungle greenery and rich culture and heritage in the mind of contemporary visitors or convey the same content of promise to the future tourists of Brunei. It claims to represent the generic attribute of all products, services, and perceptions related to tourism of Brunei and compel the formation or development of apparently discrete and separate products and services to be aligned in the line of the image created by it. When the brand is predefined an integrated vision about the destination, its attractions combined with assets and the entire reality that destination would offer for the tourists' experience become evident. This approach ignores market needs as well as wants and exclusivity of discrete and specific service produces to an extent as it primarily put the common attribute of given sets of attractions and assets into focus. A high degree of control and hegemony of destination is thus obvious.

Bhutan's mythology made her known as 'Druk Yul' or 'the land of Thunder Dragon' in the world tourism market for years. It happened without much conscious effort on branding. In Dzongkha, the language of Bhutanese, 'Druk' means 'Thunder Dragon.' Accordingly, people are called as 'Druk Gyalpo," national anthem as Druk Tsendhen, King as Druk king, and airline as Druk Air. Druk appears as a national symbol. Despite those facts contemporary Bhutan has built a new, but stronger destination image or identity recently to survive within a highly competitive market condition and crises over national identity developed out of shift from an inward-looking policy to a more liberal and investment-friendly regime, Indianization of economy, use of English in education and official works, entry of global economy and culture with the growth of international tourism, and subtle threat to territorial sovereignty when the land is locked between the two bigger and militarily powerful states. Bhutan's attempt of building neo-identity in brand form, both overtly and covertly, is nationalistic and conventional as it aims to streamline and align people, place, philosophy, product, policy, promotion, process, politics, and public relation, etc. with the state's requirement of branding and developing nationalist feeling all over nation. While building new nationalistic branding leading to the national identity, Bhutan champions the concept of gross national happiness (GNH), signifies measuring the well being of the Nation. GNH receives attention around the world as an alternative to GDP. The new brand is designed by 'Future Brand' focusing the spirit of happiness in a way the state has adopted, preached, and wanted her people to follow and implement in their life. By the new branding approach Bhutan strategically tries to put a high degree of control and hegemony over everything and every activity located at and belonged to the state; overlooks the neo-reality like development of information technology and globalization in business, trade, commerce, industry, and economy; ignores exclusivity of discrete and specific tourism services to an extent and market needs, consumers' opinions, choices or wants; influences, drives, and aligns the producers and consumers opinions and choices with the line that country wants.

Within the conventional system sometimes destination brand is derived naturally or historically without any conscious effort, a high degree of control and hegemony of destination or any organization whether governmental or nongovernmental, responsible or accountable for promotion and image building of destination. For example, based on certain universally accepted historical fact, Jerusalem has branded as 'the holy city' in the world market.

Image of a destination is also developed incidentally or by the world ignorant of branding or brand. One could cite the example of Calcutta here. 19th century Calcutta was projected as the city of palaces or London of Orient or St. Petersburg of the East. Contemporary English inhabitants of the city made Calcutta known by those phrases as the city carried a resemblance to those cities architecturally and in wealth too. With the publication of Dominique Lapierre's book on Calcutta, 'La Cité de la Joie' (Lapierre, 1985) or 'City of Joy' (Lapierre, 1992) the city began to be portrayed by the title and the story of the book. In the tourism market, Calcutta eventually becomes synonymous to the expression 'the city of Joy' by the end of the last century. Apparently, the content of the syntactic construction or expression bears positive sense, but its association with Dominique's portrayal of Calcutta in the book accomplished a negative emotional relationship between the destination and its visitors. In consequence, tourists imagined the city as a hellhole of several million people crammed to live with millions of scorpions, mosquitoes, rats, and other little creatures (Ray, 2013). Experiencing the hellhole myth visually thus become the primary aspire of the visit to Calcutta through the objective reality confronts the myth as at the end visitors are either surprised by experiencing the rich culture, heritage, and architecture of the city or disheartened by not revealing the destination as it was portrayed in the book or by author's subjective interpretation. The absence of proper plan and policy of branding or a high degree of control over tourism produces and producers ease the use of catchphrase, 'the city of joy,' arbitrarily by the stakeholders towards sell and promote the destination in the tourist market. The portrayal of Varanasi by the expression like 'the Ghat City' or 'the Holiest among sacred cities' or 'Older than history older than tradition and even older than legend' is arbitrary too. Flexibility in portraying the image by varied expressions at the same time becomes apparent. Lack of proper and scientific planning and absence of direct control by any sort of authority over branding or image building of both Calcutta and Varanasi make the tourism stakeholders to use individually convenient, preferred, varied, and arbitrarily selected or constructed catchphrases or syntactic expressions at ease to create an image for destination with the aim of reaching the market. It often builds an enormous gap between image portrayed and the objective reality of the destination. Sometimes image turns into myth and confronts the objective reality as it happened in the case of contemporary Calcutta. Use of multiple and varied images for destination promotion and marketing at a time may put the market or tourists in a dilemma to have the right perception about the destination. This would become more difficult when images confront each other.

Image building initiatives for Jaipur and Jodhpur are quite unique. Use of pink stone for construction and color to paint heritage and other edifices at Jaipur in accordance with the local traditional belief that pink is the color of hospitality and a historical surmise that the city put on pink to welcome the Prince of Wales, corroborate the logic of calling and naming Jaipur as the Pink City. Similarly, Jodhpur is known as 'the blue city.' Strolling within the old town and around Mehrangarh fort will make one understand why the destination is called 'the blue city.' The color itself relates to local heritage, myth, history, and signs of social stratification. Consistent long-term associations with the same catchphrase become possible only because both cities pursued a participative-oriented destination image building process (Colomb, & Kalandides, 2010; Van Assche, & Lo, 2011; Zenker, 2014).

Tourism destination's image building, portrayal or branding, as it is cited in the above examples, is quite often impeded by political and bureaucratic interferences, high degree of control or hegemony of state and national tourism bodies or product and producers, inability to perceive the attributes which is uniquely of destination and inherently competitive (ETC/UNWTO, 2009), arbitrary construction of expression and image content, lack of balanced representation (ETC/UNWTO 2009), disagreement between portrayed image and objective reality, negative reputation, less or no attention to market needs, consumers' choices or wants, and exclusivity of particular service, less or no interest in dynamic interaction with the potential or past visitors over destination (ETC/UNWTO 2009), and failure to understand the significance of neo-information, technological revolution, and globalization.

6.2.4.3 *IMAGE BUILDING IN E-MOBILITY ERA*

Since 80s and 90s of 20th century, we began to inhibit in e-mobility era when Web-based social communication set in motion due to growing physical alienation or loneliness with the rapid fall in traditional way of spare time usage, demand escalation for individual choices/tastes/ needs and privacy, dependency on digital resources, and an increase in variety of products and services. Information, ideas, concepts, images, people, product, money, and many more are moving to and from; within and beyond space, relations, time (Urry, 2002) through online apps and social digital media, a set of Internet-based applications build on the technological foundations of Web 2.0 in the realm of Information Communication Technologies. This set of online platform enables individuals and organizations increasingly networked with millions of others for dynamic interactions. The platform is used for promotion and presentation of tourism destination too. Government and non-government authorities including entrepreneurs and organizations concerned with tourism discretely or in combination with other components do branding and image building of destination, market and distribute varied assets, attractions, products and services under destination's name as offers for and to the market, and carry out online booking and payments etc. through web technology. Web-based tourism business communication made a destination with its tourism related information to reach more people and attract more tourists beyond time and space than it did through print media and television like audiovisual media. UNWTO (1999) too encouraged the use of the web as a crucial distribution channel and promotion tool in the tourism field. Subsequently, Worldwide online booking had increased by

37% in 2000 and by 59% in 2001 (UNWTO, 2005). While promoting and marketing a destination for tourism or building its image, tourism authorities and organizations mostly use the new technology to construct, structure and create the content of information in this regard through texts, photos and videos on the web in a way they think a destination is to be portrayed for tourism. Authenticity is often being challenged when such way of portrayal mystifies and amplifies the image, minimizes and over-amplifies the reality, and indulges pseudo image or myth and subjectivity like self-feelings and biasness over objective reality. They wanted tourists should perceive the image of a destination in the way it is portrayed on the web by them. Therefore, high degree control of tourism authorities or organizations in image building and branding continues and authenticity issues are in question even after web revolution.

Online or virtual participants remain at the core of a social or online network of dynamic interactions and participants together form an online or virtual community. Neither community would subsist, nor would communication or interaction occur without participants. A large number of individuals, groups, and organizations, spread far and wide, participate actively in virtual communication through social media. They are able to come together, generate and share varied contents such as texts, photos, pictures, and videos about ideas, concepts, information, and experiences through and on the web. Sometimes they involve in business transactions through the web too. a report (Nielsen Report, 2009) states that two-thirds of online global population use the social media or messenger media like Facebook, Twitter, YouTube, Blog, WhatsApp, etc. and above 50% of 3 billion social media users use Facebook. Social media thereby play a major role in social change, altering social patterns, economic development, political progress, cultural change, social revolution, business development as well as transactions, etc.

6.2.4.4 PARADIGM SHIFT IN IMAGE BUILDING

Social media has given opportunities to its users to obtain, exchange, create, and post information and experiences about destinations and to be in dialogue or dynamic interactions to present its image in the world tourism market. Most reports and statistics agreed that more than 70% of the users' generated contents are related to vacation travel or travel plans. A great number of social media users, not less than 50%, are inspired by those contents while deciding their travel plans (Morrison, 2015). Virtual participants' role thus becomes crucial and effective for communicating the image of a destination and its offers though authorities and organizations involved in image building and promotion of a tourism destination do not understand its relevance yet.

In e-mobility era increased web dependency marginalizes the use of traditional media or sources of information for promotion or image building. This would probably be the earliest example of a paradigm shift in the field of tourism business communication, promotion, marketing, and branding but the most crucial one is the change in established direction of information flow which empowers virtual tourist community to create, share, and communicate information and generate varied contents while in regular dialogue or dynamic interactions on social media, and reduces dependence on tourism authority and organization created promotional contents that usually are illustrated through web or traditional media.

The symbiosis between travel-tourism and ethnocultural contents is not new. It could be traced from the pre-Christian era (Ray, 2017, 2008; Swain, 2008). Travel- tour organizers, also use ethnocultural and heritage contents to pull the tourists since the days of Grand tour and Cook's tour. As the sole distinguishing element of a destination and its people, those contents become a major driver of destinations' attractiveness and competitiveness (Ray, 2017; Mckercher, 2002; Mckercher, 2003; OECD, 2009; Urry, 1990). No two destinations share exactly the same heritage and ethnocultural traits. For example, the tradition and nostalgic heritage of Rajas and Maharajas which Rajasthan still bears can differentiate it from other destinations of the world. Hence it could be the USP or UBP of the destination. This concept would be applicable to any destination. Understanding ethnocultural contents of a tourism destination mean an authentic perception or experience of lifestyle or ethnoculture and heritage of visiting destination on travel or tour. Getting such experiences in lieu of just having glimpses of attractions becomes the contemporary trend of the visits. It is apparent too in memory sharing of visits as nearly 75% of users' generated vacation travel contents in social media are either lifestyle experience on holiday or culture-heritage content of visiting destinations. So culture-heritage continues to be the major tourist puller in e-mobility era.

Ethnocultural content is embedded within tangible and intangible elements; illustrated through living cultural tradition and nonliving representations; inherited by or passed to the descending generations. The dynamicity of ethnoculture is bestowed in changes and modifications in response to the sociocultural and historical evolution of community which bears it. Visiting folk too perceive and experience this dynamicity as its reflection is found in their practice of communicating and sharing the memory of contemporary visits in social media through posting and uploading of ethnocultural and lifestyle experience related contents. Due to regular access to those contents, the meaning of culture in the mind of potential contemporary tourists turns out to be critical, varied, and complex. Thus their desire of visit for authentic cultural feel with special longing for lifestyle experience leads to shift in tourist gazes (Urry, 1990) from 'visits to cultural attractions' to 'visit undertaken for cultural experience' (Ray, 2017), collective to individual, bubble feel to authentic lifestyle experience, and conventional feel to contemporarily relevant experience of heritage-culture-history relation. As lifestyle or ethnocultural experiences of a contemporary visit are uploaded and shared through tourists' created pictures, documents, photos, videos, and any other live activity contents on the walls of Social media; the state of reality and authenticity of posted contents are established, posited and tested in relation to other experiences. This has brought the new generation tourists into conflict with tourism authorities, entrepreneurs, and organizations over urge for new product and service expectations, the creation of pseudo image or myth, over-amplification and arbitrary attempts in image building, sense of authenticity, and approaches towards the ethnocultural identity of the hosts. Tourists' search for the deeper meaning of holidays and building of authentic perception of visiting destination's image emerges out of those conflicts and dynamic interactions of virtual tourist community in the social media.

The virtual tourist community in dynamic interaction allows us to know and understand varied views or perceptions that potential guest and the hosts having about the destination. Guests' view is the external perception while the host's view is internal perception. If the image is determined by the sum of both perceptions, it would certainly be authentic; happen outside the direct control of tourism service producers, state and national tourism bodies; match the image and objective reality; resist over-amplification, creation of pseudo image or myth, and arbitrary attempts towards image building and branding. Mostly hosts cannot perceive what attribute is unique to their destination because they either over- amplify destination's attributes or overlook those as habitual or usual. Still, it is important for the overall perception of a destination. In fact, judging or perceiving an incidence while being present at the contemporary time or measuring the range and height of a mountain while being positioned on the same would not be possible. One needs a time gap or to be placed on any onward or fore point of a vertical time bar to understand the incidence or has to stand away to feel its significance and magnificence. Similarly, the guests can understand a destination and its ethnocultural features more objectively than hosts as their gaze is not habitual or usual though sometimes it is overshadowed by their individual subjectivity as it happened in case of Calcutta. Repeat visitors can capture the dynamicity of destinations' ethnocultural attributes objectively as their visits take place in different points of vertical time bar with a reasonable time gap. Tourism authorities and organizations could obtain and then transform the perceptions of guests and hosts as input towards their effort of creating an authentic and dynamic image of destination by creating interactive virtual community environment within social media domain developed by them. Virtual communities can be used to educate visitors about the destination image and brand philosophy too. Hence the sum of objective and subjective, hosts' and guests' perceptions which emerge out of dynamic interactions among virtual participants of social media provide a crucial input to the process of constructing primarily an image and then the brand of a destination. Some destinations and travel companies have already been benefited in image or brand building, promotion, and marketing endeavor by adopting virtual community perceptions about destination posted in virtualtourist.com, TripAdvisor, Travelocity and Expedia Inc., and lonelyplanet.com, etc.

6.3 CONCLUSION

In the pre-industrial state word of mouth publicity done by the visitors after a visit was the sole way of destinations' image building. It was limited within face-to-face interaction among people of visitors' usual home place. In the industrial state communication of information to a greater number of people from the larger area at a time became possible with the successive spread of print, audio, and audio-visual media. Thus interaction in person and the word of mouth publicity are marginalized. The flow of information at that state became unidirectional. In this respect, it was from creators of information like tourism authority/marketers/brand builders to reader/listener/viewers who could be the potential tourists. Consequently, high degree control of travel-tour business authorities over shaping the visitors' opinion and perception about the destination image became easier. All those shifts together made a crucial change in destination image building and then branding initiatives which were completely different from the experiences that society had in the pre-industrial state in this regard.

In the post-industrial e-mobility state, web 2.0-based online interactions led the latest and most crucial shift which left an immense impact over the greater periphery of civil society. The shift is perceived primarily in swings from traditional to deeper and new meaning of spare time usage; traditional social connectivity to individualized or personalized social connectivity marked by demand escalation for individual choices/tastes/ needs and privacy; loneliness or alienation to the pursuit of socially networked communication beyond space and time. Socio-technologically it becomes apparent in the marginalization of traditional media or sources of information, increased dependency on digital resources, and dynamicity of interaction through social networking on the web. Shift becomes evident at all spheres of society and life including tourism. It becomes crucial in tourism business communication when tourists' created and generated travel-tourism contents impact immensely over marketing, image building, branding of a destination and purchase, payment, receipt of services and products. It is also evident in online dynamic interactions as maximum numbers of posting on social media are either selfie or contents related to lifestyle, and culture-heritage experiences on vacation or travel for business and at least half of the virtual participants of dynamic interactions are motivated by those travel-related contents while doing their travel plans.

Power of creating and generating such contents is then shared between individuals and organizations, but tourists' generated travel related contents get priority over similar contents created by authorities like national tourism bodies or tourism business organizations because factual basis of vacation information including image issues of destination and authentic experience on holiday are being prioritized. The authenticity of all the travel and tourism contents, whether created by tourism authorities or generated by tourists, are to be established only by contemporary cross uploading and sharing of tourists' experience contents like documents, pictures, photos, videos or any other live activity content on the walls of Social media. Regular access to those leads to eventual shifts in the tourist gaze. This shift from 'visits to cultural attractions' to 'visit undertaken for a cultural experience," collective to individual gaze, standardized package to tailor-made tour program, bubble feel to authentic lifestyle experience, and conventional feel to contemporarily relevant experience of heritage-culture-history relation become apparent and crucial. This has created a new generation tourists. Visitors' perception about destination's image; offers; and qualitative attributes of services combined with hosts' self-perception; reflected in dynamic interactive social networks through users' generated contents, yield valuable and authentic input to the image building and then branding of a destination. Consequently, the sum of guests and hosts perceptions definitely develop an authentic insight about destination's ethnocultural perspective among all concerned for tourism, identify the culture-heritage centric brand image, match the image and objective reality, and resist over amplification; creation of pseudo image or myth; and arbitrary attempts towards image building and branding. Hence the line of difference between tourists and brand authorities in image building and branding process is blurred too when marginalization of tourism authority and organization created promotional, and branding contents to an insignificant and powerless position became inevitable. Vibrant discussions, new ideas, and constantly changing content distinguish online communities' contribution towards destinations' image building from the brand developed in the tourism authorities' Web pages. It has also enabled NTOs to stand in contrast to the entire conventional marketing paradigm where high degree of control and hegemony of either product or producers was clearly apparent, enter into a dialogue with potential visitors, perceive destinations' image reflected in virtual tourist community interactions, understand their desires, and obtain immediate feedback in the way that the destination is perceived and performs, as never before. Therefore, a paradigm shift in the initiatives of destinations' image building and branding becomes evident with the growing pursuit of posting users' created culture-heritage content related to travel on the walls of social media.

KEYWORDS

- **culture-heritage content**
- **destination branding**
- **destination image building**
- **lifestyle experience**
- **users' created content**

REFERENCES

Ashworth, G. J., & Voogd, H. (1988). Marketing the City, Concepts, Processes and Dutch Applications. *Town Planning Review, 59*, 65–79.

Ashworth, G. J., & Voogd, H. (1995). Selling the City, Marketing Approaches In Public Sector Urban Planning, Belhaven Press: London. pp. 1–60.

Baloglu, S., & McCleary, K. (1999). A Model of Destination Image Formation. *Annals of Tourism Research. 26,* 868–897.

Blain, C., Levy, S. E. & Ritchie, J. R. B. (2005). 43 Destination Branding, Insights and Practices from Destination Management Organizations. *Journal of Travel Research.* (May), 328–338.

Boo, S., Busser, J., & Baloglu, S. (2009). A model of customer-based brand equity and its application to multiple destinations. *Tourism Management. 30,* 219–231.

Buhalis, D. (2000). Marketing the competitive destination in the future. *Tourism Management. 21*(1), 97–116.

Cazes, G. (1994). À Propos Du Tourisme Urbain, Quelques Questions Préalables Et Dérangeantes. Les Cahiers Espaces, Décembre 30, 26–30.

Colomb, C., & Kalandides, A. (2010). The 'be Berlin' campaign: old wine in new bottles or Innovative form of participatory place branding?. In Towards Effective Place Brand Management-Branding European Cities and Regions; Ashworth, G., & Kavaratzis, M., Eds., Edward Elgar Publishing: Cheltenham, 173–190.

Davidson, R., & Maitland, R. (1997). Tourism Destination, Hodder & Stoughton: London.

ETC/UNWTO. (2009). *Handbook on Tourism Destination Branding: With An Introduction by Simon Anholt,* ETC/ UNWTO: Madrid.

García, Juan A., Gómez, M. & Molina, A. A. (2012). Destination-Branding Model: An Empirical Analysis Based on Stakeholders. *Tourism Management. 33*, 646–661.

Gilmore, F. (2002). A Country- Can it be Repositioned? Spain- the Success Story of Country Branding. *Brand Management. 9 (*April), 281–293.

Hailin, Q., Hyunjung Kim, L. & Holly Hyunjung, Im. (2011). A Model of Destination Branding: Integrating the Concepts of the Branding and Destination Image. *Tourism Management. 32*, 465–476.

Kavaratzis, M. (2007). City Marketing: The Past, the Present and Some Unresolved Issues. Geography Compass. *1*(3), 695–712.

Keller, K. L. (1993). Conceptualizing, Measuring, and Managing Customer-Based Brand Equity. Journal of Marketing. *57* (January), 1–22.

Kotler, P., & Gertner, D. (2002). Country as Brand, Product and Beyond, A Place Marketing and Brand Management Perspective. *Special Issue Brand Management. 9* (5), 249–261.

Philip H. Kotler, *Marketing Management: Analysis, Planning, and Control,* 8th ed., NJ: Prentice-Hall, Inc.: Englewood Cliffs, 1991.

Lapierre, D. (1992). *City of Joy; Arrow*: New Delhi.

Lapierre, D. La Cité de la Joie, (1985). Éditions Robert Laffont: Paris.

Leiper, N. (1995). *Tourism Management*, RMIT Press: Melbourne.

Mckercher, B. (2002). Testing a Cultural Tourism Typology. *International Journal of Tourism Research. 5*(1), 45–58.

Mckercher, B., & Du, Cros. (2003). Towards a classification of Cultural Tourists. *International Journal of Tourism Research. 4*(1), 29–38.

Middleton, V., & Hawkins, R.. (1998). Sustainable tourism: a marketing perspective, Butterworth- Heinemann, Oxford, pp.7–10

Morgan, N., Pritchard, A., & Piggott, E. (2002). New Zeeland, 100% Pure. The Creation of a Powerful Niche Destination Brand. *Brand Management. 9* (4–5), 335–354.

Morrison, Kimberlee. (2015). *Social Media and Travel Go Hand in Hand (Infographic) Adweek [Online] August 7.* www.adweek.com/digital/social-media-and-travel-go-hand-in-hand-infographic/ (accessed August 9, 2017).

Nielsen Report. (2009). Global Faces and Networked Places, *A Nielsen Report on Social Networking's.* www.nielsen.com/content/dam/corporate/us/…/2009/…/nielsen_globalfaces _mar09.pd (accessed July 2, 2017).

OECD (2009). The Impact of Culture on Tourism; Organization for Economic Co-operation and Development: Paris, France.

Rainisto, S. K. (2003). Success Factors of Place Marketing, A Study of Place Marketing Practices In Northern Europe And The United States. Doctoral Dissertation, Helsinki

University of Technology: Institute of Strategy and International Business, Helsinki. Disponível em URL, http://lib.tkk.fi/Diss/2003/isbn9512266849/isbn9512266849.pdf, (accessed May 15, 2017).

Ray, S. (2005). Seasonality: *An Overview. Tourism Theory and Practice. 3*(2), 7–25.

Ray, S. (2008). Cultural Tourism or Culture based Tourism: An Analytical View. *Tourism Theory and Practice. 6*(1), 7–21.

Ray, S. (2013). Journalistic Writings-Texts on Travel/ Tourism- Traditional Literature Connections. *Tourism Theory and Practice. 11*(2), 47–74.

Ray, S. (2013). How Tourism Could Make Use of History. *Tourism Theory and Practice. 11*(1), 80–119.

Ray, S. (2014). Tourism and Heritage interaction in the Post Modernity State. *Tourism Theory and Practice. 12*(1–2), 70–83.

Ray, S. (2017). Challenges towards Integration Process of Tourism and Culture in the Indian Perspective. *International Journal of Economics and Management Science. 3*(2), 23–30.

Silva, J. A., Mendes, J. & Guerreiro, M. M. (2001). A Qualidade dos Destinos Turísticos, dos Modelos *Aos Indicadores. Revista Portuguesa de Gestão.* 1, 65–81.

Swain, S. K. (2008). Cultural Tourism Destination: Craft Village–Keshna: A Case Study. *Tourism Theory and Practice. 6*(1), 51–70.

Thibaut, A. M. La ville, (1994). *Destination Culturelle, Turisme Urbain.* Les Cahiers Espaces, Décembre 30, pp. 61–64.

UNWTO (1999). *Marketing Tourism Destinations Online: Strategies for the Information Age*; UNWTO: Madrid.

UNWTO (2005). *World Overview and Tourism Topics*; UNWTO: Madrid.

Urry, John. (2002). Mobility and Proximity. *Sociology. 36*(2), 255–274.

Urry, John. *The Tourist Gaze: Leisure and Travel in Contemporary Societies*; Sage: London, 1990.

Van Assche, K., & Lo, M. (2011). Planning, preservation and place branding: A Tale of sharing assets and narratives. *Place Branding and Public Diplomacy. 7* (2), 16–126.

Zenker, S. (2014). Measuring place brand equity with the advanced brand concept map (aBCM) method. *Place Branding and Public Diplomacy. 10* (May), 158–166.

CHAPTER 7

INNOVATIVE TECHNOLOGY ADOPTION IN FILM TOURISM

AZIZUL HASSAN

PhD Researcher, Cardiff Metropolitan University, UK,
E-mail: M.Hassan15@outlook.cardiffmet.ac.uk

ABSTRACT

There are considerable evidences that innovative technologies are adopted in films. Propelled by such adoption, films support a special interest tourism type named as film tourism. This tourism type is rather well acquainted by the tourists getting benefitted from the recent entertainment industry and easier travel access. In film tourism, entertainment in its core embraces innovative technology adoption as the basic focus. This research assumes that films are such form bringing new perceptions among the viewers and thus the research searches for feature analysis of innovative technology adoption discourse in film tourism studies. However, a persistent knowledge gap is explored in this regard. To contribute to this gap, empirical instances are aligned with conceptual arguments. Bangladesh as an emerging film tourism destination is presented as a case where face-to-face interviews and personal observation remain the major data collection procedures. Logical arguments rely on these data to sustenance the basic assumption of a relatively new research paradigm for film tourism. The research shows that innovative technologies are adopted in the film industry where Bangladesh is not a yonder of this trend. Virtual environments are created in films summoning tourists to interact with the places or objects as seen in films to promote film tourism. This research determines innovative technology adoption as rather a discourse for film tourism and make calls to researchers to contribute to this relatively new research paradigm.

7.1 INTRODUCTION

Innovative technologies are adopted in films. Such adoption is rather a distinctive feature between the old day films and the modern ones. Innovative technology adoption in films is very often meant to form new perceptions about an object, place or relevant fact in a more playful and interesting way. This feature of the modern-day films is gradually harnessing immense admiration among viewers that promote film tourism. This also outlines the necessity to suggest a relatively new discourse as innovative technology adoption for film tourism. Innovative technology adoption in films suggests to concentrate, contribute, and integrate a novel discourse for film tourism. This research is based on an identified knowledge gap in film tourism literature and brings the case of an emerging film tourism destination, Bangladesh. The research then proposes that innovative technology adoption discourse can possibly result to intense tourist engagement with sights in a film. On the practical ground, this research addresses a perceived thought that innovative technology adoption creates scope to position film tourism in viable tourism markets. This adoption can also ascribe film tourism with effective market campaigns.

7.2 FILM TOURISM: AN AREA OF RESEARCH

In theory, film tourism is the touristic activities that are induced towards viewing a moving image as encompassed in television, film or pre-recorded formats in DVD, video, Blu-ray devices and the most recent digital forms (Fernandez-Young and Young, 2008). According to researchers, this can become irrelevant to differentiate film and television as such differentiation does not necessarily effect film tourism studies. This is particularly important that films were once viewed within the cinema that moved to a typical home theatre system as video, television or even DVD. Also, the continued transformation of films to many different devices outlines that film can hardly be available in any static format than continuously embracing innovative technologies. Necessarily, there are some distinctions between television programmes and films as well as the medium by means of which these images are transmitted. This is true that conceptual debates are continued many different aspects of television and film-related research studies in film tourism literature. In that specific context, the importance of the notion of film tourism can have more significance.

Film tourism is the tourism type that deals with aspects related to the film. The interactions and processes as resulted from film tourism are rather complex and unclear. The works of Hudson and Ritchie (2006a) and Evans (1997) are crucial to understanding film tourism. There are variations in defining film tourism concept. The most common differentiation is the geographical and cultural context where some researchers tend to see this as 'film tourism' while the others as 'movie tourism' and the rest others as 'film-induced tourism.' However, these terms are interchangeably similar in their meaning. Still, in the early American cinemas, 'movie' is defined as the 'moving image.' Some critics propose that 'screen tourism' is the umbrella term that is useful as it attaches both the television media and film (Connell and Meyer, 2009; Olsberg | SPI, 2007). 'Screen tourism' can possibly reduce the awkward use of individual terms and minimizes the risk of misinterpretation of an extended phenomenon. Still, film tourism tends to be a basic term for studies and research. The other relevant point is that film tourism is very often referred to as film-induced tourism where most of the film tourists are either stimulated or induced after viewing a location on film screen that acts as the facilitator to visit that place. This can be regarded as the meaning of film tourism from a narrow space. Even though some differences exist in defining film tourism, the perspective of film tourism with all other relevant terms are mostly similar in literature. However, the basic feature of film tourism can hardly be ignored in almost all cases of film truism context that, film tourism is the bridge between film, tourists and a location as viewed on the film screen.

There are multiple ways to view the film and filmic images, but mostly in recent times, innovative technologies are interpreted in this regard. Such viewing can possibly take place in virtually enhanced and multitude environments. These dissimilarities establish that film tourists tend to perceive and then attach complete filmic experiences to have comprehensive experiences about the aspects as they watch on a film.

7.3 DISCOURSES IN FILM TOURISM RESEARCH

A critical literature review on film tourism brings out parameters in film tourism research that are rather multi-dimensional in cross-disciplinary contexts. Searching the film tourism discourses from cognate sub-disciplinary contexts is important. Film tourism needs to be viewed from a more practical context than purely general. From a more conceptual understanding, the shift of film tourism as a sub-discipline is rather an important element of film tourism discourse research. Some disciplines and sub-disciplines of film tourism can help in informing the theoretical base of film tourism discourse research and can lead to contribute and create tourist engagement with film locations, themes, and images. From a cultural, geographical perspective, film tourism discourse research can have an obvious contribution towards the theoretical progression of film tourism studies. Still, more research studies are required to explore diversities in film tourism discourses. Such research area is rather a ground to involve larger respondent participation and empirical evidences encompassing the landscapes, rural and urban settings, popular cultures, arts, consumer behaviors or psychologies as outlined in films.

Some themes are developed in film tourism offering some elements for film tourism discourse research. These themes identify several elements of interests as film tourism from a consumer perspective covering profiles and characteristics of film tourists, film tourist experiences, film tourist motivation, and film tourism destinations (Kim and Richardson, 2003). On of such theme is the film tourism effects that covers social, communal and economic implications of film tourism research. The other theme discusses the film tourism business with film tourism destination marketing, film tourism consumer marketing initiatives and film tourism marketing. Another theme is studying a place from film tourism contexts attaching conflict, power relations, people, cultures, and misinterpretation (Tooke and Baker, 1997).

In general, film tourism discourses expand in two broad areas: the first discourse trails an approach that is dominantly applied for studying film tourism. This discourse is featured in many of the emerging tourism management literature. Under this discourse, film tourism research studies are placed on a practical lens identifying the incidences and effects of film tourism. Also, this domain offers marketing and management implications, challenges and opportunities as research findings summaries. Research studies with this discourse normally include empirical studies covering specific destinations in films or film production aspects. This discourse offers an activity focused, applied and real-world observations that emphasize on recording, understanding and mapping activities of film production. Outcomes of research studies having this discourse are primarily management perspective oriented (Grihault, 2003).

On the other side, the second discourse suggests a dominant focus on theories and concepts where relevant research studies outline fundamental constructs that contribute, stimulate and explain film tourism meaning, activities, responses and behavior. Researchers belonging to this category are commonly placed within a subjective or disciplinary framework of film theory or cultural geography. This discourse contributes to the understanding of film tourism that basically originates from a cultural context where in this case, Shiel (1990: 1) argues, 'cinema is the most important cultural form.' Studies deriving from relatively under-researched contexts are related to hidden symbols and meanings as attached to the cinema. Film tourism in this perspective is viewed as a practice of culture and the interrelationships between tourism and culture (Edensor, 2000).

Further, as a contribution to film tourism discourse research, film tourism becomes significant having some features that outline its progress as a valid research area. Over the last one decade or so, film tourism literature studies are enriched with considerable knowledge generation supports by new research studies and journals. These journals feature many diverse areas of film tourism including. Some of the journals that contribute to film tourism discourse research in this time period are: Tourism Planning and Development (2010); World Hospitality and Tourism Themes (2011); Tourism Review International (2009); Teoros, Revue de Recherché en Tourism (2011); Journal of Travel and Tourism Marketing (2008); Tourism Analysis (2009); Tourist Studies (2006); Tourism, Culture and Communication (2006) and few others. Contributions of the scholars and researchers in such journals on film tourism are commonly able to create awareness and encourage scholarly activities in film tourism research. Thus, the knowledge of film tourism is simultaneously developing with supports from scholars and researchers. Other than the journals mentioned, a good number of conferences and events promote film tourism research. One of the key conferences is organized in 2004 by the La Trobe and Monash Universities named as the International Tourism and Media conference. In addition, a good number of academic works as both graduate and undergraduate dissertations, PhD theses and academic essays are enriching discourses in film tourism research.

There is a gap in existing film tourism literature in innovative technology adoption in films although technologies are adopted in films. This possible area of film tourism discourse is faster becoming a turning point of a film's popularity and business return. Discourse research agenda in film tourism can be diverse and multi-faceted. In films, space, media, innovative technology adoption, place, culture, psychology, tourism are often placed symmetrically that should allure researchers to study discourses of film tourism as an area of academic and research interest. This rather emerges as an interdisciplinary inquiry area to encourage researchers to conduct meaningful research studies. Film tourism discourse research can concentrate on some specific effects, drivers and experiences of film tourist's related to innovative technology adoption in films that they watch. Innovative technology adoption discourse can support in marketing, policy formulation and planning to ensure better and fruitful engagement of film tourists with the locations or aspects shown in films. In addition, an increasing number of industry reports and online resources can help to bring out innovative technology adoption in films as an independent knowledge discourse in film tourism discourse. Thus, film tourism can become more integrated, holistic and discourse research-oriented that clearly aligns with geographical landscape and relevant theories.

7.4 FILM, CINEMA, MOVIE, AND INNOVATIVE TECHNOLOGIES

Terms as 'film,' 'cinema' and 'movie' resemble similar meaning and have a presence in the relevant literature. According to Books (2016), films are creative works of art. Cinema is a varied medium for a wide variety of academic works (Bazin, 2005). On the other side, a movie is a controversial piece of 'high' art (Kawin, 1992). However, this research prefers to use the term film for outlining the research contexts in more manageable and controlled manners. Thus, studying aspects related to film is rather essential to outline the latest phenomenon of film tourism. The unprecedented progress of innovative technology adoption is rather revolutionary in both film and television. This adoption appears in transmitting news, documentaries and entertainment features. Some researchers argue that films, in fact, turn into one of the most crucial and globally observed cultural forms (Shiel 2001; Urry and Larsen 2011).

A good number of other discourses exits in film tourism research. Some research studies in film tourism have outlined subject areas that link film tourism and destination marketing (Beeton, 2005; Walaiporn, 2008; O'Connor, 2010; Saltik et al., 2011). Also, the effect analysis of films of a tourist destination· becomes important as it is shown in films and has also been researched in few research studies (Beeton, 2005; Hudson and Ritchie, 2006; Saltik et al., 2011). In addition, some research studies have also covered the experiences of film tourists (Roesch, 2009; Sellgren, 2011). Some research studies have focused on different films as The Beach, The Lord of Rings, Sound of Music and few others (Hudson and Ritchie 2006, Warnick et al., 2005; Roesch, 2009). Some other research studies have commonly concentrated on Western and developed countries (Beeton, 2005; Walaiporn, 2008). However, O'Connor (2010) explores a lack of research covering off-location film destinations and emphasizes on carrying further research on impact assessment of films in many other countries that are popular as tourist destinations. In another approach, Roesch (2009) accepts that specific films that result in film tourism rely on best-selling novels and thus stresses on conducting more research in this area. In very few research studies, young travelers are given main attention where they are characteristically included in films tourism (Warnick et al. 2005; Sellgren 2011; Soliman 2011). In this regard, research on students' travel behavior searches for young travelers' as film tourists' behavior, motivations, and activities in film destinations (Richards and Wilson, 2003). Studies covering youth travel behavior in film tourism mainly considers films because of the effect of that particular film. Thus, the Beach film creates an effect on students to travel to Thailand (Warnick et al., 2005) or even research about the Captain Hima movie to visit Egypt (Soliman, 2011). Researchers like Sellgren (2011) tries to trace out the motivations and film tourist' behavior by conducting focus group discussions with relevant student respondents. In such focus group discussions, films as Slumdog Millionaire (India), Tomb Raider (Cambodia), the Beach (Thailand), Lost in Translation (Japan), Blood Diamond (Sierra Leone), Sex and the City (USA), Kill Bill (Japan), Crocodile Dundee (Australia), Gladiator (Italy), the Man with the Golden Gun (Thailand), Chungking Express (Hong Kong), Notting Hill (UK), and Seven Years in Tibet (Tibet) are covered.

In a separate research, Wang (1999) studies the role of specific movies in terms of destination image promotion of four United Kingdom locations. Majority of the respondents in this research are students in the UK. In this research, one movie location is mentioned as the Bodleian Library in Oxford that appears in the Harry Potter movies for the dining hall scenes. From this location's experience, Wang (1999) concludes that the key motivation of students for visiting Oxford appears as this city's global reputation for academic activities and the movie location is added as an extra sight. As Oxford is the location, more research studies can search if the rest other movies as the Goathland Railway Station, the Gloucester Cathedral and many London based spots can bring similar results. In terms of researching solely on students' travel behavior and film tourism, the key drawback is the lack of necessary data to compare diverse film tourism activities. Apart from the Bodleian Library where the Twilight is discovered or the dining scenes are filmed, lack of research about the most recent blockbusters as the Harry Potter can become important. The Indian sub-continent has a very good reputation for attracting a massive number of film tourists. In very early of the twentieth century, faster advancement of films transformed the forms of entertainment and communication. This transformation also created a popular culture in a relatively larger way than ever before. The popularity of films among the mass population was on the peak during the late 1940s. Thus, film watching turned into one of the leading cultural activities in many societies in the developing world. A gradual shift of popularity from the developed world towards the developing became on the rise.

The term 'narrowcasting' as developed by Page and Connell (2010) is rather a trend having the propagation of cable and digital channels with downloading options using the Internet technology and even the Smart-phone devices. This trend is the outcome of innovative technological advancement that replaces view on demand (VoD) services. Innovative technology adoption is serving the greater number of population in so good ways that they are never served like this before. The declining number of film admissions has resulted from increased television viewing. Television with all other digital gadgets are becoming the key form of leisure activity and means of entertainment in many countries of the world mostly in the developed countries in Europe and USA (Page and Connell, 2010). This becomes very common that, households now have more television sets and relevant digital gadgets for entertainments to watch programmes and mostly film broadcasting on the television. Television has brought films in the household that acts as a negative factor for peoples going out to watch films. These television programmes and mainly the film broadcasting has thus turned into a key facet of the daily modern human life. Apart from this, the viewing of films with innovative technologies as DVD, Blu-ray, and even Smart television is shaped as a big industry where more than 86% of the UK households own a DVD player (Vacche, 2012).

Film watching in the sub-continent including Bangladesh is a very popular cultural activity. There is a gap in adequate data on film tourism and film watching in Bangladesh. India has the largest cinema attendance in the world. In 2008, the net cinema attendance reached at over 2.9 billion admissions. This number sharply surpassed the USA admissions of 1.36 billion (European Audio-visual Observatory, 2010). On the other side, China became the third fastest and largest market of film having a sharp growth in the global box office market. China doubled its film revenue between 2009–2011 (Screen Digest, 2016). In the United Kingdom, the number of film admission was the world's sixth largest, but this country saw a decline in cinema admission during the mid-1980s. However, this has sharply risen from a record low of 72 million in 1985 to over 164 million in 2008 as parallel to the USA and the rest of Europe. Almost, 60% of the UK population tends to watch the film for a minimum once a year and 18% population make film admissions at least once or more in a month.

7.5 INNOVATIVE TECHNOLOGY ADOPTION IN BANGLADESHI FILMS: A SHIFT

Innovative technologies are adopted in the Bangladesh film industry with an enriched historical background. The historical background and origin of the Bangla films dates to as far as 1898. In this year, the Bradford Bioscope Company appeared in the scene with their arrangement of screening a film at the Dhaka Harbor's the Crown Theatre. This was the very first film released in Bangladesh. However, in filmmaking, Bangladesh started the movement in 1913. The 'New Picture House' was the very first theatre built in the then Bangladesh. On the global perspective, in the Paris De Café on 28 December 1895, the very first commercial show started by the Lumiere Brothers. In Dhaka, the Lumiere Brothers screened their first film after 6 months, and this becomes the first established bioscope in the entire Indian subcontinent. According to Cholochitro (2016), the first ever released films in Bangladesh were: 'Greek-Turkey Battle,' 'the Jump of Princess Diana,' 'the JubliMichil,' 'the Game of Snow,' and 'the French Underground Railway.' During the silent era, the first Bengali silent film was 'the Bless the World' and this film was released on 8 November 1919. This film was directed by Jotish Banerjee from the French Madden Company. At that time, Bangladesh had 80 theatres across the country. According to Mustafa (2016), the Dhaka Royal family moved forward in 1927–1928 by producing a short film named 'the Good Girl.' Following the success of this film, the Royal Family stepped out for bigger and more ventures by establishing the Dhaka East Bengal Cinematograph Society. 'The Last Kiss' was the first Bengali full-length silent film produced by the Royal Family. Khawja (2016) stated that, by the year 1947, around 80 cinemas were present in the then Bangladesh. The first Bengali initiative in the 1980s for exhibiting and producing films was named 'the Royal Bioscope Company' by Hiralal Sen in Kolkata. During this time, most of the feature films were made in Kolkata. 'Bilwa Mangal' was the first feature film in Bengali. Also, according to Hayat (2006), the Dhaka Nawab family sponsored and produced two films named 'The Last Kiss' in 1931 and 'the Sukumari' (1928–1929). Following the partition of India in 1947, the Pakistan era resulted in East Pakistan. During this time, the first newsreel was created on Mohammad Ali Jinnah's visit, and this was made by the broadcaster Nazir Ahmed in 1948. The first audible feature film with sound in East India was 'the Mukh O Mukhosh' produced by Abdul Jabbar Khan on 3 August 1956. However, complete edit, print and relevant all other works of this film were done in Lahore in the then West Pakistan. On 3 April 1957, the East Pakistan Film Development Corporation (EPFDC) was established by the then East Bengal Provincial Assembly. The first produced film by this organization was 'the Asiva' (The Life of a Village Girl) directed by Fateh Lohani.

Following a different trend of patriotism, in 1971, the documentary named 'Stop Genocide' was made on the liberation war of Bangladesh. This film directed by Jahir Rayhan was the first internationally acclaimed films of Bangladesh (Hayat, 2016). Since then, the 'Bangladeshi filmmakers are continuously making films. However, not necessarily all of them maintain the set international standard, but some of these films manage to get international recognition and awards. During the 2000s, the film industry in Bangladesh faced poor business. However, that turbulent time is gradually recovering followed by contributions of some promising filmmakers and their good quality works. Thus, the country also welcomes relatively bigger budget films to showcase the Bangladeshi culture before the world.

7.6 INNOVATIVE TECHNOLOGY ADOPTION IN FILM TOURISM DISCOURSE IN BANGLADESH

Innovative technology adoption in film tourism discourse has some empirical evidences in Bangladesh. People tend to watch films staying in their homes finding as more enjoyable than going to out personally. This is rather a significant point in researching innovative technology adoption for film tourism. Thus, film tourism is expanded from the developed world to the developing with more popularity. The rapid popularity of innovative technology adoption in film tourism is visible in the Indian sub-continent including Bangladesh. The comparatively smaller film tourism market in Bangladesh is expanding over the last few decades. Although, there are controversies and debates regarding the growth of film tourism in Bangladesh. Still, innovative technologies are adopted in Bangladesh in some well-accepted and globally awarded movies that actually contributes to promoting film tourism. This can be a positive sign of more interest creation among the possible film tourists.

Film tourism in Bangladesh has potentials. The country already has been internationally awarded for excellence but a significant lack of persists in adequate technological supports. However, the process of innovative technology adoption in films is on the rise. Also, the film industry itself or the general audience of Bangladesh have very few recognized infrastructures with global standard and class. This is the reason for which films in Bangladesh are very frequently criticized for their quality and acceptance by the general audience. This means that the lack of relevant infrastructure and innovative technological support in a real sense acts as a drawback to progress this prospective industry to an expected level of standard.

In the most recent years, the national tourism policies of Bangladesh are made favorable for tourism promotion (Hassan & Burns, 2014). On the other side conventionally, Bangladesh has a reputation for supporting different types technological innovations in many of the development areas (Hassan et al., 2015; Hassan & Rahimi, 2018; Hassan & Sharma, 2017a, 2017b; Hassan & Ramkissoon, 2017). Film cities with international standards and adequate innovative technological facilities are being implemented in Bangladesh by both the government and private entrepreneurs to meet necessities of the present time. This can propel the development of the film industry and film tourism of Bangladesh. A film city that is built for filming purpose is a studio complex equipped with all relevant, innovative technological facilities of film making. Government agencies seem to have a clear vision to implement innovative technology adoption strategies in the film industry. Thus, the development of the film city with film tourism in Bangladesh can possibly be ascribed with two main areas as technological supports and creativity supports. The film city can advance the film industry and film tourism of the country. This city can also support a filmmaker with all innovative technologies of film-making coupled with both indoor and outdoor shooting facilities. The city can also benefit a filmmaker with post-film production facilities and serve the entire film artist community of Bangladesh. Thus, the creation of a film city in Bangladesh can obviously become an approach for favoring prospective filmmakers of the country resulting in some very good quality movies with international standards. The film city with adequate and necessary technologies and equipments can obviously change the way of making films and thus serve the entire film industry with promoting film tourism. This city can also become the destination for film tourism if proper planning and policies are executed. This city can particularly help the filmmakers by offering necessary and updated services. Thus, the time length for filmmaking can be reduced significantly saving both expenses and time of the filmmakers. This city can also attract foreign peoples involved in filmmaking as well as film tourists. Innovative technology adoption is very common and becomes an essential feature of film making in these days that have effected both the making and marketing of films across the world. Innovative technological application in film making and viewing have encompassed so many gadgets and advancements that films and technological innovations have turned as inseparable these days. In developed countries, filmmaking, production, marketing and almost all areas related to a film are very intensively based on innovative technology adoption. However, in reality, a technological divide exists between the developed and developing countries making the availability and adoption of innovative and updated technologies. A film city having equipped with all necessary innovative technologies can thus be a good scope for film tourism promotion and to attract film tourists from different parts of the world. Innovative technologies thus can become a possible platform for promoting film tourism in Bangladesh.

7.7 CONCLUSION

The importance of researching innovative technology adoption in film tourism is growing in different counties across the world including Bangladesh. One of the main reasons for such popularity is the interest and participation of the general population of Bangladesh with film-related activities. The aim of this study is to analyze the context of film tourism to propose a relatively new discourse as innovative technology adoption in film tourism in relation to marketing campaigns. Findings outline positive interrelationship analysis and effect exploration of innovative technologies in promoting film tourism discourse. This research had a reference to the particular context of Bangladesh, and the research thus reaches a conclusion that there are several discourses remain valid in film tourism literature covering many different areas. However, innovative technology adoption is a promising area that is yet to attract the attention of diverse researchers. Thus, this research explores the importance of introducing technology discourse for film tourism research. The Bangladesh context has outlined that innovative technology adoption in a film city leads to promote film tourism to an expected level. This research relies more on empirical evidences than conceptual that is rather a limitation. Future research studies should have concentrated on theorizing innovative technology adoption discourse in film tourism with special reference to developing countries.

KEYWORDS

- Bangladesh
- discourse
- film
- technology
- tourism

REFERENCES

Bazin, A. (2005). What is cinema? Berkeley: University of California Press.

Beeton, S. (2005). Film-induced tourism. Clevedon: Channel View Publications.

Books, J. N. (2016). What is film? New York: Peter Lang Publishing Inc.

Cholochitro. (2016). Home. Retrieved from: http://www.cholochitro.com/index.php?option=com_content&task=view&id=33&Ite mid=49 (accessed: 9 December 2016).

Connell, J., & Meyer, D. (2009). Balamory revisited: An evaluation of the screen tourism destination-tourist nexus. *Tourism Management. 30*(2), 194–207.

Edensor, T. (2000). Staging tourism: tourists as performers. *Annals of Tourism Research. 27*(2), 322–344.

European Audiovisual Observatory. (2010). Focus: world film market trends. Strasbourg: EAU.

Evans, M. (1997). Plugging into TV tourism. Insights March D35–38. London: English Tourist Board.

Fernandez-Young, A., & Young, R. (2008). Measuring the effects of film and television on tourism to screen locations: A theoretical and empirical perspective. *Journal of Travel and Tourism Marketing. 24*(2–3), 195–212.

Grihault, N. (2003). Film tourism – the global picture. *Travel & Tourism Analyst, 5*, 1–22.

Hassan, A., & Burns, P. (2014). Tourism policies of Bangladesh – a contextual analysis. *Tourism Planning & Development. 11*(4), 463–466.

Hassan, A., & Rahimi, R. (2018). Addressing climate change effects on coastal tourism in St. Martin's Island of Bangladesh. In: Jones, A. & Phillips, M. (eds.), Disappearing destinations: Climate change and future challenges for coastal tourism. Oxfordshire: CABI.

Hassan, A., & Ramkissoon, H. (2017). Augmented reality for visitor experiences. In J. N. Albrecht (ed.), *Visitor Management*. Oxfordshire: CABI, pp. 117–130.

Hassan, A., & Sharma, A. (2017a). Wildlife Tourism for Visitor Learning Experiences: Some Evidences on the Royal Bengal Tiger in Bangladesh and India. In I. Lima and R. Green (eds.). Wildlife tourism, environmental learning, and ethical encounters. New York, NY: Springer, pp. 155–168.

Hassan, A., & Sharma, A. (2017b). Wildlife Tourism: Technology Adoption for Marketing and Conservation. In J. K. Fatima (eds.), Wilderness of wildlife tourism. Waretown: Apple Academic Press, Inc., pp. 61–85.

Hassan, A., Ahmed, M. U., & Shoeb-Ur-Rahman, M. (2015). The development, nature, and impact of medical tourism in Bangladesh. In M. Cooper, K. Vafadari and M. Hieda (Eds.),

Current issues and emerging trends in medical tourism. Pennsylvania: IGI Global, pp. 296–311.

Hayat, A. (2016). Brief history of cinema. Retrieved from: http://web.archive.org/web/20041204114 (accessed: 9 December 2016).

Hudson, S., & Ritchie, J. R. B. (2006). Promoting destinations via film tourism: an empirical identification of supporting marketing initiatives. *Journal of Travel Research, 44*, 387–396.

Kawin, B. F. (1992). How movies work. Berkeley, CA: University of California Press.

Khawja, A. (2016). Dhaka Nawab Family and Film. Retrieved from: http://www.nawabbari.com/main_arts.html?string=lastKiss.html (accessed: 09 December 2016).

Kim, H., & Richardson, S. L. (2003). Motion picture impacts on destination images. *Annals of Tourism Research. 30* (1), 216–237.

Mustafa, K. (2016). The liberation struggles of a country and a festival. Retrieved from: http://www.dhakafilmfestival.org/the-liberation-struggles-ofa-country-and-afestival/ (accessed: 9 December 2016).

O'Connor, N., Flanagan, S., & Gilbert, D. (2008). The integration of film-induced tourism and destination branding in Yorkshire, UK. *International Journal of Tourism Research, 10*, 423–437.

Olsberg, SPI. (2007). How film and TV programmes promote tourism in the UK. London: Film Council.

Page, S. J. & Connell, J. (2010). Leisure: an introduction. Harlow: Pearson.

Richards, G. & Wilson, J. (eds.) (2007). Tourism, creativity, and development. London: Routledge.

Roesch, S. (2009). The experiences of film location tourists. Clevedon: Channel View Publications.

Saltïk, I. A., Cosar, Y., & Kozak, M. (2011). Film-induced tourism: Benefits and challenges for destination marketing. *European Journal of Tourism Research. 4*(1), 44–54.

Screen Digest. (2016). China now third largest global box office market. Retrieved from: www.screendigest.com/news/2012_01_china_now_third_largest_global_box_office_market/view.html?start_ser=ci (accessed: 15th December 2016).

Sellgren, E. (2011). Film-induced tourism: the effect films have on destination image formation, motivation and travel behavior. Saarbrucken: LAP Lambert Academic Publishing.

Shiel, M. & Fitzmaurice, T. (eds). (2001). Cinema and the city: Film and urban societies in an urban context. Oxford: Blackwell, p. 1.

Soliman, D. M. (2011). Exploring the role of film in promoting domestic tourism: A case study of Al Fayoum, Egypt. *Journal of Vacation Marketing. 17*(3), 225–235.

Tooke, N. & Baker, M. (1996). Seeing is believing: the effect of film on visitor numbers to screened locations. *Tourism Management. 17*(2), 87–94.

Vacche, A. D. (2012). Film, art, new media: museum without walls?: museum without walls? New York: Palgrave McMillan.

Wang, N. (1999). Rethinking authenticity in tourism experience. *Annals of Tourism Research. 26*(2), 349–370.

Warnick, R. B., Bojanic, D. C., & Siriangkul, A. (2005). Movie Effects on the Image of Thailand Among College Student Travelers. In J. G. Peden & R. Schuster (eds.), Proceedings of the 2005 Northeastern Recreation Research Symposium, April 10–12. Bolton Landing, N.Y: U.S. Forest Service.

CHAPTER 8

DESTINATION IMAGE OF INDIA: TRENDS AND ANALYSIS

SWATI SHARMA and NARENDRA KUMAR

Amity Institute of Travel and Tourism, Amity University, Noida,
Email: ssharma3@amity.edu, nkumar14@amity.edu

ABSTRACT

Tourism image of India has been prismatic in line with its general image, and it has emerged as a must-see destination. It had an image problem and was not perceived as a holiday destination (Seth, 2006). Exploring reasons for the slow growth of tourism in India, researchers have identified that the negative image associations with the country have been one of the main reasons that hinder progressive tourism performance (Chaudhary, 2000). Traditionally India was perceived as magical and fascinating, and for most Americans, it was associated with epidemics flood and poverty. In 2001, Indian tourist's Bureau launched a campaign to portray an image of India as a peaceful, beautiful and safe country (Eli Avraham & Eran Ketter, 2012). The change in the tourism image of India was more discerning after the launch of the Incredible India campaign in 2002. The review of the literature suggests that the researches on Tourist Destination image in the past 47 years are plenty and have focused on different dimensions of the image. The perspective and concept of destination image have changed over time. In general, there are three major emphases among different researchers on the definition of the destination image. Crompton's discussed the composite structure of the image construct. In addition, others make use of attitudinal components such as affect and cognition to explain the concept of the image. A third approach is to consider the image as an overall visual or mental impression of an object, place or experience (Rezende-Parker, Morrison, & Ismail, 2003). Then in 2010 destination image was given a different connotation of subjective interpretation of reality made by the tourist (Bigné, Sánchez, & Sánchez, 2001; Gracia, 2002). However, the studies on the image of India as a tourist destination came much later than international researches and are still limited in number, and many dimensions are yet not touched. The first study on Tourism Image of India as a destination was conducted in 2000 (Chaudhary, 2000) and Ministry of Tourism, Government of India (2012) in its study emphasized need of regular monitoring of tourist satisfaction and proper control of tourist facilities. The objective of this chapter is to assess the destination competitiveness of India on different tourist attributes.

8.1 INTRODUCTION

Tourism as an industry has grown big and diverse since 1950. In 1950, 25 million tourists (overnight visitors) traveled internationally, by 1980 this number had risen more than tenfold to 278 million, while since then it has almost quadrupled. In 2012, the number of international arrivals surpassed one billion to reach 1.035 million. With over one billion international tourists traveling the world in 2012 and an estimated five to six billion traveling within their national borders, tourism is a key sector in economies worldwide, both in advanced ones and in emerging and developing countries (UNWTO, 2013). Tourism has emerged as a prevailing factor of socio-economic development (Crompton & Fakeye, 1991; Gartner, 2005). Nations across the world understood its potential and forayed into it. Consequently, as more and more areas of the world are opened and developed for tourism, the destination choices available to consumers continue to expand (Gunn, 1988; Echtner & Ritchie, 2003; Bornhorst et al., 2010; Assaker et al., 2011). This increase in the destinations choices has developed new dynamics in the industry. Tourism is no longer primarily driven by the pull of attractions but is actively pushed through marketing strategies that entail dynamic image building, seamless experience, and tourist satisfaction. Destinations (nations) ensure that they benchmark against others to top the ranks as this helps in better management of tourism along with the enhanced image. The annual Travel and Tourism Competitiveness Report of World Economic forum ranks nations across the world for their tourism readiness based on many factors namely Policy, rules and regulations, Environmental Sustainability, Safety and Security, Health and Hygiene, Prioritization of Travel Tourism, Air Transport infrastructure, Ground Transport Infrastructure, Tourism Infrastructure, ICT Infrastructure, Price Competitiveness in the Travel and Tourism Industry, Human Resources, Affinity for Travel & Tourism, Natural Resources, Cultural Resources, etc., as the parameters. The report of 2013, India is ranked at 65th out of 144 countries with high scores on natural and cultural resources, air transportation and ground transportation, and most importantly price competitiveness thus indicating the strengths for tourism. The report also identified weaknesses with low scores in quality of roads and some aspects of tourism infrastructure that are more concerned with management than the inherent appeal of attractions. The overall and parameter wise rankings of India suggest areas requiring improvisation to enhance tourism readiness of India which is possible at the level of destination managers with the participation of all the stakeholders.

The job of destination planners and managers has become very complex in today's competitive environment. They have to decide the image and position to be established on the world tourism scene given its inherent attractions, the position of competitors, choices of potential markets, resources available and the local environment. This task of establishing a frontal image requires balancing numerous back end forces. The promotional campaigns such as 'Incredible India' at the national level; "God's Own Country" for Kerala, "Padharo Mhare Desh" for Rajasthan, "Vibrant Gujarat" for Gujarat, "The Heart of Incredible India" for Madhya Pradesh, "Heaven on Earth" for Jammu & Kashmir showcase frontal images built on the plethora of attractions and facilities.

8.2 INTERNATIONAL TOURISM: REGION WISE

Global tourism can be understood through regional tourism that shows a consistent shift towards the Asia Pacific and the declining share of Europe and the Americas. Tourism in Asia reached 23.2% of overall traffic in the year 2014 compared to 0.8% in 1950, 8.0% in 1980 and 16.3% in 2000. The share of Europe has declined from 67.2% in the year 1950 to 51.39% in 2014. Table 8.1 shows the tourist arrivals in different regions of the world since 1950.

The data in Table 8.1 shows that Europe has been leading the world tourist arrivals and was followed by America. However, trend shift is seen from 2000 onwards, and in 2010 the Asia Pacific gained over America. Arrivals in the Middle East and Africa are also growing though a little slow. Tourism in Europe is still growing albeit at a slower rate compared to Asia and the Pacific.

8.3 INTERNATIONAL TOURISM: COUNTRY WISE

International tourism of countries and regions is often interlinked. Asian countries China, Thailand and Malaysia started figuring among top international tourism destinations at the time when tourism was shifting towards the Asia Pacific. Table 8.2 shows the top ten ranked countries of the world in terms of international tourist arrivals at different points of time from 1950 to 2014.

TABLE 8.1 Region Wise Tourist Arrivals of the World (1950–2014)

Year	Region Wise Tourist Arrivals (Million) and % share of total arrivals					Global Tourist Arrivals (Million)
	Africa	Middle East	Americas	Asia & Pacific	Europe	
1950	0.5(2.0)	0.2(0.8)	7.5(30.0)	0.2 (0.8)	16.8(67.2)	25
1960	0.8(1.1)	0.6(0.8)	16.7(24.2)	0.9(1.3)	50.4(73.0)	69
1970	2.4(1.4)	1.9(1.1)	42.3(25.4)	6.2(3.7)	113.0(68.0)	166
1980	7.2(2.5)	7.1(2.4)	62.3(21.7)	23.0(8.0)	178.5(62.4)	286
1990	18.9(4.3)	13.7(3.1)	109(25.0)	82.0(18.8)	304.1(69.9)	435
2000	26.2(3.8)	24.1(3.5)	128.2(19.0)	110.1(16.3)	388.0(57.5)	674
2010	50.3(5.3)	60.9(6.4)	149.7(15.7)	204.0(21.4)	475.3(50.0)	949
2012	52.3(5.0)	52.6(5.0)	162.1(15.6)	232.9(22.5)	534.8(51.6)	1035
2013	56.0(5.1)	52.0(4.7)	169.0(15.5)	248.0(22.8)	563.0(51.7)	1087
2014	55.7(4.9)	51.0(4.5)	181.0(15.9)	263.3(23.2)	581.8(51.39)	1132
2020 (e)	85(6.2)	101(7.4)	199(14.6)	355(26.1)	620(45.5)	1360

Source: Adapted and modified UNWTO (World Tourism Barometer, 2013).

It is evident from the data in Table 8.2 that France and the USA have maintained the top positions while others have ranked behind the new entrants such as China, Russia, Malaysia, and Thailand. These countries from Asia are the drivers behind the shift in tourism towards the Asia Pacific.

8.4 INDIAN TOURISM

8.4.1 INTERNATIONAL TOURISM IN INDIA

The Indian tourism and hospitality industry has emerged as one of the key drivers of growth among the services sector in India. According to the second tourism satellite account (TSA), 2009–10, tourism GDP accounted for 3.7% of GDP in terms of direct impact and 6.8% of GDP when indirect effects were included. This contribution to GDP had increased by 59% overall as compared to findings of TSA report (2002–2003) when tourism value added was 2.78% of GDP in terms of direct impact, and 5.83% of GDP when indirect effects were included. Tourism in India is driven by domestic tourism through its international inbound tourism has seen faster growth compared to global trends, and its outbound has grown very fast in recent times. Foreign Tourist Arrivals (FTA's) in India since year 2001

TABLE 8.2 Top 10 Countries in Terms of International Tourist Arrivals (1950–2014)

Rank	Years						International arrivals 2014	Revenue Ranking-2014	International Tourism recipets 2014 (US $)
	1950	**1970**	**1990**	**2000**	**2010**	**2014**			
1	USA	Italy	France	France	France	France	83.7 Million	USA	$177.2 Billion
2	Canada	Canada	USA	USA	USA	USA	74.8 Million	Spain	$65.2 Billion
3	Italy	France	Spain	Spain	China	Spain	65.0 Million	China	$56.9 Billion
4	France	Spain	Italy	Italy	Spain	China	55.6 Million	France	$55.4 Billion
5	Switzerland	USA	Austria	China	Italy	Italy	48.6 Million	Italy	$45.5 Billion
6	Ireland	Austria	Mexico	UK	UK	Turkey	39.8 Million	UK	$45.3 Billion
7	Austria	Germany	Germany	Mexico	Turkey	Germany	33.0 Million	Germany	$43.3 Billion
8	Spain	Switzerland	UK	Canada	Germany	UK	32.6 Million	Thailand	$38.4 Billion
9	Germany	Yugoslavia	Canada	Germany	Malaysia	Russia	29.8 Million	Australia	$32.0 Billion
10	UK	UK	China	Austria	Mexico	Mexico	29.1 Million	Turkey	$29.5 Billion
Total Tourist Arrival	**25 Million**	**166 Million**	**436 Million**	**683 Million**	**880 Million**	**1132 Million**			

Source: Adapted and modified from UNWTO website.

is given in Table 8.3. It can be observed that the FTAs in India have been increasing from all regions during the year 2012. The percentage share in FTAs in India during 2012 was the highest for Western Europe (24.23%) followed by North America (18.07%), South Asia (22.07%), South East Asia (8.93%), East Asia (7.12%), Eastern Europe (5.50%), West Asia (5.39%), Africa (0.36%), Australasia (0.05%) and Central & South America (0.91%). The arrivals in India from Western Europe were also the highest during the years 2011 and 2010 (India Tourism Statistics, 2014). Table 8.3 shows international tourist arrivals to India over time compared to arrivals globally and in the Asia Pacific region.

TABLE 8.3 Share of India in International Arrivals in World and in Asia – Pacific Region 2001–2015

Year	International Tourist Arrivals (in millions)		Foreign Tourist Arrival in India (in millions)	Percentage share and rank of India in international tourist arrivals of the world		Percentage share and rank of India in international tourist arrivals of Asia & Pacific region	
	World	Asia & Pacific region		% Share	Rank	% Share	Rank
2001	683.4	114.5	2.54	0.37	51st	2.22	12th
2002	703.2	123.4	2.38	0.34	54th	1.93	12th
2003	691.0	111.9	2.73	0.39	51st	2.44	11th
2004	762.0	143.4	3.46	0.45	44th	2.41	11th
2005	803.4	154.6	3.92	0.49	43rd	2.53	11th
2006	846.0	166.0	4.45	0.53	44th	2.68	11th
2007	894.0	182.0	5.08	0.57	41st	2.79	11th
2008	917.0	184.1	5.28	0.58	41st	2.87	11th
2009	883.0	181.1	5.17	0.59	41st	2.85	11th
2010	940.0	204.4	5.78	0.61	42nd	2.83	11th
2011	983.0	217.0	6.29	0.64	38th	2.90	9th
2012	1035	233.6	6.58	0.64	41st	2.82	11th
2013	1087	249.8	6.97	0.64	42nd	2.93	11th
2014	1135	263.5	7.68	0.68	41st	2.97	12th
2015	1186	279.2	8.02	0.67	37th	2.99	11th

Source: Adapted and modified from India Tourism Statistics (2012).

8.5 DESTINATION IMAGE

The review of the literature suggests that the researches on Tourist Destination image in the past 47 years are plenty and have focused on different dimensions of the image. The perspective and concept of destination image have changed over time. In general, there are three major emphases among different researchers on the definition of the destination image. Crompton's discussed the composite structure of the image construct. In addition, others make use of attitudinal components such as affect and cognition to explain the concept of the image. A third approach is to consider the image as an overall visual or mental impression of an object, place or experience (Rezende-Parker, Morrison & Ismail, 2003). Then in 2010 destination image was given a different connotation of subjective interpretation of reality made by the tourist (Bigné, Sánchez and Sánchez, 2001; Gracia, 2002).

However, the studies on the image of India as a tourist destination came much later than international researches and are still limited in number, and many dimensions are yet not touched. The first study on Tourism Image of India as a destination was conducted in 2000 (Chaudhary, 2000) and Ministry Of Tourism, Government of India (2012) in its study emphasized need of regular monitoring of tourist satisfaction and proper control of tourist facilities.

8.5.1 CONSTRUCTS OF TOURIST DESTINATION IMAGES

The researches on images used constructs to define and measure images (Gunn, 1972; Hunt, 1975; Mayo, 1975, Crompton, 1979; Gartner, 1989, 1993; Chon, 1991, 1992; Baloglu & Brinberg, 1997; Baloglu & McCleary, 1999; Gallarza, Saura, & Garcia, 2002; Pike & Ryan, 2004; Tasci, Gartner, & Cavusgil, 2007; Molina et al., 2010). The frameworks provided for measurement of images (Fakeye & Crompton, 1991; Gallarza, Saura, & Garcia, 2002) were questioned for the incompleteness of constructs in other studies. Echtner and Ritchie (1993) suggested use of multiple dimensions for measuring destination image and to reduce the complexity of the concept: (1) destination image has to be viewed as having attribute-based and holistic components, (2) these components have functional (tangible) and psychological (abstract) characteristics, (3) images range from common and functional to unique, and (4) a combination of structured and unstructured methodologies should be used in order to identify the destination image. This multidimensional approach of the destination image viewed attributes as a whole combination than as individual attributes.

This was based on a previous framework by the authors in 1991 that suggested three different sets of components in the destination image formation process (Figure 8.1).

1. Attribute-holistic set – comprised of those perceptions of individual attributes and the more holistic impressions of the place
2. Functional-psychological set – comprised of elements that are observable and measurable and those that are less tangible
3. Common-unique set – comprised of the destination images ranging from common to unique.

In a later work, Echtner and Ritchie (2003) commented that in most studies there is not a concrete indication of the authors are considering attribute-based or the holistic components of the image, or even both. Ahsen (1977), Aaker (1991), (1996), Kapferer (1991), Aaker & Joachimsthaler (2000), and Gracia (2002) viewed image to be made up of a large number of elements and attributes making it very complicated to define.

8.6 TOURIST IMAGE OF INDIA

Tourism image of India has been prismatic in line with its general image, and it has emerged as a must-see destination. It had an image problem and was not perceived as a holiday destination (Seth, 2006). Exploring reasons for the slow growth of tourism in India, researchers have identified that the negative image associations with the country have been one of the main

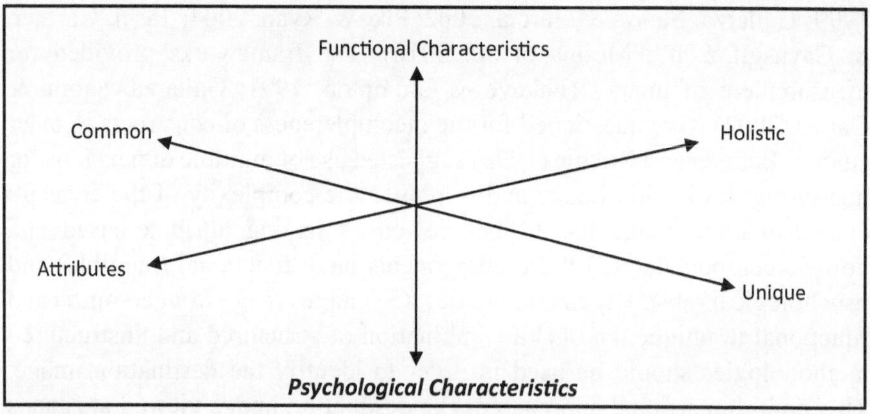

FIGURE 8.1 Dimensions of destination image (*Source*: Echtner & Ritchie, 1991).

reasons that hinder progressive tourism performance (Chaudhary, 2000). Traditionally India was perceived as magical and fascinating, and for most Americans, it was associated with epidemics flood and poverty. In 2001, Indian tourist's Bureau launched a campaign to portray an image of India as a peaceful, beautiful and safe country (Eli Avraham, EranKetter, 2012). The change in the tourism image of India was more discerning after the launch of the Incredible India Campaign in 2002. In 2003, Conde Nast Traveler selected India as one of the top ten preferred destinations among discerning traveler. In 2004, in a survey of lonely planet India was placed among the top five favorite destinations (Introduction to Tourism management Ksitiz Sharma). In 2007, India was ranked as world's no one preferred holiday destination by Conde Nast Traveler (Kevin Lane Keller, M. G. Parameswaran, Isaac Jacob). A study by Bandyopadhyay and Morais (2005) highlighted the dissonance between nationally projected images of India and the image India portrays. This study analyzed two marketing campaigns of India Eternally yours (1997) and incredible India (2002) and found the focus on five themes; cultural diversity, cultural richness, personal enlightenment, natural beauty, and royal treatment. C. Michael Hall, Stephen Page (2012) observed that India has not been able to build a tourism image that sells. Its product development and marketing have been sluggish in its long history of tourism. Seth (2007) stressed that the Indian tourism product underwent massive diversification by adding new areas of adventure, medical, rural, and Buddhist tourism to its age-old allurement of cultural and spiritual products. The need for diversification of tourism basket of India especially rural tourism was emphasized by the government (Incredible India report, 2006–07). Ministry of tourism of Govt. of India highlighted this in its report stating that a large number of efforts have been made towards the development of adventure activities such as trekking, rock climbing, mountaineering, aero sports, trekker huts, wildlife viewing, water sports, etc. for foreign tourist in various states of India along with a number of initiatives towards cruise tourism in conjunction with the Ministry of shipping to offer tourists a large array and variety of tourism activities (Incredible India Report, 2012–13).

The research studies on the holistic image of tourism are scarce though scattered researches are available on select tourist places in India (Das et al., 2007; Dwivedi, 2007; Dwivedi et al., 2009) that are survey based. These studies analyze the constructs of images on the basis of perception and experiences of tourists. Chaudhary (2000) found India to be highly rated for its rich art forms and cultural heritage but irritants like cheating, begging, unhygienic conditions, lack of safety dampen the spirits of tourists. India can be positioned on the world map only after these hygiene factors are improved along with other motivators. The study by Swain & Mohan (2006) raised a number of issues, facilities, amenities, and quality of services at the destinations in their study on the image of incredible India. Safety has been a serious concern for tourists as found by Agrawal (2007) as a greater percentage of tourists were not satisfied with the safety in India which could have been, in the case of non-visitors due to their lack of experience or poor information from various news, articles, or some other unforeseeable reasons which might have left a negative impact in their mind regarding the safety in India. Nandan (2010) focused on the need of creating India as a business hub and to promote business tourism in India. Chand (2012) found no significant differences in the perceived image of India as a tourist destination among visitors from four countries; USA, UK, Canada, and Germany. However, this study found demographic factors influencing the perceived image of India as a tourist destination and suggested cultural segmentation while developing their tourism strategies.

Khan (2013) found that the projected image of the country on the national destination marketing organization website, 'Incredible India" is understated in its content and appears shallow in contrast to the intensity of the perceived image conveyed by experienced tourists on the travel blog virtualtourist.com as tourists report equally strong positive and negative images of the country. Off late India has emerged as one of the World's most cost-efficient and fastest growing medical tourism destination. The uniqueness of India is its ability to offer holistic medical services such as Unani, yoga, meditation, Ayurveda, and homeopathic treatments and despite the advantage of having low medical costs, the Indian government is taking a different approach in promoting its medical tourism industry by highlighting its wellness elements. (Wong, 2014). Kaur (2016) found natural attraction to be positively perceived across all destinations and infrastructure emerged as an area in need of dire attention. The other dominant attributes found to mark destinations' positioning were – local cuisine & food outlets, hotels & restaurants, famous handicraft, and parking facilities. The studies on tourism image of India present a complex picture having unique attractions but missing facilitation in an age when destinations across the world are creating seamless and hassle-free experiences.

8.7 NEED OF THE STUDY

This study covers limited albeit important area of tourist image of India in the context of brand incredible India. Therefore, in this study, the focus of the researcher will be on; understanding the concept of Tourist destination image of Incredible India from the viewpoint of inbound tourists, factors contributing to the formation of Image of India, components of the image and the buying behavior of important inbound tourist market segments (demographically and geographically).

8.8 FINDINGS

The answer to the objective has been achieved with the help of Expectation experience matrix (EEM) and Strength weaknesses opportunities threats (SWOT) analysis based on responses of tourists on tour related attributes. EEM Importance-Performance analysis (IPA) is a direct measurement technique which emerged from the earlier work of Martilla and James (1977) variant of IPA only and maps customers' expectations and experiences that can be plotted on a grid divided into four quadrants. Each quadrant is created on the basis of the mean scores of the expectations-experience ratings and the median value of responses. The placing of variables in the grid is then assessed for sound management decisions. Attributes that are rated high in expectation and high in experience score suggest that service providers keep up the 'good' work. In contrast, attributes having low expectation rating and a low experience rating suggest that these areas may offer only little advantage and should, Therefore, be on a lower priority. Attributes that are rated high in expectation and low in experience are the missed opportunities, and service providers need to concentrate here and pay particular attention for improvement. Another variant of IPA is EEM (Experience Expectation Matrix) that is used to analyze expectation experience of customers' needs and desires in a 2×2 matrix and results are displayed and interpreted into four quadrants as under (Table 8.4, Figure 8.2).

Figure 8.2 indicates that all the 47 variables of the study are located in the 3 quadrants and quadrant 2 of 'possibly overkill' does not contain any variable.

Quadrant 1 – Low priority – contains safety and security of women, safety and security within cities, safety & security during travel, bars and discotheques, quality of life, hygiene and cleanliness at tourist sites, modern medical facilities, beaches, domestic connectivity, water connectivity, climate/weather & promptness in travel services. Safety and security of the women comes out as an issue of highest concern followed by safety and security within cities. Domestic connectivity is also an area to be focused

TABLE 8.4 EEM Matrix

Quadrants	Implications
Quadrant I Missed opportunity	Attributes in this quadrant are very important to respondents, but performances are fairly low. This suggests that these attributes are the missed opportunities.
Quadrant II Keep up the Good work	Attributes are very important to respondents and performance is also appreciable here.
Quadrant III Lower priority	Attributes rated as having low importance and low performance. Both performance and importance is low.
Quadrant IV Possible over kill	This contains attributes of low importance but where performance is really high. Respondents are satisfied with the performance. But putting efforts in this area could be unnecessary.

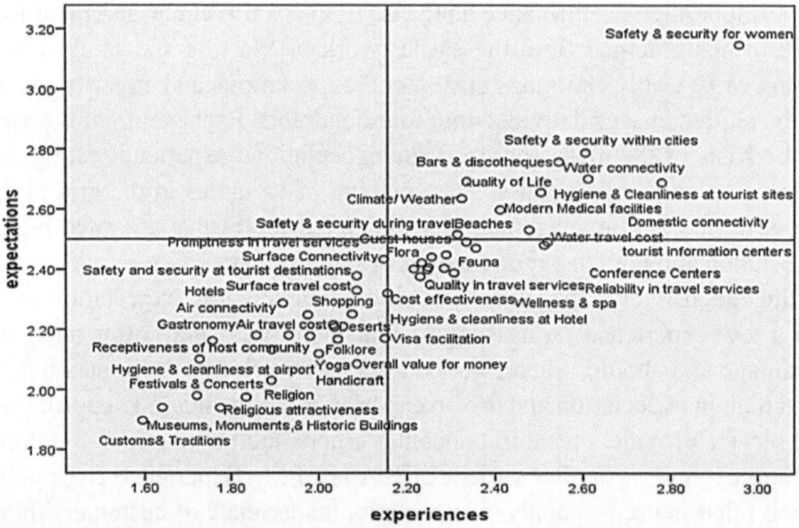

FIGURE 8.2 Expectation-Experience Matrix.

which presently lacking the priority and falling under the low-performance category. Quality of life in India is falling in this quadrant highlighting it just opposite to what India is marketed for.

Quadrant 2 – Possible overkill – No variable found place here. This suggests that no attribute has been promoted and used to such extent that tourists develop fatigue from it. This can also be treated as a blessing at this point of timing giving scope for promotion and image building around all the attributes deemed appropriate by destination managers.

Quadrant 3 – Keep up the good work – contains the variables of surface connectivity, safety and security at tourist destinations, surface travel cost, hotels, shopping, air connectivity, air travel cost, gastronomy, deserts, hygiene and cleanliness at airport, folklore, yoga, festivals and concerts, religion, religious attractiveness, museums/monuments/historic buildings, customs and traditions, handicrafts, overall value for money and receptiveness of host community. This contains an interesting mix of tradition and modernity. Heritage, museums, religion are placed with hotels and airports. Safety and security at tourist destinations are considered good. India is also high on overall value for money. All these attributes can be highlighted further through the Incredible India Campaign. Gastronomy and Receptiveness are falling in the quadrant which is found the statistically significant component.

Quadrant 4 – Missed opportunity – Guesthouses, flora, fauna, tourist information centers, quality in travel services, the cost-effectiveness of medical facilities, wellness and spa, hygiene and cleanliness at the hotel, visa facilitation, water travel cost and reliability in travel services fall in this category. Tourist info centers can be made good selling points with more work on them. Visa facilitation is found to be on the borderline with Quadrant 3 and 4 showing the improvisation and movement towards keeping up the good work. The initiatives taken by the Ministry of tourism like E-visa and Visa on arrival schemes could be a strong reason for that. Another point to be noted here is that quality & reliability of travel services is falling in this zone indicating the lack of professionalism at the customer service level. The attributes under missed opportunities need to be converted into strengths.

On the basis of EEM matrix SWOT analysis has been done to get insight into the areas for building the positive tourism destination image of India (Table 8.5).

The variables found in Keep up the work good work can be counted as strengths, variables under missed opportunities could be tapped as opportunities and variables in low priority quadrant could be taken as a threat.

TABLE 8.5 SWOT Analysis of India As a Destination

STRENGTHS	WEAKNESSES
History, culture, heritage, traditions, religion, folklore, gastronomy, museums, monuments, festivals and handicraft.	Tourist information centers, quality in travel services, the cost-effectiveness of medical, wellness and spa, hygiene and cleanliness at the hotel, visa facilitation, water travel cost and reliability in travel services.
OPPORTUNITIES	**THREATS**
Tourist information centers, quality in travel services, the cost-effectiveness of medical, wellness and spa, hygiene and cleanliness at the hotel, visa facilitation, water travel cost and reliability in travel services.	Safety and security of women, safety and security within cities, safety and security during travel, bars and discotheques, quality of life, hygiene and cleanliness at tourist sites, modern medical facilities, beaches, domestic connectivity, water connectivity, climate/weather and promptness in travel services.

8.9 A WAY AHEAD FOR DESTINATION IMAGE OF INDIA

The findings of the study suggest that following action areas as a way forward for academicians as well as practitioners. Tourism destination image of India has remained rooted in its ancient attractions through a few tourism facilitators such as airport and hotels have added value to it. There is a need to position India with ancient attractiveness as its core product and modern facilities as peripherals to make it a composite product in line with the Kotler's concept of product (Kotler, 1969).

A few factors that can be considered such as hygiene, safety, and security of women, safety, and security at tourism destination and cleanliness are absent, and it can be presumed from Herzberg's classical study that motivators will not work unless hygiene factors are in place. This also supports the need of Swacch Bharat spirit in the context of tourism (Herzberg, 1968). Visa is a borderline factor on tourism expectation experience matrix, and the current efforts for facilitating the tourist visits to the country may be continued. Low cost of water travel has emerged as the single attribute that was found to be statistically significant. Its possibility for tourism can be further explored and can be synced with the government of India plans of waterways (Ministry of transport, 2015).

8.10 CONCLUSION

Incredible India campaign needs a revisit in terms of its less reach though it was found to be effective in its impact. WTTC has also suggested the need for Incredible India Campaign 2. Both images of India and Incredible India campaign shall be monitored on a continuous basis with the help of well-designed tools as suggested by MOT report 2103. Thus feedback obtained shall be used for corrections in the image as well as Incredible India Campaign. Tourist image of India monitored through tourist surveys shall be decomposed on different factors such as source country, destination choices, media exposure, etc. This will provide better insight into the different dimensions of the image. Measurement matrices need to be developed to study the link between the Incredible India Campaign and tourists' image of India. These matrices shall be continuously improvised through further studies.

KEYWORDS

- **destination branding**
- **destination image**
- **EEM**
- **Incredible India Campaign**
- **marketing**
- **SWOT**

REFERENCES

Aaker, D. A. & Joachimsthaler, E. (2000). *Brand Leadership: Building Assets in the Information Society,* Free Press, NY.

Aaker, D. A. (1991). *Managing Brand Equity: Capitalizing on the Value of a Brand Name,* The Free Press, NY.

Aaker, D. A. (1996). *Criando e administrandomarcas de sucesso*, Futura, São Paulo.

Ahsen, A. (1977). Eidetics: An overview, *Journal of Mental Imagery, 1*(1), 5–38.

Aksoy, R., & Kiyci, S. (2011). A Destination Image As a Type of Image and Measuring Destination Image in Tourism (Amasra Case). *European Journal of Social Sciences, 20*(3), 478.

Alastair M. Morrison. (2013). *Marketing and Managing Tourism Destinations*, Routledge, p. 291.

Anholt, S. (2010), *Places, Identity, Image and Reputation,* Palgrave MacMillan UK ISBN 978-0-230-25128-1.

Anholt,S. (2006). T*he Anholt City Brand Index: How the World Views Its Cities*, 2nd Edition. (Volume 2, Issue 1, pp. 18–31), Palgrave Macmillan UK.

Anholt, S. (2006). *The Anholt City Brands Index: How the World Views Its Cities*, 2nd Edition.

Arturo Molina, Mar Gómez, & David Martín-Consuegra (2010). Tourism marketing information and destination image management. *African Journal of Business Management 4*(5), 722–728.

Assaker, G., Vinzi, V. & O'Connor, P. (2011). Examining the Effect of Novelty Seeking, Satisfaction, and Destination Image on Tourists' Return Pattern: A Two Factor, Non-Linear Latent Growth Model. *Tourism Management, 32,* 890–901.

Baloglu, S., & Brinbecg, D. (1997). Affective images of tourism destinations. *Journal of Travel Research, 4,* 11–15.

Baloglu, S., & McCleary, K. (1999). A Model of Destination Image Formation. *Annals of Tourism Research, 26*(4), 808–889.

Beerli, A. & Martín, J.D. (2004). Tourists' characteristics and the perceived image of tourist destinations: a quantitative analysis. *A Case Study of Lanzarote, Spain, Tourism Management, 25*, 623–636.

Bigné, E., Sánchez, M. I. & Sánchez, J. (2001). Tourism image, evaluation variables and after purchase behavior: inter-relationship, *Tourism Management, 22*, 607–616.

Bornhorst, T., Ritchie, J. & Sheehan, L. (2010). Determinants of Tourism Success for DMOs & destinations: An Empirical Examination of Stakeholders' Perspectives. *Tourism Management, 31,* 572–589.

Buhalis, D. (2000). 'Marketing the Competitive destination of the future.' *Tourism Management, 21,* 97–116.

Buhalis, D., & Cooper, C., (1998). Competition or co-operation: The needs of Small and Medium-sized Tourism Enterprises at a destination level, in E., Laws, Faulkner, B., and Moscardo, G., (ed.), *Embracing and Managing Change in Tourism*, Routledge, London.

Chaudhary, M. (2000). India's image as a tourist destination: a perspective of foreign tourists, *Tourism Management, 21*, 293–297.

Choi, M. W., Chan, A., & Wu, J. (1999) 'A qualitative and quantitative assessment of Hong Kong's image as a tourist destination, ' *Tourism Management, 20*, 361–365.

Chon, K-S. (1991). Tourism destination image modification process. Marketing implications, *Tourism Management, 12* (1), 68–72.

Chon, K.-S. (1990). The role of destination image in tourism: A review and discussion. *Revue de Tourisme, 45*(2), 2–9.

Dobni, D. & Zinkhan, G. M. (1990). In search of brand image: a foundation analysis, *Advances in Consumer Research, 17,* 110–119.

Echtner, C. & Ritchie, J. B. (2003). The Meaning and Measurement of Destination Image. *The Journal of Tourism Studies, 14*(1), 37–46.

Echtner, C. M. & Ritchie, J. R. B. (1991). The measuring and measurement of destination image, *The Journal of Tourism Studies, 2* (2), 2–12.

Eli Avraham & Eran Ketter, (2012). *Media Strategies for Marketing Places in Crisis*, Routledge, pp. 160.

Fakaye, P. C. & Crompton, J. L. (1991). Image differences between prospective, first-time and repeat visitors to the lower Rio Grande Valley, *Journal of Travel Research, 30* (2), 10–16.

Fayos-Sola, E., (1996). Tourism policy: a midsummer night's dream?, *Tourism Management, 17*(6), 405–412.

Fürsich, E., & Robins, M. (2008). Visiting Africa: Constructions of Nation and Identity on Travel Websites, *Journal of Asian and African Studies, 39*(1–2), 113–152.

Gartner, W. C., & Hunt, J. D. (1987). An analysis of state image change over a twelve-year period (1971–1983). *Journal of Travel Research, 2,* 15–19.

Goodwin, M. (1993). 'The City as Commodity: The Contested Spaces of Urban Development,' in Kearns and Philo (eds). 145–162, Pergamon, Oxford.

Graham Hughes, & Chris Fil, (2007). Marketing communications 2007–2008, Routledge, – Business & Economics, pp. 45.

Gunn, C. A. (1988). *Vacation Scape, Designing Tourist Regions*, 2nd Edn., New York: Van Nostrand Reinhold.

Holbrook, M. B. (1978). Beyond attitude structure: toward the informational determinants of attitude, *Journal of Marketing Research, 15,* 545–556.

Hunt, J., D. (1975). Image as a Factor in Tourism Development, *Journal of Travel Research, 13,* 1–9.

Kapferer, J. N. (1991). As Marcas, Capital da Empresa, CETOP, Portugal.

Kearns G., & Philo C. (eds) (1993). Selling Places: The City as Cultural, Capital, Past, and Present, Oxford: Pergamon.

Kevin Lane Keller, Parameswaran, M. G. & Isaac Jacob, (2011). Strategic Brand Management: Building, Measuring, and Managing Brand Equity, Pearson Education India, pp. 286.

Kotler, P. & Certner, D (2004). El paíscomomarca y comoproducto, *Gestión de Negocios, 5*(3), 54–62.

Lawson, F., & Baud-Bovy, M. (1977). *Tourism and Recreational Development.* London: Architectural Press, ISBN 0843601663, 9780843601664.

Lin, C-H., Duarte, B., Kerstetter, D. L. & Hou, J-S. (2007). Examining the Role of Cognitive and Affective Image in Predicting Choice Across Natural, Developed, and Theme-Park Destinations, *Journal of Travel Research, 46*, 183–194.

LopesSérgio Dominique Ferreira (2011). Destination image: Origins, Developments and Implications, PASOS. *Revista de Turismo y Patrimonio Cultural. 9*(2) 305–315, ISSN 1695–7121.

Lubbe, B. (1998). Primary Image as a Dimension of Destination Image: An Empirical Assessment, *Journal of Travel and Tourism Marketing, 7* (4), 21–43.

Matos, C. (2012). Globalization and the mass media. In: *Encyclopedia of Globalization.* Oxford: Wiley-Blackwell.

Mazursky, D. & Jacoby, J. (1986). Exploring the development of store images, *Journal of Retailing, 62* (2), 145–165.

Meler, M., & Ruzic, D. (1999) 'Marketing identity of the tourist product of the Republic of Croatia,' *Tourism Management, 20,* 635–643.

Metaxas T., (2003). 'The image of the city as a 'good': The creation of a city's promotional package through a strategic framework analysis of City Marketing procedure' in Beriatos E. et al. (eds.). Sustainable Planning and Development, Wessex Institute of Technology and Dept. of Planning and Regional Development (Univ. of Thessaly), pp. 427–438.

Michael Hall, C. & Stephen Page (2012). Tourism in South and Southeast Asia Routledge, pp. 227–228.

Moutinho, L. (1987). Consumer behavior in tourism, *European Journal of Marketing, 21* (10), 5–44.

Oppermann, M. (1996). 'Convention destination images: analysis of association meeting planner's perceptions,' *Tourism Management, 17* (3), 175–182.

Park, C. W., Jaworski, B. J., & Maclnnis, D. J. (1986). Strategic brand concept-image management. *Journal of Marketing, 50*(4), 135–145.

QuHailin, Kim Lisa Hyunjung, & Im Holly Hyunjung, (2011). A model of destination branding: Integrating the concepts of the branding and destination image. Elsevier, *Tourism Management 32,* 465–476.

Reynolds, T. J., & Gutman, J. (1984). Advertising is image management. *Journal of Advertising Research, 24*(1), 27–37.

Russel, J. A. & Prat, G. (1980). A description of affective quality attributed to environment, *Journal of Personality and Social Psychology, 38,* 311–322.

Ryan, C. & Cove, J. (2007). Structuring Destination Image: A Qualitative Approach, *Journal of Travel Research, 44* (2), 143–150.

Seth Pran Nath, (2006). Successful Tourism: Volume II: Tourism Practices, Sterling Publishers Pvt. Ltd, 1 December 2006, p. 376.

Stern, E. & Krakover, S. (1993). The formation of a composite urban image, *Geographical Analysis, 25* (2), 130–146.

Tasci, A. D. A., & Gartner, W. C. (2007). Destination Image and Its Functional Relationships, *Journal of Travel Research, 45*(4), 413–425.

Um, Seoho & Crompton, J. L. (1990). Attitude determinants in tourism destination choice. *Annals of Tourism Research, 17,* 432–448.

Vogt, C. A., & Andereck, K. L. (2003). Destination perceptions across a vacation. *Journal of Travel Research, 41*(4), 348–354.

Walmsley, D. J. & Young, M. (1998). Evaluative images and tourism: the use of perceptual constructs to describe the structure of destination images, *Journal of Travel Research, 36* (3), 65–69.

Whynne-Hammond (1985). Elements of Human Geography, Allen & Unwin, University of Michigan ISBN 0049100815, 9780049100817.

Wicks, B. E. & Schuett, M. A. (1991). Examining the role of tourism promotion through the use of brochures, *Tourism Management, 12*(4), 301–312.

CHAPTER 9

PRO-POOR TOURISM: A STUDY OF GOVERNMENT STRATEGIES IN THE STATE OF UTTAR PRADESH, INDIA

ALOK KUMAR

Associate Professor, Department of Tourism Management and Hospitality Studies, IPHC, Mekelle University, Ethiopia, Africa, Mobile: 251-985024149; E-mail: kumaralok1975@yahoo.co.in

ABSTRACT

Poverty alleviation is a major concern for all developing countries including India. It can be achieved through higher sectorial growth and ensuring that the poor have a stake in that growth. There is evidence of tourism contributing to the economic growth of countries. Tourism development often provides the dual advantages of generating employment and income while promoting cultural heritage and traditions leading to sustainable development. Today tourism is one of the largest service industries in India. An out of the box approach to tourism development is proposed, through "pro-poor tourism," which focuses on unlocking opportunities for the underprivileged hitherto relegated to the bottom of the economic pyramid. With this vision, the paper aims to examine pro-poor tourism as a vehicle for development in the state of Uttar Pradesh, India. It highlights the current status of tourism in the country with reference to pro-poor tourism approaches as an effective instrument of poverty alleviation. The paper aims at gaining an insight into the actual and potential roles and motivations of various stakeholders in this endeavor of tourism for poverty alleviation.

9.1 INTRODUCTION

Tourism is widely recognized as one of the largest and rapidly growing industries in the global economy. The diverse nature of the tourism industry, which incorporates many different sectors and types of jobs, means accurate statistics on employment are very difficult to ascertain. Therefore, Tourism to a developing country has always been considered as a source of foreign exchange earnings, which are essential for economic development. The role of international tourism in generating economic benefits has long been recognized in many developing countries (Jenkins, 1991; WTO, 1994). Therefore, tourism as a sector of developing countries' economies deserves priority on the part of the governments and the business community (United Nations, 1999). The tourism industry has grown rapidly worldwide, resulting in economic and social benefits; this industry may play a vital role in enhancing a country's trade performance.

According to the World Travel and Tourism Council, India will be a tourism hotspot from 2009–2018, having the highest 10-year growth potential. As per the Travel and Tourism Competitiveness Report, 2009 by the World Economic Forum, India is ranked 11th in the Asia Pacific region and 62nd overall, moving up three places on the list of the world's attractive destinations. It is ranked the 14th best tourist destination for its natural resources and 24th for its cultural resources, with many World Heritage Sites, both natural and cultural, rich fauna, and strong creative industries in the country. India also bagged 37th rank for its air transport network. The India travel and tourism industry ranked 5th in the long-term (10-year) growth and is expected to be the second largest employer in the world by 2019 (www.rgcresearchjournal.org, 2012).

With proper interventions, such economic benefits can play a crucial role in the process of poverty alleviation. According to the United Nations, about 25,000 people die every day of hunger or hunger-related causes, yet there is plenty of food in the world for everyone. The problem is that hungry people are trapped in severe poverty. Considering that the tourism sector is touted as the largest economic sector in terms of the number of people employed or other criteria, its contribution to job creation and income generation for local communities is a very important element for overall development. Poverty reduction requires strategies on a variety of complementary fronts and scales, but a prerequisite of significant progress is pro-poor growth which benefits the poor. Tourism can be one source of such growth (Stephanie Chok et al., 2007). This is especially the case for developing countries and households with lower incomes. Tourism policies with poverty reduction focus provide an attractive package for local governments to achieve increased job creation and income generation goals. Pro-poor tours are nature-based and designed for those that want to experience the unique scenery, flora, fauna, history, and culture of each area.

The state of Uttar Pradesh is home to some of India's most iconic tangible and intangible assets and is India's biggest tourist draw, attracting 155 million domestic and 1.8 million international tourists out of 748 million domestic and 6.2 million international tourists visiting India in 2011, contributing a total of US$ 6 billion to the state's economy. Paradoxically, despite such staggering numbers, Uttar Pradesh remains India's third most lagging state, with a 37.7% poverty rate. Moreover, the economic and service delivery benefits of the tourism sector have had a little trickle-down effect in the state in general, and in the communities adjacent to cultural assets in particular. Many of the state's most recognized assets are home to or surrounded by some of the poorest communities, and the assets themselves fact great threats as a result of, for instance, limited institutional capacities and lack of coordinated management (Uttar Pradesh Pro-Poor Tourism Draft, 2015).

The Uttar Pradesh Pro-Poor Tourism Development Project (the Project) aims to address this paradox by unlocking the potential of the unique heritage of Uttar Pradesh through pro-poor tourism development for inclusive growth and by poverty reduction in asset-rich but lagging areas of the state. Ultimately, the Project is expected to improve living standards, create job opportunities and promote thriving local enterprises among the poor communities living in the Buddhist Circuit, the Braj region and the city of Agra, while protecting the state's unique heritage through a restructured tourism sector (UP Tourism, 2015).

9.2 OBJECTIVES OF STUDY

This chapter aims to examine pro-poor tourism as a vehicle for development in the state of Uttar Pradesh, India. It highlights the current status of tourism in the country with special reference to the state of Uttar Pradesh applying pro-poor tourism approaches as an effective instrument of poverty alleviation. The chapter intends to gain an insight into the actual and potential roles and motivations of various stakeholders in this endeavor of tourism for poverty alleviation.

9.3 REVIEW OF LITERATURE

Pro-poor tourism is broadly defined as tourism that generates net benefits for the poor (PPT Partnership, 2004a). 'Net benefits' means benefits outweigh costs (Ashley et al., 1999). This encompasses economic as well as environmental, social, and cultural dimensions (Ashley et al., 1999; UNESCAP, 2003; WTOOMT, 2002). PPT arises from the belief that tourism can and should contribute to pro-poor economic growth. Pro-poor growth is 'growth that enables the poor to participate in actively and significantly benefit from economic activity' (PPT Partnership, 2004b: 2).

Pro-poor tourism is set up in developing countries as a means to improve the local economy for local people. It enhances the linkages between tourism businesses and poor people; so that poverty is reduced and poor people are able to participate more effectively in tourism development. The aims of pro-poor ranges from increasing local employment to involving local people in the decision-making process. Any type of company can be involved such as a small lodge or a tour operator. The most important factor is not the type of company or the type of tourism, but that poor people receive an increase in the net benefits from tourism (Tour Directory, 2015).

People may be defined as poor due to their economic circumstances; however, they may be socially and culturally rich. These resources are often of interest to tourists which creates opportunities to create a source of income for these communities. By using this tourism 'wealth,' tourists and tourism businesses are able to assist the economically poor by stimulating local economic development and increasing livelihood opportunities.

Local people can gain from tourism businesses and tourists through:

1. employment in tourism businesses or businesses which supply the industry;
2. sales of goods and services to tourism businesses;
3. direct sales of goods and services to tourists;
4. contributions from tourists to the maintenance of their natural and cultural heritage;
5. traveler and business philanthropy (e.g., donations);
6. Infrastructure gain when tourism brings electricity, telecommunications, roads or potable water.

We use "pro-poor tourism" (PPT) as shorthand for the opportunities which enable the economically poor to engage in and benefit from tourism. PPT is tourism which provides *net* benefits for the economically poor; the term 'net benefits' is used because there are often losses of livelihood opportunities associated with tourism, such as loss of seashore and fishing access, agricultural land or access to natural resources in national parks. Some gain, others lose. Reporting net benefit ensures that the negatives are not ignored and the benefits are counted and demonstrated. The idea is not enough; the idea has to deliver clear, tangible gains (PPTP, 2004b).

9.3.1 DEFINITION OF POVERTY

The United Nations defines extremely poor people as those who live on less than the equivalent of one US dollar per day. This is measured in purchasing power parity (PPP), which is based on the cost of a similar basket of goods in different countries, expressed in United States dollars. Merely poor people are those who live on less than US$2 per day. There are about 1.3 billion people in the developing world (21% of the world population) in extreme poverty, which live on less than US$1 per day. More than 2.7 billion live on less than US$2 per day. In India even now in spite of all the development during the past five decades, nearly 35% of the population was getting less than $1 (PPP) a dying 1999–2000 (World Development report 2007). Amartya Sen (1977) defined poverty as "the poor is not an economic class, nor convenient category to use for analyzing social and economic movements. Poverty is the common outcome of a variety of disparate economic circumstances, and policy to tackle poverty must of necessity go beyond the concept of poverty" (www.rgcresearchjournal.org, 2012).

Pro-poor tourism (PPT) is tourism that results in increased net benefits for poor people. PPT is not a specific product or niche sector but an approach to tourism development and management. It enhances the linkages between tourism businesses and poor people; so that tourism's contribution to poverty reduction is increased and poor people are able to participate more effectively in product development (Stephanie Chok et al., 2007).

PPT is tourism that generates net benefits for the poor PPT is not a specific product or sector of tourism, but an overall approach. Rather than aiming to expand the size of the sector, PPT strategies aim to unlock opportunities for economic gain, other livelihood benefits, or engagement in decision-making-for the poor (Ashley, 2001).

PPT is mostly determined as tourism that results in increased net benefits for poor people. PPT is not a specific product or niche sector but an approach to tourism development and management. It enhances the linkages between tourism businesses and poor people. According to PPT view, by the help of the pro-poor understanding, tourism's contribution to poverty reduction is increased, and poor people are able to participate more effectively in product development (PPT, 2011). PPT is used to refer to interventions that specifically focus on addressing poverty which moves beyond the "trickle-down" theory (Walter and Mandke, 2002: 3). Before a critical analysis of pro-poor discourse is realized, earlier development efforts tried to be discussed on the past initiatives of global agencies.

Tourism is one industry that has continued to grow despite the global economic slowdown, accounting for more than 9% of global GDP in 2011 (WTTC, 2012a). The beneficiaries of this growth have, however, largely been multinational corporations, with local communities often seeing little growth in income. Ashley, Boyd & Goodwin (2000), estimate that on average around 55% of tourism expenditure goes outside the destination country. Meanwhile, local communities frequently raise concerns about the negative impacts of tourism on the environment and on local community access to land. Perceptions of tourism as a threat to local culture are also widespread. Hotels and resorts undoubtedly have significant impacts on the local economies and societies within which they operate. One of the key issues in assessing the impact of tourism is whether local communities are in a position to take advantage of the opportunities that tourism might offer. These opportunities can only be accessed with the support and motivation of tourism operators (Stephanie Chok et al., 2007). Pro-poor tourism initiatives need to ensure that the aim of recruitment from poor communities is in the interests of building the skills and capacity of people for further employment rather than simply the building of the company brand via corporate social investment. The main cash injection into the local economy from tourism is from the wages of local staff; Therefore, the opportunity for local employment is important for the poor. Tourism enterprises that show a commitment to genuine job creation within poor communities can make a real difference and bring a range of benefits.

Treating staff well ensures that the organization is able to secure the best staff through:

- lower staff turnover.
- lower training costs.
- higher staff morale.
- better quality service.
- more authentic experience for guests/customers.
- positive company public image.
- employing local people with a strong identity and sense of place (culturally rich but economically poor) can contribute to a sense of place and provide an enterprise with is competitive advantage point of difference.

9.3.2 OVERVIEW OF STATE OF UTTAR PRADESH (INDIA)

The literal meaning of the state of Uttar Pradesh is "Northern Province." It is located in the northern part of India. The state is commonly known as U.P. by its abbreviation. The state was created on 1 April 1937 as the United Provinces with the passing of the States Reorganization Act and renamed Uttar Pradesh (State of Uttar Pradesh) in 1950. Lucknow is the capital of Uttar Pradesh, and the city of Kanpur is regarded as its economic and industrial capital. Uttarakhand, a new state was carved from the mountainous Himalayan region of Uttar Pradesh on 9 November 2000. The boundaries of U.P meet with Rajasthan in west, Haryana, and Delhi in the northwest, Uttarakhand and the country of Nepal is in its north, Bihar in the east, Jharkhand to the southeast, and Madhya Pradesh to the southwest. It the fifth largest Indian state with an area of 93,933 square miles (243,290 km) which constitutes of 6.88% of the total area of India. With a population of over 200 million (2011; Census), it is the most populous state in India. Hindi is the official language of the state. It was ruled by powerful empires in ancient and medieval India. It boasts of two major rivers of the country 'Ganga and Yamuna' which meet at Prayag (Allahabad); where famous Kumbhmela takes place. The historical, natural, and religious tourist destinations of the state are Taj Mahal at Agra, Varanasi, Piprahwa, Kaushambi, Ballia, Shravasti, Kushinagar, Lucknow, Chitrakoot, Jhansi, Allahabad, Meerut, and Mathura (Travel Guide, 2015).

Uttar Pradesh is a land of ancient history, language, and culture. The several luminaries were born on this land. The Hindu epics 'Ramayana, Mahabharata and Puranas' are believed to be written over here. Varanasi and Allahabad have been Centre of Hindi and Urdu literature since past (Tour Directory, 2015). The state has been rich in linguistic traditions. The languages like Hindi, Urdu, Sanskrit, Hindustani, Braj Bhasha, English, Awadhi, Bagheli, Bhojpuri, Bundeliand Kannauji are spoken here(Uttar Pradesh Tourism Report, 2015).

Uttar Pradesh is the second largest Indian state by the economy, with a GDP of INR 708,000 crore. Agriculture and service industries are the core sector of the state's economy. The state of Uttar Pradesh is unique in itself but still it is the third poorest state in India. It has got over 50 million people living below the poverty line. The state has got per-capita incomes at US$445 per annum compared to the national average of US$922 per annum. It is one of the lowest in India (Uttar Pradesh Pro-Poor Tourism Draft, 2015).

9.3.3 PRO-POOR TOURISM PROJECT IN UTTAR PRADESH

The convergence of high tourism growth in poverty-stricken countries has generated a large amount of interest in tourism as a poverty alleviation strategy (Sofield et al., 2004; UNESCAP, 2003; WTO-OMT, 2004a). This is sometimes termed 'pro-poor tourism' (PPT). Touted as an 'alternative' approach to mainstream tourism development models, PPT strategies are directed towards generating net benefits for the poor (DFID, 1999).

The pro-poor tourism project in the state of Uttar Pradesh (India) was prepared with meaningful participation and through consultation of many stakeholders. These stakeholders included were central, state and local government officials to representatives of the private sector, experts, religious bodies, visitors, pilgrims, and local communities. It has shared concerns of both producers and decision-makers. The project is framed in such a way to reap the benefit from the project activities, the construction, and operation of the infrastructure and other activities which would be supported under the Project. The consultations from Stakeholder started from its concept stage. The Project is expected to support the state's plans for restructuring its tourism sector in a pro-poor manner with a view to increasing benefits to local communities and improving the management of its tourism destinations. The pro-poor approach adopted; also aims at utilizing the impact of key sectors in the thrust area, such as accessibility/connectivity, environmental preservation, asset management and business development on local communities. The aims would be achieved by promoting an integrated area-based approach in which tourist destinations and their ancient heritage structure will be revamped, promoted, served, managed and linked geographically. It will be implemented on the area-based approach. The area-based approach will allow for improved understanding of a given area's potential and linkages with other industries. It will further strengthen the skills of the local population, and promote small-scale entrepreneurship, especially amongst poor women and youth (Uttar Pradesh Pro-Poor Tourism Draft, 2015).

9.3.3.1 OBJECTIVE OF PRO-POOR PROJECT IN UTTAR PRADESH

The forecast of high tourism growth in developing nations, where widespread poverty exists, has led to considerable interest in tourism as a tool for poverty alleviation. Powerful bureaucratic and business alliances have been forged to expand this programme. International development agencies are also turning to tourism as a way of alleviating poverty. Generally, tourism is too often regarded as a panacea without an attendant recognition that, like any other industrial activity, tourism is highly political. As a global industry, tourism operates within a neo-liberal market economy which presents severe challenges to meeting pro-poor and sustainable development objectives (Stephanie Chok, Jim Macbeth & Carol Warren, 2007). The Uttar Pradesh Pro-Poor Tourism

Development Project has been planned to contribute towards improving living conditions and increasing income opportunities for the poor. This would be done by the vehicle of enhanced tourism development in the Buddhist Circuit, Braj region, and Agra.

9.3.3.2 ROLES AND RESPONSIBILITIES OF KEY STAKEHOLDERS

The international financial institutions like the Asian Development Bank and the World Bank are also supporting Pro-Poor tourism projects in Uttar Pradesh. It is part of their broader 'pro-poor growth' objectives (ADB, 2002; Sofield et al., 2004). The State Department of Tourism will be the agency responsible for the overall management and coordination of the project at the state level project. The Development Authorities such as the Archeological Survey of India (ASI) and the State Department of Forests (DoF) will be the key implementing Agencies at the destination level. The committees comprising the services and infrastructure users will be established to be the first line of operational oversight and for proper functioning and maintenance of the project-financed facilities for operation and maintenance at the tourism product level (World Bank Report, 2015).

9.4 RESEARCH METHODOLOGY

The study employed exploratory and descriptive tools to collect and analyze data. The research is exploratory and entirely based on the usage of secondary data which includes Journals (both print and online), Periodicals, Reports, and Books, etc. Use of internet sites and annual reports of local governments and organizations have been made throughout the work especially in the literature review.

9.5 LIMITATIONS OF THE STUDY

Although this research paper throws light on Pro-Poor Tourism, it also has a few limitations viz. secondary data was used, and study is exploratory in nature. Use of case study approach or interviews with the stakeholders could help in better understanding of the subject and would have generated more information but was not conducted because of practical constraints in the field setting. During the course of this work, it was realized that the ambit of the Uttar Pradesh Pro-Poor tourism project could be made broader based through innovative tourism strategies viz. environmental tourism and wildlife tourism, handloom tourism, adventure tourism, etc. Further research is desirable in order to increase more contributions in this area and to validate the findings of the study with secondary data. Hence, this study can act as a base for future research.

9.6 RESULT AND DISCUSSION

The tourism sector is growing at a fast pace, and there is immense potential for generating employment and earning a large amount of foreign exchange besides giving a fillip to the overall economic and social development of the country. This chapter emphasizes upon tourism as an important sector in the development of the economy with special emphasis on pro-poor tourism approach. After an in-depth study to this end, this Project will provide Uttar Pradesh with the necessary policy framework, capacity building and infrastructure development strategies to help structure its tourism economy keeping the pro-poor approach as the dominant theme. The Project is expected to meet its objective of: (i) increasing accrual of benefits to the state's residents especially its poor through the provision of jobs and public services to some of the most underprivileged communities in the state, living in heritage-rich but economically lagging areas; (ii) enhancing the management of the State's unique natural and cultural assets as endogenous sources of inclusive growth; and (iii) enabling productive private investments and associated job creation opportunities especially to the youth and women.

9.7 CONCLUSION

Tourism and Poverty are a much-needed addition to the literature on tourism development in developing countries. This chapter infers that opportunities exist to improve the impact of tourism on local communities through a targeted approach to recruitment and training. Importantly, the approaches can be mutually beneficial for tourism organizations. But, globally, there is a lack of convincing empirical evidence to support the claim that tourism benefits the poor. On a global scale, PPT advocates must be explicit about the fact that stakeholders include winners and losers working on a vastly uneven playing field.

It is important to consider what forms of tourism development are likely to emerge from such an asymmetrical political landscape, despite the best of intentions. Tourism is highly political in its nature owing to the profound impact of values of powerful stakeholders which greatly shape the outcomes.

The Pro-Poor tourism Project Uttar Pradesh is aligned with India's 12th Five-Year Plan, which stresses upon the pro-poor tourism approach as a priority is aimed at increasing the net benefits to the poor and by ensuring that tourism growth contributes to job creation, particularly for poor female and young workers, and ultimately leads to poverty reduction. The Government of Uttar Pradesh shares this vision and is currently updating its 2002 Tourism Development Policy, prioritizing the Buddhist Circuit, Agra and the Braj region as drivers of the socio-economic development of the state. To this end, this chapter will provide Uttar Pradesh with the necessary policies, approaches, capacities, and infrastructure to help structure its tourism economy using the pro-poor approach.

KEYWORDS

- **employment**
- **India**
- **poverty**
- **pro-poor tourism**
- **Uttar Pradesh**

REFERENCES

A Critical Theoretical Evaluation on Pro-Poor Tourism and Poverty Alleviation, (2012). *9*(17), 115–133.

Ashley, C., Bennett, O. & Roe D. (1999). Sustainable Tourism and Poverty Elimination Study. Deloitte and Touche, International Institute for Environment and Development and Overseas Development Institute, London. on WWW at http://www.odi.org.uk/pptourism/dfid_report.pdf.

Ashley, C., Boyd, C. & Goodwin, H. (2000). Pro-poor Tourism: Putting Poverty at the Heart of the Tourism Agenda. Overseas Development Institute Natural Resource Perspectives.

Ashley, C., Roe, D. & Goodwin, H.(2001b). Pro-Poor Tourism Strategies: Making Tourism Work for the Poor, Overseas Development Institute.

Asian Development Bank (ADB) (2002). Report and Recommendation of the President to the Board of Director on Proposed Loans to the Kingdom of Cambodia, Lao People's Democratic Republic, and the Socialist Republic of Viet Nam for the Greater Mekong Subregion: Mekong Tourism Development Project. Asian Development Bank, Manila. http://www.adb.org/Documents/RRPS/CAM/rrp_35282-cam.pdf. (accessed on 29 May 2005).

Department for International Development (DFID) (1999) Tourism and poverty elimination: Untapped potential. DFID, UK. http://www.propoortourism.org.uk/dfid_summary.PDF (accessed on 29 August 2004).

http://rgcresearchjournal.org/volume-I,issues-I,april-june2012/commerce and management/ pro-poor tourism approach in India-statusandimplication.pdf.

Jenkins, C. L. (1991). "Tourism policies in developing countries," in Medlik, S. (Ed.), Managing Tourism, Butterworth- Heinemann, Oxford, pp. 269–78.

PPTP (2011). What is Pro-poor Tourism? http://www.propoortourism.org.uk/what_is_ppt.html.

Pro-Poor Tourism Partnership (PPTP) (2004b). Tourism and poverty reduction – making the links. Pro-Poor Tourism Info-Sheets, Sheet No.3. http://www.propoortourism.org.uk/info_sheets/3%20info%20sheet.pdf . Pro-poor tourism: improving livelihood opportunities for local people, http://www.propoortourism.info/intro.html.

Sofield, T., Bauer, J., Delacy, T., Lipman, G. & Daugherty, S. (2004). Sustainable Tourism–Elimination Poverty (ST–EP): An Overview. Queensland, Australia: CRC for Sustainable Tourism.

Stephanie Chok, Jim Macbeth & Carol Warren (2007). Tourism as a Tool for Poverty Alleviation: A Critical Analysis of 'Pro-Poor Tourism' and Implications for Sustainability, Current Issues in Tourism, 10(2–3), 144–165.

Tour Directory (2015). Retrieved from http://www.ecotourdirectory.com/pro-poor-tourism.htm on 16 January 2015 at 10.00 Hrs.

Travel Guide (2015). Retrieved from https://www.ixigo.com/travel-guide/uttar-pradesh on 15 January 2015 at 15.05 Hrs.

UNESCAP (2003). Poverty Alleviation through Sustainable Tourism Development. UN, New York. http://unescap.org/ttdw/Publications/TPTS_pubs/Pub_2265/pub_2265_fulltext.pdf.

United Nations (1999). Facilitation of Travel in the Asian Region; ESCAP Tourism Review No. 20, New York, p. 3–5.

Uttar Pradesh (2015). Pro-Poor Tourism Development Project Environmental and Social Management framework Draft.

Uttar Pradesh Tourism (2015): Retrieved from http://uptourism.gov.in/pages/top/experience/top-experience-literatur on 15 January 2015 at 15.15 Hrs.

Uttar Pradesh Tourism Report (2015). Retrieved from http://uptourism.gov.in/pdf/english_080515.pdf on 15 January 2015 at 15.30 Hrs.

Walter, J. & Mandke, P. (2002). "Exploration of the National Policy Issues Related to the Use of Tourism Development in Poverty Reduction in Southeast Asia," Conference Proceedings: Tourism in Asia, Hong Kong: Hong Kong Polytechnic University.

World Bank Report (2015). Retrieved from http://wwwwds.worldbank.org/external/default/WDSContentServer/WDSP/IB/2015/05/01/090224b082e2d4bc/RenderedPDF/India000Uttar00ronmental0assessment.pdf.

World Travel and Tourism Council (WTTC) (2012a). Travel and Tourism: Economic Impact 2012. Retrieved on Jan 2013, from http://www.wttc.org/site_media/uploads/downloads/world2012.pdf.

WTO-OMT (2002). Tourism and Poverty Alleviation. Madrid, Spain: WTO-OMT.

CHAPTER 10

THEORY AND PRACTICES OF ECOTOURISM FOR SUSTAINABLE TOURISM

G. ANBALAGAN

Assistant Regional Director,
Indira Gandhi National Open University (IGNOU),
Regional Centre, Madurai, Tamil Nadu, India, 625018

ABSTRACT

Tourism is one of the major businesses in any country and also contributing to the economic development of the nation besides exploring the culture, tradition and natural resources. Ecotourism is a major step towards attaining sustainable practices in the tourism sector and promoting the economical development with equitable resource mobilization. Many countries during the recent decades have witnessed the importance of tourism to attract the foreign visitors and thereby generating considerable income generation with the involvement of local resource exploration. Environments are always a changing phenomenon and because of the vast developments in the tourism sector and modern practicing styles of human being large quantum of wastes are disposed in the natural resources and thereby conservation of nature is a major challenge in the developing nations. Increased consumption of natural resources, deforestation, industrial developments, exploitation of wildlife and improper dumping of wastes in the natural places are the major challenges to the ecotourism besides varied climatic conditions. Lack of knowledge on the protection of the environment among the tourists is also one of the reasons for increasing waste disposal in the tourist areas. The governments need to train knowledge and value-oriented manpower for the servicing sectors of the tourism area so as to maintain ecological balance and to minimize the impact of pollution on the natural environment.

The present chapter describes about the theory and major practices for ensuring ecotourism and strategies needed to promote healthy and sustainable development in the tourism sector in order to build inclusive and ecologically sound tourism areas for the future.

10.1 INTRODUCTION

Tourism has attained a major shift during the technological era with the massive developments of Internet-based access to information and worldwide practice of connection with nature for leisure, joy and peaceful recreations. Many countries have understood the importance of tourism to showcase the culture and traditions among youngsters to boost the connectivity of nature with human settings as many technological innovations have slowly created a gap between the human and nature in the recent years. Ecotourism is important concern among many nations to develop ecological sensitivity among human beings, and it helps to build inclusive knowledge on the importance of an environmentally friendly approach to conserve the forests and wildlife towards healthy and sustainable future. Overconsumption and deforestation activities in many regions slowly devastated the growth of greeneries and a massive conservation approach is a dire need to safeguard the ecology and environment to protect the mother earth for the safe consumption and healthy life of our future generations. Many nations use the term ecotourism to conserve the natural resources by exploiting tourism in such places and at the same time promotes sustainability for a futuristic approach which envisages the importance of tourism for the development of the ecosystem and thereby generating potential income for safeguarding the natural areas. Tourism activities are interlinked with sustainability in almost all kinds and scales of tourism and environment (Clarke, 1997). Most of the school students are encouraged for ecotourism as part of the conservation activities as well to tune their sensitivity for the conservation of nature and environment through exclusive strategies of ecotourism. Such activities not only help to understand the importance of nature and wildlife and explores their knowledge to induce several environmental conservative practices to combat the climate change, global warming, deforestation, extinction of rare species of flora and fauna, wildlife conservation, river pollution, and so many environmentally connected important problems. Often it is very late the governments' initiatives towards conservation of natural areas through ecotourism as most of the tribes in the hilly and mountain forests able to conserve the important species through their traditional activities and sense of knowledge.

10.2 ECOTOURISM

Ecotourism is the conceptual phenomenon practiced after the 1980s to promote sustainable tourism related practices in order to conserve the nature and elicit the problems on environmental concerns. Due to the modern practices and rapidly changing lifestyles among human beings all over the world, it is pertinent to promote and practice healthy and safe practices in all forms of life to conserve the Mother Nature from the evils of pollutants as the environmental issues are alarming across many countries. There are a variety of issues surround us which either directly or indirectly affects the natural environment and pollute the natural resources, in turn, affects the human and wildlife. Climatic change in the recent years is the major example in these context since heavy vehicular emissions, air pollutants, industrial emissions, burning of fossil fuels, smokes, burning of crackers, accumulation of excess obnoxious gases in the atmosphere and several other causes the pollution of the atmosphere which affects the ozone layer. The ozone layer is very important to prevent the ultraviolet radiation passes on the earth from the rays of the sun which cause several adverse impacts on the human and animals and can cause severe skin cancer problem. Furthermore, the excess concentration of carbon dioxide gases in the atmosphere causes global warming effect which leads to melting of ice caps in the polar region and thereby increases the sea water levels. Excess sea water levels damage the seashore and even engulf the land slowly and hence the flooding affects the natural lands. Therefore, to curb the menace of air pollution, increasing greeneries, and forest conservation practices are much needed to conserve the nature. The excess greening on the earth absorbs the carbon dioxide from the air and thereby reduces the pollution in the atmosphere. Air pollution is the major cause in the recent years to maximize the impact of respiratory diseases among human beings and depletion in the rainfall and several other causes. Indeed the green trees act as a protector of the Mother Nature and enormously contributes to the decrease of air pollution.

Sustainable tourism has become much concern which reflects the traditional concepts and cultural principles in any region, and moreover, sustainable tourism does not necessarily contribute to those of sustainable development (Hunter, 1995). Majority of the country like India where the cultural traditions and natural mountains, forest covers, and adventurous spaces are more common become helpful in developing the ecotourism practices. Most of the tourists try to spend their leisure time with family at the nature bound areas to enjoy the moments as these natural areas like forests, beaches, mountains, waterfalls, etc., provides space for healthy activities like swimming, adventure walk, trekking, gliding, and sightseeing, etc. Ecotourism is an entirely different form of tourism practice as it not only simply visiting the natural places and it has close linkage with the nature and helps the human beings to aware the importance of nature and its ethical practices like traditional medicinal plants which are helpful in combating several diseases. Visiting the natural farmland containing the medicinal plants helps to breathe the natural form of air with the mix of aroma arouses from medicinal plants helps to increases the power of health. Likewise, taking a bath from a waterfall where the water flow through several forms of medicinal plants and varieties keeps the body healthy and gives full energy in the form of mental peace. In the southern part of India, many tribal peoples provide spa and body massage to the tourists at their locally serene green villages with the herbals and medicinal extracts which provides positive energy in the mind and body. Many tourists prefer to visit such serene villages of tribes to enjoy the natural massage and healthy air to breathe. The tribal peoples can able to sell their locally available products like honey, beans, herbals, fruits, and veggies popular in such areas with the tourists to earn income. The cultural dance of tribes and exotic beauty of the local villages attracts the tourists and the resorts formed by the government, and private firms are helpful to enjoy the ecotourism in most of the regions. Ecotourism is a type of marketing tactics in the competitive edge of tourism in the global arena and helps to connect the man with nature in many ways to enjoy the pleasure and serene beauties of nature. Nature always favor to the human beings as the expanse of land and sea exploited to cater the needs of an exponential number of peoples seeking pleasure from mother nature and which totally clear that tourism is ultimately depend on the availability of natural areas.

Ecotourism helps the peoples located in the ecological tourism region for their livelihood improvement and increasing the potential of earning through so many allied activities with the tourists by the way of selling the local products and helping in the activities related to tourism. It is evident from several tourist areas that the tribes are associated with tourism departments in extending support during the trekking, sightseeing in the forests, nature walk, and adventurous sports, etc. In view of degrading environment and deforestation activities, mining, poaching and extinction of important species the forest ecosystem is facing a lot of problems, moreover the tribal peoples living close to forest areas also facing several issues with the livelihood improvement. Many researchers coined import theoretical perspectives on ecotourism, and they are as follows:

Rajasthan in India is one of the states with many cultural traditions and tourism attractions and tourism activities in Rajasthan state act as dynamic practice in uplifting the social status of the peoples (Batra 1990), the peoples need much orientation on the social barriers and orthodox, a traditional system to facilitate foreign tourists. The foreigners prefer to visit Rajasthan for various reasons, and the monuments, palaces built by the erstwhile kings of the state makes attraction.

The difference between the concept of sustainable tourism and sustainable development cannot be overruled, and at the same time, sustainable development needs more constructive approaches for the improvement in changes as well as practices in the field level. Sustainable tourism should be consistent with the tenets of sustainable development (Stabler & Goodall, 1996) so as to develop supporting policies and conceptual frameworks. The theoretical concept of sustainable tourism has foundation on the understanding of the belief in which it has built upon; however, there has been consistent and fundamental failure to build a theoretical linkage between the concept of sustainable tourism and its parental paradigm (Lanfant & Grabum, 1992). Ecotourism has been nurtured to develop sustainable development in the field of tourism although there was a pitfall in attaining theoretical goals in which it has emerged, and more so the ecotourism theory has not been successfully put into practice all over the years due to many ground-level reasons. Even though the prescriptions and guidelines have emerged to promote successful implementation, there has been a lack of widely utilized methods to identify potential progress of the ecotourism in sites (Lindberg & Hawkins, 1993). The International Union for Conservation of Nature (IUCN) on National Parks and Protected Areas (CNPPA) defines ecotourism as environmentally responsible travel and visit to relatively undisturbed natural areas to enjoy and appreciate nature which helps to promote conservation of nature and provides socio-economic benefit to the local people involved in tourism activities (Ceballos Lascurain, 1996). The 'earth summit' in 1992 has laid the foundation for the need to enforce principles of sustainable development within the wider economic and social processes highlighted the need for sustainability and potential of the tourism for advancing the goals for developments (Berry & Ladkin, 1997; Pigram & Wahab, 1997).

Empowering the tribal population is one of the major challenges in any country as they mainly depend on the forest and wild related activities for their livelihood. In order to improve their economy and social participation, the ecotourism is one of the keys in several ways to utilize their talents for tourism practices. The tribes in northeastern states of India is the major example in this regard as they use bamboo for making of several antiques, furniture, hunting, and toys, etc. Majority of the tribes in the forest areas much aware of the important produces from forests such as honey, fruits, medicinal plants, and its roots, etc. furthermore they dwell in naturally made protective places and well aware of animal activities and safeguarding practices from wild animals.

In recent years, many employees of corporate industries especially working IT sectors prefer to spend their holidays in the resorts close to beaches, forest areas and other tourism places where the natural scenic beauty, spa, entertainment sightseeing, trekking, nature walks, bathing in the waterfalls and aeration at medicinal gardens as most of the urban areas become polluted and the air pollution is common cities. In order to stay with the nature connection and enjoying the natural beauties, the tourists prefer such type of stay for rejuvenating breathe to purify the body. On the other hand, many tourism agencies performing a vital role in making such natural arrangements for the tourists and advertising in the online social media based portals to attract the peoples which help to earn much profit through tourism in the natural places of importance such as forests, sea beaches, and waterfalls, etc.

10.3 TOURISM AND SUSTAINABLE DEVELOPMENT

The number of International travelers is expected to increase from 698 million to 1.6 billion by 2020, according to the World Wildlife Fund (WWF). Furthermore, many countries earn major economy through the foreigners visiting to enjoy tourism activities as the rich natural heritage, medical facilities, archaeological attractions, and wildlife, etc. increases the visit of foreigners. In India, the tourism facilities are vast and diverse due to rich cultural diversity and natural landscapes including forest areas. The increase of tourists favors the income generation to the national development and employment opportunities to many peoples. Sustainable practices alone can persist in a long-term approach since many natural water bodies including groundwater pollution are more common due to manmade activities and other industrial activities. Many naturalists and conservationist in India and abroad much worried on the loss of biodiversity and extinction of rare species due to pollution and hunting activities. The forest dwellers in many areas cut trees of important economic nature and involved in the hunting/poaching of animals for medicines and other reasons.

According to the world tourism industry, growth of national parks and protected areas are the major concern to infuse effective environmental education to the students as young minds conceive a lot of ideas towards conserving the nature when they exposed to connect with the nature and inculcated the effective environmental education. Thus helps to build sustainable practices from the younger stage with motivational aspects to attain sustainability and to safeguard the important areas of nature for the future dynamism. The environmental education at school level necessitates the importance of ecotourism so as to inculcate the importance of nature and environment and combating the ill effects of pollution through sustainable practices at various levels. Many research studies support the importance of nature-based connectivity to study the environmental education at field level for training the younger generation with potential knowledge and skills needed for combating the issues. Sustainability cannot be possible unless the effective practices towards conservation are in place as most tribal people living in forest areas lack knowledge. Non Governmental Organizations (NGO) involves in protection and conservation of the environment, and they do mobilize the tribes in forest region to conserve the wildlife and trees through motivations and explaining the need of sustainable approach for the future consumption of resources.

Developing the young entrepreneurs at the ecotourism areas to produce natural products famous in such areas to sell with the tourists is one of the importance aspects for their livelihood improvement and also to prevent the migration of rural peoples into urban areas seeking employment since many urban areas in particular cities in India are overpopulated and facing the crisis of water consumption, scarcity of water, loss of wetlands, peri-urbanization, lack of facilities for managing municipal solid wastes. Even in urban areas, the development of eco-parks, green buildings, and waste management facilities can help to attract the peoples to enjoy nature walks since in the modern life many individuals facing work pressure and psychological disturbance due to various issues. The development of ecological parks, water theme parks and conservation areas in the urban areas are much helpful to the birds and other small species of animals as the increasing urbanization, cellular towers, lay of roads, cutting of trees, improper waste dumping, and water shortage, etc. are major issues in the depletion of bird species. The zoological parks in many countries have tourist attractions since the kids visiting zoos likely to develop the kindness on animals and the importance of wildlife for the sustainable environment. In the context of neoclassical economics and more traditional ecological perspectives, the growth of economy or resource substitutions approach towards development diametrically oppose the eco-centric and deep ecology approach which represents the virtual rejection of the sustainable exploitation of natural resources (O'Riordan, 1981a, 1981b; Turner, 1993). Therefore, the sustainable development always supports combining developmental theories with the sustainability.

The loss of environmental conditions naturally degrades the approach towards sustainability and more over the tourism facilities in the ecological sensitivity areas require much attention and constructive approach with the involvement of communities at local level to attain long-term success. The programmes on ecotourism support the partners in the effective management of tourism plans for land use, construction of buildings, guidelines for visitation, developing tools for resource managers and to offer technical assistance through studies and research on environmental impacts assessments. The African countries provide ample number of avenues for ecotourism due to enormous wildlife in the region, and many tourists prefer to enjoy the sightseeing of wild animals. Wildlife conservation, on the other hand, has given much priority by the respective government to ensure the protection of wild animals from hunters and other illegal activities. The phenomenal growth of tourists visiting such region is the major example for the increasing trend of ecotourism and wildlife conservation. Climate change and environmental pollution are the major threats to the loss of ecosystem as increased rains, floods and disasters rapidly affect the natural ecosystem.

While exploring the sustainability concept, there are several issues that emerge adherence to the environmentalism and conservation of nature. The use of natural resources for the socio-economic well being of the peoples and effective distribution of cost benefits must ensure the sustainable socio, economic and environmental conditions (Fox, 1994). Many countries while increasing the facilities for the ecotourism practices fails to develop equitable environmental conditions to sustain in a long-term approach. Globally environmental detritions are increasingly contributed to the loss of biodiversity and environment, and on the other hand, effective conservation practices through appropriate participatory approach are very much needed to equalize the losses. Lack of planning, insolvent measures and rapid consumption of natural resources affects the growth of ecotourism, and suitable alternative models need to be developed in the areas like waste management, land consumption, resource utilization and conservation of wildlife. In order to conserve the non-renewable resources, an effective mechanism needs to be implemented at ground level to minimize the waste and energy. Further, the consumption of freshwater, soils and natural ecosystem like lake, river, etc. due care is absolutely necessary to promote ecotourism without much affecting the nature and restrictions need to be practiced within the ambit of sustainable yield so that the imbalances between the resource consumption and conservation practices properly managed. Climate-based tourism activities like winter or summer to the natural areas like hills, mountains have a profound increase in the modern age due to an increase of tourists in such regions. During the peak periods of tourist attraction and visits, there has been a huge accumulation of wastes which can increasingly damage the surrounding environment and wildlife. The tourist areas are facing the crisis of water, accommodation facilities, and handling the movement of vehicles which disrupts the entire surrounding environment in the places like Shimla, Ooty, Darjeeling, Kashmir, etc. in Indian continent. The tourists face increased expenses in availing facilities in such tourism areas at the time of peak periods due to congestion of vehicles and an increased number of tourists for attractions. There need proactive measures to safeguard the environmental facilities and effective waste management techniques to combat the problems of pollution and to protect the well being of nature and also to prevent the diseases that arise from waste handling.

Sustainability in the modern world of developments is a mega challenge until the proper handling mechanisms are set in place as land development and encroaching the agricultural lands for developing tourist attractions has severely affected the agriculture land usage. The increased consumption of water at the tourism places need proper recycling and re-use facilities to conserve its use through agricultural lands adjacent to the areas. The abnormal use of plastic carry bags poses a serious threat to wildlife as it prevents the penetration of rainwater in the soil and takes thousands of years for decay in the soils. The food waste thrown with plastic bags is normally consumed by the wild animals which create high ill-health effects on them. Handling of plastic wastes is the serious mega problem in the tourism areas as the quantum waste generation is comparatively higher than other types of wastes and lack of recycling of plastic wastes facilities are the major threat to the environment.

The rich wildlife at Ghana has attracted international tourists, and the country has adopted several programmes for popularizing tourism through conservation approach. Canopy walkway is one of the awards winning tourism activity in such region where the tourists enjoy a spectacular view of the rainforest ecosystem at Kakum National Park of Ghana. The birds and wildlife can be observed through canopy walkway including the animal activities at nature through rural people help of Ghana (Source: IGNOU Study Materials on Ecotourism). The concept of sustainable development arose in the broader sense of international concern on the issues of ecologically important areas. Further, in the tourism sector, a possible solution to the environmental and social degradation of the resources and moreover tourism is a resource-dependent industry (Cooper, 1995; Murphy, 1998).

Absolutely ecotourism can make a bigger profit to the nation for the developmental activities and helps to earn potential income for generating fund towards conservation of nature. On the other hand, there are a lot of threatening issues like destroying resources through the ecotourism if proper planning is not meticulously emphasized. The polar region like Antarctica faces the dump of waste materials through visitors which threatens the life of the animals in the ice caps. Furthermore, the climatic change around the globe already altered the polar region through the slow melting of ice caps and the same pose serious threat to the polar bears.

Many rivers in India are facing the problem of pollution due to ritual activities and dumping of wastes besides industrial and other pollutants entry into the natural stream. Though naturally, rivers have self-purification potential for a certain extent, the excess dumping of wastes and intrusion of wastewater creates a hurdle to the self-purification process and damage the aquatic animals.

10.4 MEASURES TO IMPROVE ECOTOURISM

The appropriate government needs to perform a vital role in the development of ecotourism with a sustainable approach to conserve nature. Mass awareness among the people about the importance of a sustainable approach to conserve the environment and preventive measures to address the ill-health effects of pollution needs to be addressed through local volunteers and non-governmental organizations to sensitize the society. The problems and prospects of the ecotourism need to be explained in a much better way through the involvement of experts and local planners for the developmental activities and conservation of the environment since the waste management needs greater significance. The displacement of human settlements and construction of hotels, resorts, and recreation points are to be meticulously planned without affecting the nature, and the rehabilitation and resettlement of displaced human society need to be taken care with necessary provisions and facilities. Restrictions on the indiscriminate construction activities and careful execution of policies for planning and developments need to be practiced to conserve the local environment so that the massive construction at the hill slopes, coastal regions, and mountains can be controlled to some extent. Tourism can undoubtedly create positive economic returns for destination countries; foreign exchange earnings, employment growth, tax revenues, and can substantially stimulate the economy overall (Inskeep, 1991). The sustainable approach to tourism planning is based on the achievement of two prerequisites: a strategic and long-term orientation in tourism planning and multiple stakeholder participation in the planning process (Simpson, 2001).

Training the local youths with the key objectives of environmental education to perform as tourist guides with foreign languages which will help the tourists to enjoy their activities smoothly and observe the local places easily. The rural peoples at ecotourism areas may be provided options to sell their local products of importance at the shops and stalls with the tourists to earn income for livelihood expenses, and thus the government can provide self-employment opportunities to the needy peoples.

10.5 CONCLUSION

Ecotourism attained significant improvement in many countries as foreign travelers prefer such type of tourism for leisure and to connect with nature besides enjoying the sightseeing on wildlife, birds, flora, and fauna. Developments for tourism activities are inevitable at the same time careful planning and execution of regulatory mechanism can prevent the indiscriminate use of natural resources which can be helpful to the conservation strategies and improving the local facilities like the ecological park, water conservation, health and sanitation facilities to the rural peoples. Nature has several good things to the human beings, and at the same time, excess consumption of natural resources and harming the nature threatens the human society in the form of flood, climate change, water scarcity, etc. Hence, meticulous planning and careful execution of developments for the ecotourism to generate economical resources are very much important. Several traditional practices are slowly disappearing due to technological development, and machine-based work in the modern era and the increasing population, on the other hand, prefers to procure information in a digital way. The tourism sector with the advancement of digital communication and internet-based media sharing practices attained steady shape to cover anyone from any part of the world. Digital transformation in the current world explicit the avenues for financial transaction online, banking services, and information sharing practice. However, the technologies for the tourism sector need to be practiced for a sustainable way to attain proper utilization. Sustainability can be achieved with the real conceptual practice of equitable utilization of resources from nature and addressing the gap between optimal consumption and resource availability in order to maintain balance and save our mother earth from the evils of pollution and deterioration of natural resources. Theoretical perspectives on sustainability can be helpful to better utilization of ecotourism for the conservation of nature and human community if followed through pre-determined approach with careful consideration of available use of resources.

KEYWORDS

- **ecotourism**
- **environment**
- **pollution**
- **sustainable development**
- **wildlife**

REFERENCES

Batra, G. S. & Chawla, A. S. (1994). *Tourism Management: A Global Perspective*. Deep and Deep Publications, New Delhi, 10–40.

Berry, S. & Ladkin, A. (1997). Sustainable tourism: A regional perspective, *Tourism Management, 18*(7), 433–440.

Ceballos-Lascurain, H. (1996). *Tourism, Ecotourism, and Protected Areas*. Gland, Switzerland: IUCN.

Clarke, J. (1997). A Framework of Approaches to Sustainable Tourism. *Journal of Sustainable Tourism, 5,* 224–233.

Communities in eGyanKosh. http://egyankosh.ac.in/ (accessed on 16 March 2019).

Cooper, C. (1995). Strategic planning for sustainable tourism: the case of the offshore islands of the UK. *Journal of Sustainable Tourism, 3*(4), 191–209.

Fox, W., (1994). Ecophilosophy and Science, *The Environmentalist, 14*, 207–213.

Hunter, C. (1995). On the Need to Re-Conceptualize Sustainable Tourism Development. *Journal of Sustainable Tourism, 3,* 155–165.

Hunter, C. (1997). Sustainable tourism as an adaptive paradigm. *Annals of Tourism Research, 24*(4), 850–867.

Inskeep, E. (1991). *Tourism Planning: An Integrated and Sustainable Development Approach*, John Wiley & Sons, New York.

Lanfant, M. F., & Graburn, N. H. (1992). International tourism reconsidered: the principle of the alternative. Tourism Alternatives: Potentials and Problems in the Development of Tourism, 88–112.

Lindberg, K., & Hawkins, D. E. (1993). Ecotourism: a guide for planners and managers .

Murphy, P. E. (1998). Tourism and sustainable development, in W.F. Theobald (ed.), Global Tourism, Butterworth Heinemann, Oxford, pp. 173–190.

Neto, F. (2003). A New Approach to Sustainable Tourism Development: Moving Beyond Environmental .

O'Riordan, T. (1995). Linking the environmental and social agendas. *The Environmentalist, 15*(4), 233–239.

Pearce, D. W. & Turner, R. K. (1993). Defining Sustainable Development, in D. W. Pearce (ed.), *Blueprint 3: Measuring Sustainable Development,* pp. 3–14, Earth Scan, London.
Ross, S., & Wall, G. (1999). Ecotourism: towards congruence between theory and practice. *Tourism Management, 20*(1), 123–132.
Simpson, K. (2001). Strategic planning and community involvement as contributors to sustainable tourism development. *Current Issues in Tourism, 4*(1), 3–41.
Stabler, M., & Goodall, B. (1996). Environmental auditing in planning for sustainable island tourism. *Sustainable Tourism in Islands and Small States: Issues and Policies, 1,* 170–196.
The International Union for Conservation of Nature (IUCN). https://www.iucn.org/ (accessed on 16 March 2019).
Wahab, S. & Pigram, J. J. (eds.). (1997). *Tourism, Development, and Growth: The Challenge of Sustainability*, Routledge, London.
World Wild Life. https://www.worldwildlife.org/ (accessed on 16 March 2019).

CHAPTER 11

INTERNAL BENCHMARKING MODEL FOR DESTINATION PERFORMANCE EVALUATION

TONEY K. THOMAS

Taylor's University Lakeside Campus, No.1 JLN Taylor's, 47500, Subang Jaya, Malaysia, E-mail: tonythomas@taylors.edu.my

ABSTRACT

Destination performance evaluation has become an increasingly important task for effective destination management. However, it is perceived as a complex task due to the inclusion of diverse subsectors, business components and performance elements of the tourism and allied sectors. Although destination performance evaluation is a tedious task, it is essential for enhancing destination image, positioning, and ultimately achieving market leadership. The positioning of a tourist destination in terms of its performance is inevitable for building external competencies. There are many advantages in evaluating the performance of tourist destinations, such as, to understand the efficiency of tourism resources, the strength of resource availability, marketing strategies and promotions, and to strengthen policies. Destination performance evaluation also facilitates the enforcement of successful policies and minimizes and eliminates unsuccessful policies and processes. Traditional destination performance is evaluated by using the satisfaction index of the visitors. However, such attempts will only limit to evaluate the perception of the visitors. Destination benchmarking is a comprehensive and holistic approach that can be adopted for destination performance evaluation in a most effective and critical way. Nevertheless, there are several benchmarking models which exist at different levels, such as: (a) Internal Benchmarking, (b) External Benchmarking, (c) Generic Benchmarking, (d) Competitive Benchmarking, (e) Functional Benchmarking, and (f) Relationship Benchmarking. Rare attempts are made to consolidate destination performance indices to develop a destination performance model. This chapter is an attempt to develop an internal benchmarking model and adoption methods to execute it.

11.1 DESTINATION BENCHMARKING

Benchmarking destination is relatively a new exercise in tourist destinations. It is an important task to position a tourist destination in the international market (Thomas, 2011). In the current scenario, new destinations are entering into the leisure and recreation market. This leads to new demand segments arising from emerging economies, with new tourism business models mushrooming in travel and related service sectors. The new (tourism) generation are significantly different from the existing traditional tourists as such diverse traveler groups are formed. Also, looking on the future demand and growing market of baby boomers, as well as constructing tourism in the society by the concept 'tourism for all' – including people with different abilities – this paves the way for destinations to be more competent than their competitors or neighboring destinations. The emerging core destination management activity is the periodic monitoring of visitor satisfaction and the regular resource monitoring of other involved parties that are vulnerable to damage caused by tourism (Mathias, & Klaus, 2004). Benchmarking is a tool to build competency and enhance quality, while also to turn a critical weakness into a strength. Therefore, destination benchmarking can be considered as a diagnostic tool that compares a destination's tourism performance relative to other destinations in a comprehensive manner using appropriate methods and techniques. Depending on the purpose, benchmarking can be classified as: External Benchmarking (identifying the performance of one destination to another); Internal Benchmarking (measuring own performance of the destination in order to specify areas that need to be benchmarked (Zairi, 1992). Generic Benchmarking (identifying the best practices or the best performing business in the industry and improve one's own performance by adopting good practices by others or guidelines established by professional national or international organizations. Competitive Benchmarking (a comparison of one destination with its direct competitors only); Functional Benchmarking (a comparative performance matching that attempts to seek world-class excellence by comparing business performance not only against competitors but also against the best business operating in similar fields and performing similar activities or having similar problems, but in different industries) (Kozak, 2004) and Relationship Benchmarking (similar to External Benchmarking, in this process the benchmarker already has a relationship in advance of a benchmarking agreement) (Kozak, 2004).

The above typologies interrelated with each other. It is important to analyze 'own' destinations' performance before external benchmarking with another destination. Without internal benchmarking, external benchmarking is not possible. Consequently, internal benchmarking can be set as the basic criteria to perform other types of benchmarking which is suitable for a product or service.

11.2 WHO HOLDS THE STAKE?

Tourism is an extraordinary 'inter-,' 'intra-' and 'multi-disciplinary' phenomena. Sometimes, a major role of tourism development is performed by an unrelated allied sector, or tourism development can be downshifted by an unrelated allied sector. Therefore, it is truly a complex task to set a performance standard unless there is a strong 'meta-governance' system that prioritizes tourism as a development tool for a specific 'state.' A new destination will have a great challenge in overcoming issues of 'tourism ownership'; it is difficult to specify who owns tourism in a country as it is a complex network of processes and functions. In these processes, there are people or the 'destination community,' local residences, tourism entrepreneurs and tourists, where there are direct human interfaces happening. Also, there are regulatory bodies such as national and regional tourism organizations, NGOs involved directly in the process to set functions of the tourism system in a specific geographic setting. Furthermore, there are tourism resources including culture, heritage and natural resources which are tangible items that spur tourist visitations. A destination's performance is determined by the total performances of the stakeholders. It is, indeed, difficult to consider several subsectors in order to evaluate the performance of a tourist destination. However, a closer estimation is possible by finding the most involved performance components in tourism development. In order to 'model' internal performance evaluation of a tourist destination, in this study, four performance core components have been identified. They are: (a) Tourist Satisfaction Index; (b) Community Impact Assessment; (c) Managerial Action (Tourism Policy); and (d) Stakeholders' Perception (Tour Operators). Service quality has been a traditionally accepted evaluation point for any service as word-of-mouth is an important promotional tool for service consumption; in fact, tourist satisfaction index has been considered as a precedence in this context. Therefore, tourist satisfaction indices are subdivided by 'cognitive indices.' Meanwhile, one main characteristic of tourists at a tourist destination is 'temporality' – the term 'temporality' is important, where a tourist performs his/her role in a new social, cultural and geographic setting temporarily, and what experience is sought at that point of time is the creation of reflection. Tourists travel to a tourist destination with expectation. Thus, during the reflective phase, if the expectation exceeds the service performance, a positive disconfirmation is the result. Therefore, in this study, specifically for the satisfaction component, the expectancy disconfirmation theory has been adopted. The cognitive satisfaction involves: (a) Cognitive Product Satisfaction; (b) Cognitive Service Satisfaction; (c) Cognitive Emotional Satisfaction; and (d) Cognitive Dissonance. According to Oliver [4], the cognitive approach leading to tourist satisfaction is the consumer's response to the congruence between performance and comparison standard (Bosque, & Martin, 2008). A cognitive-affective approach to tourist satisfaction explains satisfaction formation as influenced by the individual's cognitive judgments (Bosque, & Martin, 2008). According to Oliver (1993), satisfaction is influenced by cognitive evaluations such as expectations and disconfirmation. Expectations are defined as the individual's belief about how a product is likely to perform in the future (Bosque & Martin, 2008). The review of the tourist's expectation using a simulation theory (Sherif, 1961), expectancy disconfirmation falls under interactive theories which focus on explaining tourist satisfaction between situational and personal facets (Sirgy, 2009). The expectation level then becomes a standard against which the product is judged (Yuksel, & Yuksel, 2001). Individuals encounter a psychological conflict when they perceive discrepancies between performance and prior belief (Bosque & Martin, 2008). Studies also revealed that under these circumstances, satisfaction would be led by expectations (Bosque & Martin, 2008; Churchill Jr., & Surprenant, 1982). A positive disconfirmation is expected to generate positive behavioral reactions from customers such as return, repeat purchase recommendation, and positive word-of-mouth (Neal, & Dogan, 2008).

11.3 DESTINATION COMMUNITY INVOLVEMENT IN TOURISM DEVELOPMENT

Destination community can be a direct or indirect stakeholder of the tourism industry. It's also suggested that a lack of community involvement has been highlighted as one of the main factors leading to a high rate of tourism plan failure (Burns, 1999; Yuksel, & Yuksel, 2001). The impact of tourism on the community is 'bipolar.' The success or failure is based on which 'pole' dominates the impact. The less negative impact with 'high' positive effect can be a better balance. "A general appreciation of tourist activity appears to be characteristic of most travel destination residents, incidents of negative reaction or resistance can sharply decrease traveler satisfaction and severely damage a community's hospitality" (Pearce II, 1980). If the tourist is a foreigner, the inhospitable reactions can be exacerbated and are reflected in the form of jealousy, xenophobia, and may be manifested in the form of disinterest, to rudeness, to physical hostility (Pearce II, 1980). The rationality principle of community receptivity on tourism development is associated with the 'social exchange theory.' This perspective fulfills the rationality principle to measure economic impact through the examination of social exchange. In the economic context, increased income injection to the destination community is not only a parameter of social sustainability, but also, it is the crucial ability to retain the income generated from tourism in the community.

11.4 THE POWER RELATION

"It has become widely recognized that planners and entrepreneurs must take the views of the host community into account if the industry is to be sustainable in the long term" Community, development authorities, planners and policy makers alone can exert direct control over the destination's tourism development (Jamal, & Getz, 1997). Most narrow interpretations show that individuals often rely on coalitions with other private or public individuals or agencies (Reed, 1997). As such, the community tourism development is a continuous structured and collective decision and action by individuals and agencies in emergent tourism settings is required, where interests are not collectively organized (Reed, 1997). Since economic development is one of the key outcomes for the nation through tourism development, frequently non-economic impacts are not valued; instead, this must be incorporated into the policy-making process (Lindberg, & Johnson, 1997). The main issues observed in the context of tourism development is that different parties involved in the decision making for tourism development have their own vested interests, and these interests may conflict with other party's priorities and may cause more developmental issues and ultimately, adversely impact tourism development. Whatever the causes, the ultimate suffering goes to the destination communities. Establishing and optimizing a balance by different stakeholders will make a difference in increasing sociocultural and economic impacts. Therefore, it is confirmed that community leadership is heterogeneous, being drawn from a number of power basis (Reed, 1997). Improper and imbalanced distribution of power causes losses to everybody: residents, visitors, and the tourism industry (Keogh, 1990).

11.5 NEED OF COMMUNITY IMPACT ASSESSMENT

Tourism's impact on the community is quite different from that of other economic sectors. The success of sustainable tourism development depends on the active support of the local populations (Gursoy, & Rutherford, 2004). Positive social capital development is the outcome of the costs and benefits of tourism development in the performance scorecard of the destination community (Thomas, 2017). Many researchers suggest that tourism-dominated/interested communities should plan their evolution more systematically, thereby taking into account residents' attitudes and perceptions about its growth at the outset (Reid, Mair, & George, 2004). The impact of such growth must be monitored over a long period of time, and their effects on daily life are assessed regularly (Reid, Mair, & George, 2004). To investigate residents' perceptions of tourism's impact, and to test the relationship between these perceptions, the role of tourism in the local economy, persons who benefit from it, engage with it, and community attachment, all have to be assessed (Andereck et al., 2005). It has also been noted that economic benefits are often seen to largely improve the quality of life of residents; sociocultural factors may not always be as positive as it seems (Andereck et al., 2005; Liu, Sheldon, & Var, 1987). Tourism development has an effect on sociocultural characteristics of residents such as habits, daily routines, social lives, beliefs, and values. These factors may, in turn, lead to psychological tension (Dogan, 1989). Studies also explain that, in an area with high levels of tourism, there is often an increase in population as a result of new residents relocating from outside areas (Andereck et al., 2005; Hair, Black, & Babin, 2010). The rapid growth of tourism without planning and proper destination management directly causes loss of resident identity and local culture. From the negative perspective, adverse effect on traditional family values, cultural commodification, crime, drug abuse, degradation of morality, increased use of alcohol, openness of sex, increased prostitution, gambling, crowding of public facilities and resources and declining resident hospitality are the major ones (Snepenger, & Johnson, 1991; Kousis, 1989; Andereck et al., 2005; Hair, Black, & Babin, 2010. There is a possibility that development policy clashes with local requirement when development plans only focus on meeting the needs of the tourists (Andereck et al., 2005). Measuring resident reaction is important in the context of these negative effects which not only affect the quality of life, but commercial tourism ventures may also be hampered or terminated by excessive negative resident sentiments toward tourism development (Williams, & Lawson, 2001).

11.6 THE SERVICE PROVIDERS (TOUR OPERATORS)

The role of tour operators and services associated to their functional areas are an important consideration on destination performance. An organization is characterized by its relationship with various groups and individuals, including employees, customers, suppliers, governments and community members (Freeman, 1999; Sautter, & Leisen, 1999). The effort to understand the complex and elaborate set of interactions among tourists, tour operators, governments and local communities has led to the development of tourist behavior (Smallman, & Moore, 2010). Tourists' decision-making processes are complex, involving many sub-decisions, occurring continuously, even from prior to the decision of 'where to go' through to 'what are we going to do now that we are here' and beyond (Smallman, & Moore, 2010). The dominance of intangible factors in tourism is problematic (Sirakaya, & Woodside, 2005; Smallman, & Moore, 2010). The implementation of specific codes of conduct by tour operators has grown in popularity in recent years. This is a clear indication of tour operators and supply chains have an important role in the image and development of tourists' destinations (Thomas, 2017). Owing to the importance of tourism and sustainable development, tour operators are the main point to educate the tourists. Today, organizations and planning bodies must be more careful to take a hard look at the various types of persons and groups which affect or are affected by the tourist service (Sautter, & Leisen, 1999). The role of tour operators is to act as a negotiation point to create a platform for responsible tourist visitation. Tourism planners need to consider the interest of the different stakeholders in tourism business as defined by their roles and serve them with regards to the particular development initiative (Sautter, & Leisen, 1999). Tourism planners need to be concerned about the perspectives of the stakeholders' group, while recognizing that their interests are exclusively touristic (Sautter, & Leisen, 1999). Since tour operators' business involves direct link with all supply chains, they can be considered as the main evaluation point for a destination's performance.

11.7 THE PUBLIC SECTOR

To view the effectiveness of tourism development at the destination level, tourism policy implementation and effectiveness can be an important measurement point. Tourism policy is a road map of tourism development, which contains the priorities and action plan showing the direction in which the tourism industry needs to grow. Tourism policy development is considered with great importance as a managerial activity by the public sector organization representing the interest of the country's social and economic goals. Tourism policy development and implementation is "a function of political and administrative action, rather than economics or business" (Richter & Richter, 1985; Airey, & Chong, 2010). Tourism policy is used to reflect a broader understanding of the political, societal and human context of public sector-led decision making and policy conceptualization is a holistic function of interactions and processes in negotiation with all the parties directly or indirectly impacted from tourism development (Airey, & Chong, 2010). The choices of selecting the thrust areas of tourism development depend as much on political as on cultural consideration (Richter, & Richter, 1985). The policy development process has to focus on the factors that affect policy formulation and implementation as well as the subsequent effect of the policy (Howlett, Ramesh, & Perl, 2003; Airey, & Chong, 2010). This is with the consideration that the major performance components in the tourism policy is important as the tourism sector consists of political, economic and social facets. Studies also describe that tourism in a place is also typified by a diverse number of organizations and issues (Elliot, 1997; Airey, & Chong, 2010). Therefore, tourism policy decision must be collective and holistic and should be made with the involvement of government agencies, non-government organizations, and businesses (Airey, & Chong, 2010). Coordination of tourism policy is important for its effective implementation, because the policy is a fragmented power structure, where authority and power get diffused (Jansen-Verbeke, & Spee, 1995; Bramwell, & Sharman, 1999; Airey, & Chong, 2010). Due to many expected benefits of involving stakeholders in the tourism policy development and implementation process, conceptualization, realization, and implementation of tourism policy are a collaborative role of different stakeholders. This kind of collaboration is very important as it enhances the coordination of policies and related actions, and promotes consideration of economic, environmental, and social impacts of tourism. Furthermore, stakeholder's collaboration avoids the cost of resolving adversarial conflicts among stakeholders in the long term (Healey, 1997). The stakeholder involvement in policy-making also adds value by building on the new knowledge, insights, and capabilities of stakeholders in the tourist destination (Bramwell, & Sharman, 1999; Roberts, & Bradley, 1991).

11.8 COLLABORATIVE POLICY MAKING

Some recent evaluation of tourism policy implementation effectiveness shows that stakeholders' involvement in developing and practicing tourism policy is inevitable. Stakeholders such as tour operators and travel companies are the frontline segment of the tourism industry that experiences the outcome of tourist satisfaction, issues, and opportunities. In this way, stakeholders can solve many issues in relation to tourism development (Bramwell, & Sharman, 1999). Since tourism policy consists of multiple segments, to protect the interest of the beneficiaries of each segment and to simplify and solve the issues associated with them, inter-organizational collaboration is important, which has been explained in the inter-organizational collaboration theory, which explains tourism development as forming a basis to develop an evolutionary model of partnerships in destinations, and described the "importance of inter-collaboration to assess factors constraining or promoting the effectiveness of such partnership" (Bramwell, & Sharman, 1999; Wang, & Xiang, 2007). Stakeholders and community collaboration also help to overcome power imbalances by involving all stakeholders in a process that meets their needs (Bramwell, & Sharman, 1999; Jamal, & Getz, 1997). It is also suggested that a stakeholder who is impacted by the actions of other sectors has a right to get involved, not only to moderate those impacts, but also to develop the resources and skills required for involvement (Bramwell, & Sharman, 1999). Based on these facts, tour operators can represent to explain the effectiveness of tourism development of a country, which is directed by the public sector and reflected through the tourism policy of a country or state.

Local communities are also an integral part of the tourism policy development. "Planning should draw on the webs of relations found in local areas and build the capacities of stakeholders so that they can have a more direct influence on their own lives" (Bramwell, & Sharman, 1999; Healey, 1997). However, it is observed that often, tourism planning and policy development is structured as a top-down approach and hence, results in public opposition. The study shows that tourism development and resident attitude are correlated. Local communities expect a higher level of economic benefits, but many of the social and environmental consequences are perceived as negative impacts on the destination (Keogh, 1990).

11.9 MODELING INTERNAL BENCHMARKING

Organizations should first begin with internal benchmarking, followed by external benchmarking and generic benchmarking for product or service positioning (McNair, & Leibfried, 1992; Zairi, 1992; Kozak, Bigné, & Andreu, 2005). In order to conceptualize an operational model for internal destination benchmarking, a descriptive and positivist approach in including the destination performance core components are incorporated. Based on the pillars of destination performance identified, the following conceptual framework has been proposed for the destination benchmarking model (Figure 11.1).

11.9.1 METHODOLOGICAL APPROACHES

There are tangible and intangible evaluation components involved in internal destination benchmarking, while services, products, and resources are also an integral part of destination performance. A methodological plurality is indeed important in designing the methodological process. Therefore, a mixed methodological approach has been adopted in the benchmarking

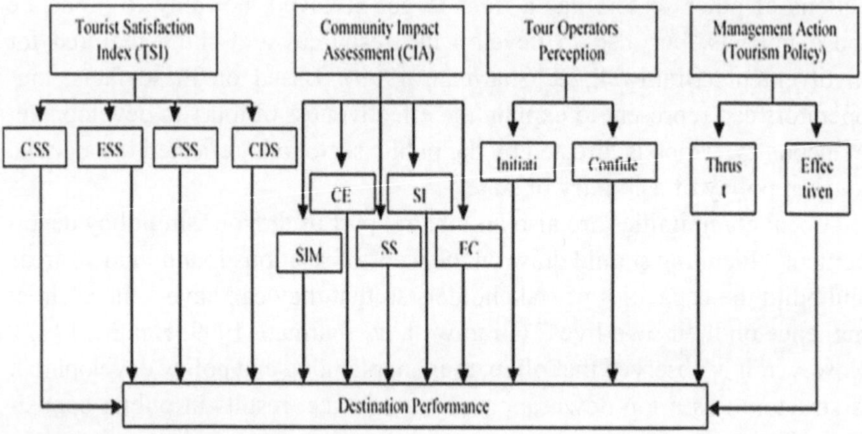

FIGURE 11.1 Conceptual framework.

process of the proposed model. In an argument, an interpretive paradigm has been adopted in several studies related to the stakeholders of tourism such as tour operators and hotel-related sectors. The managerial actions are mainly reflected through policies and actions, which is readily available, and with documentation information accessible through tourism organizations, its implementation effectiveness can be evaluated through onsite inspection and project progress through public sector organizations. A satisfaction survey has been successfully adopted through a positivist paradigm, where successful models have been adopted in tourism and related services using 'SERVQUAL (Service Quality) IPA (Important-Performance Analysis) and HOLSAT, KANO's model of customer satisfaction (Parasuraman, Zeithaml, & Berry, 1985; Tribe, & Snaith, 1998; Alegre, Rambla, & Valiente, 2009). In service satisfaction, specifically in the case of tourist satisfaction, the models argues that satisfaction is viewed from the relationship of performance of holiday attributes against expectations of the performance of those attributes, as from the basic nature of tourism phenomena, tourists' visit a place with substantial amount of dreams and expectations and tourism industry is 'unusually' selling dreams and expectations. Therefore, in a positivist approach, for Tourist Satisfaction Index, an expectancy disconfirmation model perfectly fits the context of destination benchmarking. The findings of the results (both positivist and interpretivist approach) can be possibly triangulated, as it is indeed a separate segment of the models and can providest and alone findings. A general assumption can be created that a satisfied community will bring better holiday satisfaction effects to the tourists; a successful policy brings better community impact to the destination; that the tourism industry is dependent on the local community's hospitality, and Therefore, it should be developed according to the host community's needs and desires (Tribe, & Snaith, 1998; Andriotis, 2005). The community attitude as essential for visitor satisfaction and repeat visitation (Sheldon, & Abenoja, 2001). The measurement of the host community's perceptions of tourism development plays a vital role in the future success of the tourist destination (Andriotis, 2005). The support of tourism can be measured by the perception of the local population, which can dictate the extent of the host community's acceptability of tourism (Andriotis, 2005). A positive word-of-mouth brings more tourists to the destination that supports major stakeholders. A successful tourist destination provides better 'meta-governance' structure though prioritizing tourism to the destination as it brings prosperity and positive economic impacts. Therefore, the result of the mixed paradigm can be interpreted through the 'triangulation technique,' and a triangulation effect can be resulted from the destination benchmarking exercise. It is also debated that "each destination may have different attributes." Tourists satisfied in one destination may differ from those satisfied in other destinations based on each destination's specific attributes. As a result, a "collective satisfaction scale cannot be used across destinations" (Andriotis, Agiomirgianakis, & Mihiotis, 2007). Hence, the variables in the constructs need to be re-adjusted based on the destination profile prior to using the model for destination performance evaluation.

11.9.2 THEORETICAL FOUNDATION

There are several theoretical bases for destination benchmarking as it consists of the integration of multiple models in one frame. The most suitable theoretical support for community-impact assessment identified for the proposed model is 'social exchange theory,' whereas a community, receptivity is in direct correlation with the benefit sought from tourism development. The support of tourism can be measured by the perception of the local population, which can dictate the extent of the host community's acceptance of tourism (Andriotis, 2005). The social exchange theory is considered as a major conceptual sociological approach to the study of tourism and community relationship where the objects offered for exchange have value, are measurable, and there is a mutual dispensation of rewards and costs between actors (Ap, 1992; Andriotis, 2005). The interpretation of the social exchange theory as supporting the community members "balance the costs and benefits of tourism development and their support for tourism depend on the outcome of this cost-benefit equation" (Lindberg, & Johnson, 1997; Andriotis, 2005).

The basic concept of the definition of the stakeholder theory is how an organization shapes its business concept. Since tourism is a product that consists of many components and elements, shaping the tourism business involves multiple parties. The descriptive stakeholder theory explains how managers and stakeholders actually behave and how they view their actions and roles (Fontaine, Haarman, & Schmid, 2006). Friedman (2006) explains the normative stakeholder theory by describing how managers or stakeholders should act and view the purpose of organizations, based on some ethical principles (Friedman, & Miles, 2006; Fontaine, Haarman, & Schmid, 2006). In tourism, the normative stakeholder theory is emphasized due to the influence of normative principle on the stakeholder to enhance the sustainable development of the industry. The theory stating that stakeholders consist of customers, employees, local communities, suppliers and distributors (Friedman, & Miles, 2006; Fontaine, Haarman, & Schmid, 2006). The application of the principle in tourism stressing the "plurality of organizational interest groups and the political nature of organizational goal setting and policy implementation" (Fontaine, Haarman, & Schmid, 2006). This approach focuses on "policy communities" which consists of people who interact with networks. In this networking process, it is considering the dynamics of "complex relationships" by examining them as they shift and change (John, & Senbet, 1998; Fontaine, Haarman, & Schmid, 2006). Since tourism policy is a partial outcome of managerial action that contributes at various levels of development, hence, a reciprocal relationship is required to be established when tourism explores a geographical region, that is, the contribution of tourism policy to stakeholders and destination community and vice-versa as the importance of people in the process and their support which contributes to a social conceptualization (Stevenson, Airey, & Miller, 2008).

11.9.3 CONSTRUCT DEVELOPMENT

11.9.3.1 TOURIST SATISFACTION INDEX

Cognitive-emotional satisfaction scale – Cognitive Emotional Satisfaction (CES) encompasses feelings of the tourist when visiting a tourist destination (Dick, & Basu, 1994; McMullan, & O'Neill, 2010). Cognitive Product Satisfaction Scale (CPS) – CPS represents tourism product dimensions of tourism-related activities which eventually measure tourist's consumption experience on major tourism products. Cognitive Service Satisfaction Scale (CSS) – CSS represents tourist's service dimensions, which measures tourist's experience on different services in the tourist destination. Cognitive Dissonance Scale (CDS) – CDS represents the mixture of favorable and unfavorable statements on which respondents are asked to rate their point of agreement or disagreement (Fallon, & Schofield, 2003).

11.9.3.2 COMMUNITY IMPACT ASSESSMENT

Economic Condition (EC) – (Perceived economic benefits of tourism development) – Community groups engaged in exchange transaction are keen to support tourism development. Community tourism is treated by the resident community as a means of income that supports the community

(Emerson, 1962). Perceived economic costs of tourism development – Tourism is considered to be destructive to community life (Reid, Mair, & George, 2004). Issues consist of the increased price of goods and services, crime rates, traffic issues and noise (Gursoy, & Rutherford, 2004). Based on the social exchange theory, locals who utilize the resource base view impacts negatively due to the belief that tourism may create sometimes unacceptable costs to the community even though tourism acts as a tool for economic gain, but an imbalance in development can cause social annoyance and vice versa, social acceptance and support (Gursoy, & Rutherford, 2004).

Social Issues (SI) – (Perceived sociocultural costs of tourism development) Tourist's presence deteriorates the quality of life of a destination community, through the overuse of resources and exploitation through mass sales which eventually causes crowds (Sirakaya, & Woodside, 2005). The continuous host-guest encounter may cause unacceptable cost for the destination community. Perceived social and cultural benefits – Tourists and tourism development supports the communities' native culture, provide better infrastructure access in the native area and improved standard of public facilities (Gursoy, & Rutherford, 2004). Good establishment of synergy between the community and tourists with the cultural exchange through understanding and the positive impact of cultural identity will enrich the sociocultural status of the destination community (Gursoy, & Rutherford, 2004).

Social Image (SI) – (Community attachment) – Proud of being a member and level of satisfaction of one being a member of the community during tourism development is an essential indicator for social sustainability. In a developed economy, social image overrules the social benefits of tourism development (Gursoy, & Rutherford, 2004). In a developing economy, the context is the reverse, and social benefits, particularly economic benefits overrule the social image as the destination community considers income from tourism as the primary focus of tourism development.

Social Services (S) – (Tourism related infrastructures are shared services) – The local community can use these infrastructures for their normal, everyday life. However, the destination community reacts positively or negatively based on the level of access to these shared services in their day-to-day life (Gursoy, & Rutherford, 2004).

Conservation Effort (CE) – CE deals with the impact of tourism on the sensitive areas of the destination community. There is a possibility that local development policy becomes increasing favoring the needs of the tourists, often without regard for the environment, local culture and heritage. If tourism development supports to conserve these, there will be a positive receptivity that can be observed in the destination community (Andereck et al., 2005).

Community Perception (CP) – (Community support for tourism development) – This construct explains the receptivity of the local community to support or oppose tourism development in their local areas.

11.10 MODEL TESTING

The proposed model was pilot tested to confirm its validity and fitness in a specific context. The destination, Kerala, had been chosen to run the model for internal benchmarking. The specific traits of this destination are collectively described as: Kerala's literacy rate is comparable to the most advanced regions of the world with 93.91% (Economic Report. (n.d)). In terms of poverty, Kerala has the lowest poverty rate in India. In Human Development Index (HDI), Kerala ranks the highest with 74.0 life expectancy rate. The growth of the Kerala economy is driven by the growth in the tertiary (services) sector. In Kerala, the service sector, including tourism, contributes 63.8% of the state's GDP. Trade, hotels, and restaurants (primarily tourism sector) are the largest contributors and accounts for 22% of the state's economy. The National Geographic Traveller ranked it as one of the 10 paradises in the world. In a recent survey of 700 brands in India, Kerala emerged as the only major tourism brand. Kerala is considered as India's only super brand, and the fastest growing tourist destination in India and tourism is a key driver of Kerala's economy (Tourism Update. (n.d)). Kerala is emphasizing on local, community-based tourism (Community Tourism. (n.d)). Greater emphasis on tourism development, the (6s) framework: (1) Swagat (hospitality); (2) Soochna (information); (3) Suvidha (facilitation); (4) Suraksha (security); (5) Shyyog (cooperation); and (6) Sanrachna (infrastructure) and finally emphasis of tourism in the Kerala tourism policy to support the Vision 2025 set by the state government of Kerala (Kerala Tourism Plan. (n.d)). On a darker side, literatures are debating on the 'issues related to infrastructure'; there are rampant harthal (strikes) and safety and security related issues in Kerala (Kerala Harthal. (n.d)). All the above aspects are taken into consideration in setting the constructs and variables specifically to evaluate 'Tourist Satisfaction' and 'Community Impact Assessment.' In order to collect samples for Tourist Satisfaction Survey and Community Impact

Assessment, the survey instruments were designed based on the expectation disconfirmation paradigm with a five-point Likert scale. Questionnaires were distributed in the most prominent tourist districts of Kerala, which are Ernakulam, Thiruvananthapuram, and Alappuzha. These three districts receive 75% of the total tourists' arrivals in Kerala. Destination communities in these destinations are also engaged in various Micro, Small and Medium Enterprises (MSMEs).

The main purpose of the research was to check the model fit, based on the recommended framework in this research. Structural Equation Modeling (SEM) was employed to conduct the main analysis. This model explains the interrelationship among constructs (Hair, Black, & Babin, 2010). The structural equation modeling shows the relationship between latent constructs using paths and Goodness of – Fit Index (GFI) measures the fit function of the conceptual model. The techniques adopted for analyzing the Tourist Satisfaction Index and Community Impact Assessment are the quantitative framework of the study within the proposed model. In both cases, the result shows that the model reasonably fits to confirm the model. In the case of descriptive data analysis, there were no significant variation found for the case of tourist expectation and experience. However, the mean score of fourteen items in the scale shows a mean score of less than 2.5, such as variety of entertainment, friendly people, entertainment, organized tours, roads, physical conditions, of shopping cleanliness, restrooms, error-free services, speed of service, staff knowledge, and destination cleanliness. The items that score above three in the mean values are good place, variety of restaurants, value for money, variety of attractions, easy to get around, tourist attraction, tourist information, staff behavior, affordability, and ease of access. The choice of destination selection and meeting the expectations also show positive receptivity. In the context of community impact assessment, fifteen items out of seventeen show a lower mean value. Items such as nature destruction, heritage, and cultural issues, own job creation, issues of subsidies, noise pollution, littering, crowd, drugs and alcohol use, crime rates, land use issues, commodity price, travel facilities, freshwater, electricity, and safety are of concern by the destination community. The receptive factors of tourism development are economic prosperity and job creation. In the context of policy implementation effectiveness, it is criticized that basic infrastructure development, waste management, arrival facilities, accommodation, and environmental friendliness are the main issues; however, there are several receptive factors such as visitor experience, provision of unique experience, marketing effectiveness, safety, and security (by tourists) where the community have concerns on this component but have positive perceptions. The stakeholders' analysis reveals that some significant initiatives are in action that benefits tourism development – these include the benefits, prospects of tourism business and talent and skills are receptive. However, they also note the issues of tour operations due to the 'Hartal' (specific) that reflects the lack of coherence and issues of appropriate 'meta-governance' structure in the state for effective coordination of tourism. By incorporating the entire model in a triangulated method, Kerala tourism performance overall rated in the medium performance. In terms of the level of integration of tourism performance in the state, it can thus be rated as 'type 1' destination which is characterized by "very limited or integration between the attribute and other resources. No significant relational capabilities are developed, and synergies that the resources may generate are not exploited. Level of resource management against sustainable development criterion is very low. Resource management is short term, and the aim is maximum exploitation of the most valuable resources to achieve the highest possible performance." Such a description in the outcome of the research is described as:

- Limited integration with the political situation and tourism; hartal has an evil effect on tourism development.
- Synergy between support services, follow up for maintenance required to be enhanced. The tour operators as the ambassadors and image builders seem to lose confidence in the performance. Improvement needed for the effective and efficient manpower distribution
- Improvement needed for the enforcement of policies, the right to travel is restricted, and is unable to provide what is promised.
- Lack of confidence of travelers restricts the expected increase of tourists visit.

11.11 CONCLUSION

Many of the results of this study reinforce the benchmarking model based on four performance core components of tourism. It is clear that the model tested major components, subcomponents, and elements of the tourism industry. Tourism phenomena consist of several components and elements; however, for evaluation and benchmarking purpose, measurable indicators can be selected based on the weight of those indicators influencing on overall destination development. The indicators proposed in this model has been customized for the context of Kerala tourism – such adjustments are based on the sociocultural, economic, political and geographic context of the tourist destination. In order to use the proposed model in a different context, a prior destination profiling required to be done to adjust the variable that are relevant to the context. There are also no clear studies conducted on destination profiling based on the benchmarking result, and this is for the appropriate categorization of a tourist destination. The result of the study can be used to conduct further research on external, generic and competitive benchmarking with other tourist destinations. Based on the result of the current research, it is also important to conduct a study on finding the reasons for the weak performance of the index core components and constructs, which will significantly help to enhance the functionality of the tourist destination. A destination benchmarking exercise is not common in tourism, either due to its complexity or that it is not perceived as an important exercise in the current scenarios of tourism development. Internal benchmarking is also a systematic follow up of destination development, where the growth of the destination is regulated and supported by policies, procedures, and plans. The result of this benchmarking will directly help policy makers and planners in shaping destination development strategies. Benchmarking, as a diagnostic tool, plays an important role in 'correction,' 'enhancement,' 'elimination,' 'exclusion,' and 'maintenance' of development-related aspects in a tourist destination. It is also important to take into consideration that destination benchmarking is not a 'time off' procedure; rather it is a continuous and systematic exercise to perform that eventually gives destination performance results positively. Tourism is more prominent in these days, irrespective of the destinations' development status or economic condition. More and more people are involved in leisure trips, more and more destinations are emerging, and new patterns of tourist travel is evident. Regional tourism has rapidly growing, particularly in emerging countries. Middle-class tourists are growing in numbers in the Asia Pacific region which holds a significant number of potential travelers. Competition – in terms of luxury brands – are slowly overshadowed by low-cost destinations, which in future will be able to attract middle-income groups. These expected transformations need to be understood from the position of one's 'own' performance, and internal destination benchmarking is the solution to deal with most of the above transformational trends to equip the industry with the changing needs of destinations based on market demands and national needs.

In terms of furthering the destination performance exercise, it is possible to include 'generic' performance related aspects such as internationally guided and accepted best practices promoted by associations, organizations, and NGOs. Such indicators can be included in the performance evaluation framework. Since tourism is not a seasonal phenomenon, a better result can be derived by conducting sampling in a longer span of time. In this way, a cross-sectional collection of samples can be derived throughout a set period, which will help to include more appropriate 'subjects' in the sample frame. Advancing the level of internal benchmarking, microelements that have a certain chance in destination development, indirectly related as well as unrelated sectors influenced by tourism, can be included to derive more effective outcomes.

KEYWORDS

- **community impact assessment**
- **destination benchmarking**
- **destination community involvement**
- **destination performance evaluation**

REFERENCES

Airey, D., & Chong, K. (2010). National policy-makers for tourism in China. *Annals of Tourism Research, 37*(2), 295–314.

Alegre, M. A., Rambla, X., & Valiente, O. (2009). L'elecció de center escolar a Catalunya. Elements per a undebat.

Allen, L. R., Long, P. T., Perdue, R. R., & Kieselbach, S. (1988). The impact of tourism development on residents' perceptions of community life. *Journal of Travel Research, 27*(1), 16–21.

Andereck, K. L., Valentine, K. M., Knopf, R. C., & Vogt, C. A. (2005). Residents' perceptions of community tourism impacts. *Annals of Tourism Research, 32*(4), 1056–1076.

Anderson, E. W., & Sullivan, M. W. (1993). The antecedents and consequences of customer satisfaction for firms. *Marketing Science, 12*(2), 125–143.

Andriotis, K. (2005). Community groups' perceptions of and preferences for tourism development: Evidence from Crete, *Journal of Hospitality & Tourism Research, 29* (1), 67–90.

Andriotis, K., Agiomirgianakis, G., & Mihiotis, A. (2007). Tourist vacation preferences: The case of mass tourists to Crete. *Tourism Analysis, 12*(1–1), 51–63.

Ap, J. (1992). Residents' perceptions on tourism impacts. *Annals of Tourism Research, 19* (4), 665–690.

Ap, J., & Crompton, J. L. (1998). Developing and testing a tourism impact scale. *Journal of Travel Research, 37*(2), 120–130.

Bosque, D. R., & Martin, S. H. (2008). Tourist satisfaction: A cognitive-affective model. *Annals of Tourism Research, 35* (2), 551–573.

Bramwell, B., & Sharman, A. (1999). Collaboration in local tourism policymaking. *Annals of Tourism Research, 26* (2), 392–415.

Burns, P. M. (1999). An introduction to tourism and anthropology. Psychology Press.

Churchill Jr., G. A., & Surprenant, C. (1982). An investigation into the determinants of customer satisfaction. *Journal of Marketing Research,* 491–504.

Community Tourism. (n.d). Retrieved from http://www.business-standard.com/india/news/kerala-to-develop-community-based-tourism/417098/.

Dick, A. S., & Basu, K. (1994). Customer loyalty: toward an integrated conceptual framework. *Journal of the Academy of Marketing Science, 22*(2), 99–113.

Dogan, Z. (1989). Forms of adjustment sociocultural impacts of tourism. *Annals of Tourism Research, 16*, 216–236.

Economic Report. (n.d). Retrieved from http://www.spb.kerala.gov.in/~spbuser/old/html/Economic%20Review%202010/CHAPTER%20-%2011.pdf.

Elliot, A. J. (1997). Integrating the "classic" and "contemporary" approaches to achievement motivation: A hierarchical model of approach and avoidance achievement motivation, *Advances in Motivation and Achievement. 10*(7), 143–179.

Emerson, R. M. (1962). Power-dependence relations. *American Sociological Review*, 31–41.

Fallon, P., & Schofield, P. (2003). First-timer versus repeat visitor satisfaction: the case of Orlando, Florida. *Tourism Analysis, 8*(2), 205–210.

Fontaine, C., Haarman, A., & Schmid, S. (2006). The stakeholder theory, *Edlays Education, 1*, 1–33.

Freeman, R. E. (1999). Divergent stakeholder theory. *Academy of Management Review, 24*(2), 233–236.

Friedman, A. L., & Miles, S. (2006). *Stakeholders: Theory and Practice*, Oxford University Press.

Gursoy, D., & Rutherford, D. G. (2004). Host attitudes toward tourism: An improved structural model. *Annals of Tourism Research, 31*(3), 495–516.

Gursoy, D., & Rutherford, G. D. (2004). Host attitudes towards tourism: An improved structural model. *Annals of Tourism Research, 31*(3), 495–516.

Hair, J. F., Black, W. C., & Babin, B. J. (2010). *Multivariate Data Analysis: A Global Perspective.*

Healey, P. (1998). Collaborative Planning in a Stakeholder Society, *Town Planning Review, 69*(1), 1.

Healey, P. (1997). *Collaborative Planning: Shaping Places in Fragmented Societies*. UBC Press.

Howlett, M., Ramesh, M., & Perl, A. (2003). *Studying Public Policy: Policy Cycles and Policy Subsystems,* (Vol. 3). Oxford: Oxford University Press.

Jamal, T. B., & Getz, D. (1997). Visioning for sustainable tourism development: Community-based collaborations, *Quality Management in Urban Tourism*, 199–220.

Jansen-Verbeke, M., & Spee, R. (1995). A regional analysis of tourist flows within Europe, *Tourism Management, 16*(1), 73–80.

John, K., & Senbet, L. W. (1998). Corporate governance and board effectiveness. *Journal of Banking & Finance, 22*(4), 371–403.

Keogh, B. (1990). Public participation in community tourism planning. *Annals of Tourism Research, 17*, 449–465.

Kerala Harthal. (n.d). Retrieved from http://www.thehindu.com/todays-paper/tp-national/tp-kerala/article3600714.ece, 8th July 2011.

Kerala Tourism Plan. (n.d). Retrieved from http://www.visionkerala2025.in/VisionKerala2025.pdf

Kousis, M. (1989). Tourism, and the family in a rural Cretan community. *Annals of Tourism Research, 16* (3), 318–332.

Kozak, M. (2004). Introducing Destination Benchmarking: A Conceptual Approach. *Journal of Hospitality and Tourism Research, 28*(3), 281–297.

Kozak, M., Bigné, E., & Andreu, L. (2005). Satisfaction and destination loyalty: A comparison between non-repeat and repeat tourists. *Journal of Quality Assurance in Hospitality & Tourism, 5* (1), 43–59.

Lindberg, K., & Johnson, L. R. (1997). The economic values of tourism's social impacts. *Annals of Tourism Research, 24*(1), 90–116.

Lindberg, K., & Johnson, R. L. (1997). Modeling resident attitudes toward tourism. *Annals of Tourism Research, 24*(2), 402–424.

Liu, J. C., Sheldon, P. J., & Var, T. (1987). Resident perception of the environmental impacts of tourism. *Annals of Tourism Research, 14*(1), 17–37.

MacLeod, N. (2006). Cultural tourism: Aspects of authenticity and commodification. Cultural *Tourism in a Changing World: Politics, Participation and (Re) Presentation,* 177–190.

Mathias, F., & Klaus, W. (2004). Destination Benchmarking: An indicator systems potential for exploring guest satisfaction. *Journal of Travel Research, 42,* 212–225.

McMullan, R., & O'Neill, M. (2010). Towards a valid and reliable measure of visitor satisfaction. *Journal of Vacation Marketing, 16*(1), 29–44.

McNair, C. J., & Leibfried, K. H. (1992). *Benchmarking: A Tool for Continuous Improvement,* John Wiley & Sons.

Mok, C., Slater, B., & Cheung, V. (1991). Residents' attitudes towards tourism in Hong Kong. *International Journal of Hospitality Management, 10* (3), 289–293.

Neal, D. J., & Dogan, G. (2008). A Multifaceted analysis of tourism satisfaction. *Journal of Travel Research, 47* (1), 53–62.

Oliver, R. L. (1980). Conceptualization and measurement of disconfirmation perceptions in the prediction of consumer satisfaction. *Refining Concepts and Measures of Consumer Satisfaction and Complaining Behavior*, 2–6.

Parasuraman, A., Zeithaml, V. A., & Berry, L. L. (1985). A conceptual model of service quality and its implications for future research. *The Journal of Marketing*, 41–50.

Pearce II, A. (1980). Host Community acceptance of foreign tourists: Strategic Considerations. *Annals of Tourism Research, 7* (2), 224–233.

Pizam, A., & Pokela, J. (1985). The perceived impacts of casino gambling on a community. *Annals of Tourism Research, 12*(2), 147–165.

Reed, G. M. (1997). Power relations and community-based tourism planning. *Annals of Tourism Research, 24*(3), 566–591.

Reid, D. G., Mair, H., & George, W. (2004). Community tourism planning: A self-assessment instrument. *Annals of Tourism Research, 31*(3), 623–639.

Richter, K. L., & Richter, L. W. (1985). Policy choices in south Asian tourism development. *Annals of Tourism Research 12*, 201–217.

Richter, L. K. (1989). *The Politics of Tourism in Asia,* University of Hawaii Press.

Roberts, N. C., & Bradley, R. T. (1991). Stakeholder collaboration and innovation: A study of public policy initiation at the state level. *The Journal of Applied Behavioral Science, 27*(2), 209–227.

Sautter, E. T., & Leisen, B. (1999). Managing stakeholders a tourism planning model. *Annals of Tourism Research 26*(2), 312–328.

Sheldon, P. J., & Abenoja, T. (2001). Resident attitudes in a mature destination: the case of Waikiki. *Tourism Management, 22*(5), 435–443.

Sherif, M. H. C. J. (1961). *Social Judgment: Assimilation and Contrast Effects.*

Sirakaya, E., & Woodside, A. G. (2005).Building and testing theories of decision making by travelers, *Tourism Management, 26*(6), 815–832.

Sirgy, J. M. (2009). Toward a Quality-of-Life theory of leisure travel satisfaction. *Journal of Travel Research, 49*(2), 246–260.

Smallman, C., & Moore, K. (2010). Process studies of tourists' decision-making. *Annals of Tourism Research, 37*(2), 397–422.

Snepenger, D. J., & Johnson, J. D. (1991). Political self-identification and perceptions on tourism. *Annals of Tourism Research, 18*(3), 511–515.

Stevenson, N., Airey, D., & Miller, G. (2008). Tourism policy making: The policymakers' perspectives, *Annals of Tourism Research, 35*(3), 732–750.

Thomas, T. K. (2011). Benchmarking Tourist Destination-Concept to Practice in Developing Economy. *International Journal of Culture and Tourism Research, 4*(1), 129–142.

Thomas, T. K. (2015). Measuring Community Impact Assessment for Internal Destination Performance Evaluation in an Exploring Tourist Destination. *ATNA-Journal of Tourism Studies, 10*(1), 53-71.

Tosun, C. (2002). Host perceptions of impacts: A comparative tourism study. *Annals of Tourism Research, 29* (1), 231–253.

Tourism Update. (n.d). Retrieved from http://www.keralatourism.org/rt-keralaupdate.php.

Tribe, J., & Snaith, T. (1998). From SERVQUAL to HOLST: holiday satisfaction in Varadero, Cuba, *Tourism Management, 19*(1), 25–34.

Türker, A. N., & Öztürk, A. S. (2013). Perceptions of residents towards the impacts of tourism in the Küre Mountains National Park, Turkey. *International Journal of Business and Social Science, 4*(2).

Wang, Y., & Xiang, Z. (2007). Toward a theoretical framework of collaborative destination marketing, *Journal of Travel Research, 46* (1), 75–85.

Williams, J., & Lawson, R. (2001). Community issues and resident opinions of tourism. *Annals of Tourism Research, 28*(2), 269–290.

Yuksel, A., & Yuksel, F. (2001). The Expectancy-Disconfirmation paradigm: A critique. *Journal of Hospitality and Tourism Research, 25*(2), 107–132.

Zairi, M. (1992). The art of benchmarking: Using customer feedback to establish a performance gap. *Total Quality Management, 3*(2), 177–188.

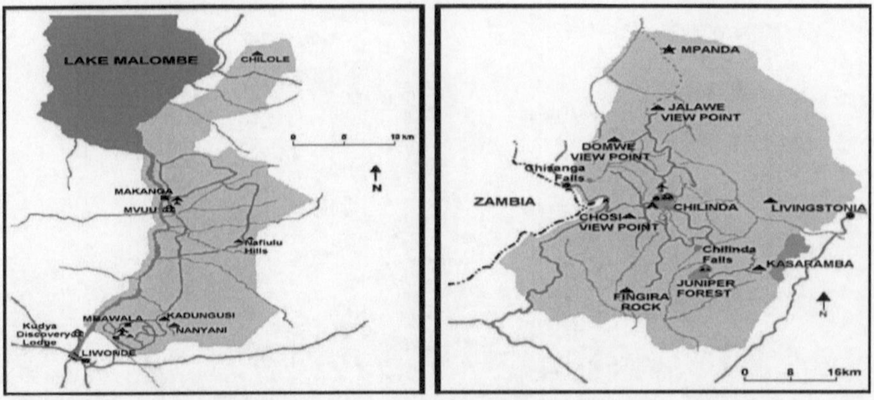

FIGURE 3.1 Showing Location of the Study Area. Map A: *Liwonde* Wildlife Reserve. Map B: *Nyika* Wildlife Reserve. Source: Safari Guide Africa (2017).

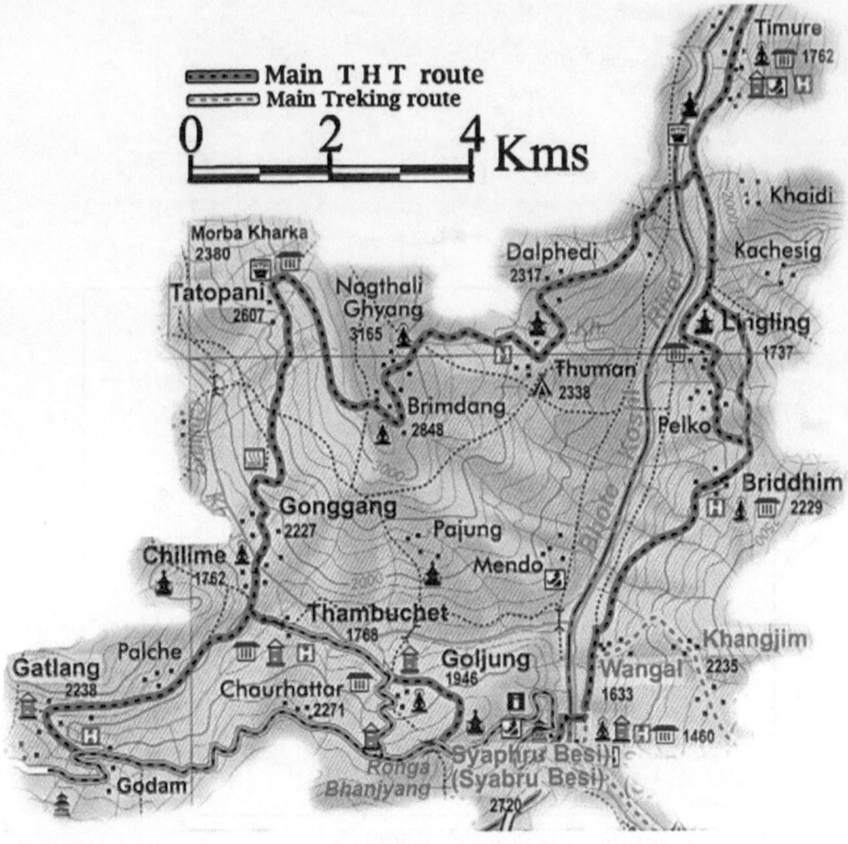

FIGURE 5.1 Tamang Heritage Trail location map.

FIGURE 13.1 Garhwal district.

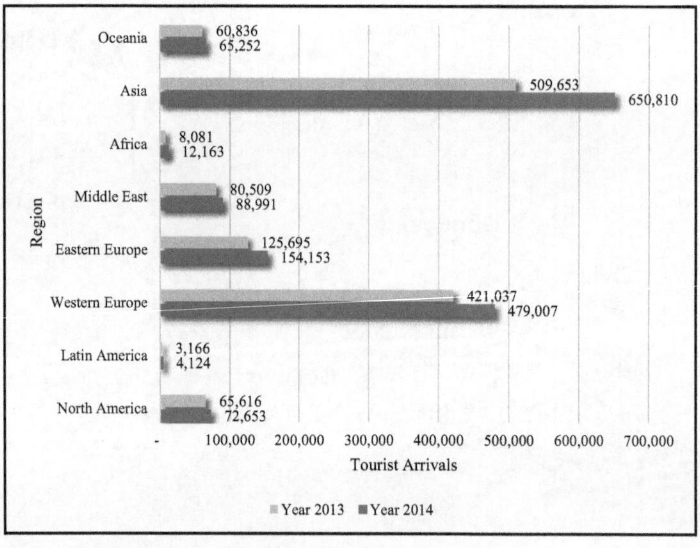

FIGURE 16.1 Tourist arrivals statistics 2013–2014
(*Source*: **Sri** Lanka Tourism Development Authority, 2014).

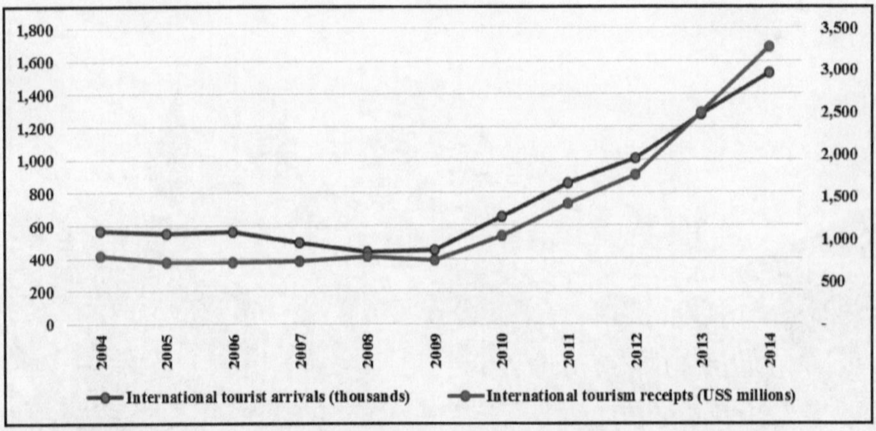

FIGURE 16.3 Tourists arrival and receipts data for Sri Lanka 1995–2013 (*Source*: World Economic Forum –Travel and Tourism Report 2015).

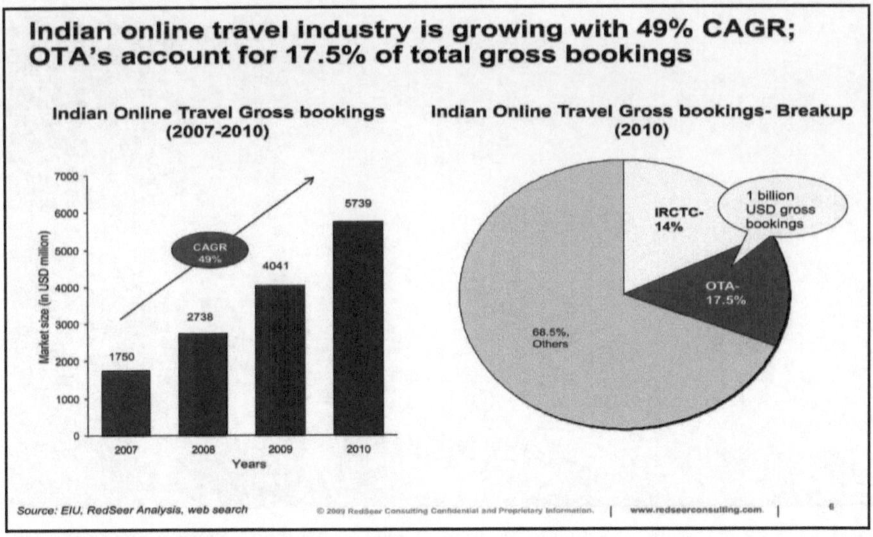

FIGURE 17.1 Indian online travel industry.

FIGURE 18.1 Saint Mark's Basilica in Venice in Assassin's Creed II. *Source*: Figure adapted from an external source (https://br.pinterest.com, 2016).

FIGURE 18.2 The city of Rome in Assassin's Creed Brotherhood. Source: Figure adapted from an external source (http://www.push-start.co.uk, 2016).

CHAPTER 12

SUSTAINABLE FOOD PRACTICES FOR HOLISTIC TOURISM DEVELOPMENT

ALOK KUMAR

Associate Professor, Department of Tourism Management and Hospitality Studies, IPHC, Mekelle University, Africa, Mobile: +251-985024149; E-mail: kumaralok1975@yahoo.co.in

ABSTRACT

Tourism is one of the world largest industry in which hotels are the major stakeholders. Hotels cater to accommodation and food to the tourists. In tune with sustainable tourism, hotels are also going green. They are following environment-friendly practices to get a competitive edge. The present sustainability idea originated in the 1970s, and it took 1987 to Brundtlant commission to outline the holistic concept of sustainable development in its report. The goal of Sustainable Development is clearly secure economic development, social equity, and environmental protection. Food is one of the basic necessities of tourists. Food is symbolic of a destination and an important motive for tourism. We are more concerned about sustainable tourism but without providing sustainable food; tourism cannot be sustainable. Indian Vedas has understood the importance of food & sustainability thousands of years back. Bhagavad Gita in its seventeenth chapter categorized food in its own way. The ancient sustainable practices focused on safety, quality and waste prevention of food leading to a healthy mind and soul. Safe food and eco-friendly hotels are the emerging international practices in the hotel industry, and these could be vital for sustainable tourism. This chapter is a maiden attempt to present the concept of sustainable food and explain emerging practices and law that can ensure sustainable food.

12.1 INTRODUCTION

Tourism is recognized as one of the world's largest industries and continues to expand at a rapid rate (UNWTO, 2006). Over 922 million people traveled in 2008, and it is estimated international tourism will increase to 1.1 billion by 2020 (UNWTO, 2008). Tourism is also a profitable economic endeavor, making it an economically enticing industry and a form of development for many countries around the world. However, the benefits of tourism are often accompanied by many negative environmental and social impacts. The World Travel and Tourism Council identified nine priority areas for action by national tourism organizations and industry-based associations or organizations. These included:

1) assessing the capacity to bring about sustainable tourism planning for sustainable tourism development,
2) planning for sustainable tourism development, and
3) measuring progress in achieving sustainable development.

Tourism is primarily a business. As the WTO states: "It is important to ensure that ecotourism is a good, economically sustainable business and that profits are generated from it. If there are no prospects for profit, then private entrepreneurs will not invest, and there will be no benefits to distribute for local communities or conservation purposes" (Vereczi 2002).

The hotel industry forms a large part of the tourism industry, and if not managed properly, it has the potential to be detrimental to the social and natural environments within which it functions. Hotels are also resource-intensive, and in order to reduce their impact, it is imperative to go green!" The consumers using hotel services are conscious about environmentally friendly practices in India. They patronize the hotels that have adapted green practices though not compromising on service quality. The consumers would prefer to use lodging that follows these practices but are not willing to pay extra for these services. Indian hotels have a competitive advantage over similar products if they follow green practices. The hotels would have to invest in environmentally friendly practices and look at long-term gains. The government needs to acknowledge and institutionalize the practice by instituting rewards and offering benefits in taxes (Manaktola & Jauhari, 2007).

Food is one of the basic necessities of tourists which hotels cater to. Cuisines and gastronomy are an important attraction which tourists are looking at a destination. Food is symbolic to a destination and an important motive for tourism leading to new concept-gastronomic tourism. Even within a city, people love eating out. There has been a drastic change in the lifestyle of common folk in this modern age; Youngsters such as college goers and professionals are compelled to take at least a meal outside the home. This has brought the revolution into the hotel industry, and it is certain that food is going to focal point in the era to come. So, tourism is going to be food-centric. As we are more concerned about sustainable tourism but without providing sustainable food; tourism cannot be sustainable. Indian Vedas has understood the importance of food and sustainability thousands of years back and expressed "Annam Brahma." "Food is the supreme Brahman" is the famous statement of the Taittriyoanisad. The foods tradition in India developed over a period of more than five thousand years. Hence, the Indian tradition of food is far superior to western tradition, since it is holistic studied thoroughly not only from the point of spiritual emancipation but also from the point of view of creating a peaceful and healthy society in which every human being is enlightened, sober, peace-loving and healthy (Lakshmithathachar, 2000). Bhagavad Gita in its seventeenth chapter clearly categorized food into:

- Satvika food: It promotes mental vigor and causes knowledge.
- Rajasik food: It is bitter, sour, saltish, excessively hot, pungent burning. Such foods are consumed by persons who have got Rajas in the ascendancy. It causes pain, grief, and disease.
- Tamsik food: It is stale, tasteless, putrid and rotten. It is left over and unclean foods. This will cause lethargy and sleep.

So, the focus was on safety, quality and waste prevention of food leading to healthy mind and soul as several texts in Indian Ayurveda testify. This ancient Indian practice is the essence of sustainable food. Food safety management system, as well as food legislation ensuring quality & reducing wastages, are mandatory requirements for ensuring sustainable food. Reduction of food wastages will ensure availability of wholesome food to the large population of the world facing a drastic shortage of food.

12.2 NEED AND OBJECTIVE STUDY

The brief review of existing literature on related areas points the absence of comprehensive and intensive study touching on these aspects of hoteliering as yet. However, in regards to other industries, a lot of scientific and detailed studies have been undertaken by researchers and institutions. It has been unfortunate that hotel business has not been able to draw the attention of researchers to any noticeable extent barring few by hotel association and institution to draw the attention of the government towards their problem.

The paucity of data is a bottleneck for refraining researchers from undertaking the in-depth study. Safe food and eco-friendly hotels are the emerging international practices in the hotel industry, and these could be vital for sustainable tourism. Further focus on practices and law that can ensure sustainable food is a maiden attempt to explain this concept.

12.3 SUSTAINABLE DEVELOPMENT: AN OVERVIEW

Sustainable development is a holistic concept based on a simple principle. As outlined in the 1987 Brundtland report, the concept involves "development that meets the needs of the present without compromising the ability of future generations to meet their own needs" The sustainability idea originated in the 1970s (WCED, 1987). The first international meeting that tackled the impact of human activities on the environment and eventually their impact on the human race was the 1972 UN Stockholm Conference on the Human Environment. In 1987, the Unsponsored Brundtland Commission released Our Common Future, a report that captured widespread concerns about the environment and poverty in many parts of the world. The Brundtland report said, in part: "Economic development cannot stop, but it must change course to fit within the planet's ecological limits." It also popularized the term Sustainable Development (SD).

"Sustainable tourism development meets the needs of present tourists and host regions while protecting and enhancing the opportunity for the future. It is envisaged as leading to management of all resources in such a way that economic, social, and aesthetic needs can be fulfilled while maintaining cultural integrity, essential ecological processes, biological diversity, and life support systems" (WTTC, 1998).

The 1992 UN Conference on Environment and Development (also known as Earth Summit 1992 or the Rio Summit) again brought environment and development issues to the forefront. It produced the Rio Declaration on Environment and Development, which added the principles of intergenerational equity, citizen participation, and empowerment of women, youth, and indigenous peoples. Its accompanying "Agenda 21: Program of Action for Sustainable Development" identified ways by which various stakeholders can operationalize the actions called for by the document on a wide range of issues. Attended by leaders from government, public interest groups, nongovernmental organizations, as well as private business, Earth Summit 1992 generated a high level of public awareness of and engagement in global environmental protection for the health and well-being of future generations.

Furthermore, it highlighted the value of international cooperation in global issues, such as environmental degradation. As with Stockholm, the 2002 World Summit on Sustainable Development in Johannesburg, South Africa, identified poverty as the main cause of environmental degradation and social problems. As such, measures to eradicate poverty were deemed essential to a sustainable future. These measures called for greater access to basic health services, clean water, and sanitation facilities, and education. It also called for greater participation by women and indigenous peoples in economic activities. Other proposed policies involved giving poor people access to basic rural infrastructure and credit facilities, and the application of new environment-friendly technologies (UN, 2002).

Eber (1992) as cited in Bramwell (1998) states that tourism can contribute to sustainable development when: "… It operates within natural capacities for the regeneration and future productivity of natural resources; recognizes the contribution that people and communities, customs and lifestyles make to the tourism experience; accepts that these people must have an equitable share in the economic benefits of tourism; and is guided by the wishes of local people and communities in the host areas.

12.3.1 THE PURPOSE OF SUSTAINABLE DEVELOPMENT

The goal of Sustainable Development is clearly secure economic development, social equity, and environmental protection. As much as they could work in harmony, these goals sometimes work against each other. The rapid development of good living, travel, and the consumer society has often resulted in less protection to the environment and to some groups of the world population.

12.3.2 PRINCIPLES OF SUSTAINABLE FOOD

People and businesses adopting a sustainable approach to food should:

1) Use local, seasonally available ingredients as standard, to minimize energy used in food production, transport and storage.
2) Specify food from farming systems that minimize harm to the environment, such as certified organic produce.
3) Limit foods of animal origin (meat, dairy products, and eggs) served, as livestock farming is one of the most significant contributors to climate change, and promote meals rich in fruit, vegetables, pulses,

whole grains, and nuts. Ensure that meat, dairy products, and eggs are produced to high environmental and animal welfare standards.
4) Exclude fish species identified as most 'at risk' by the Marine Conservation Society, and choose fish only from sustainable sources – such as those accredited by the Marine Stewardship Council.
5) Choose Fairtrade-certified products for foods and drinks imported from poorer countries, to ensure a fair deal for disadvantaged producers.
6) Avoid bottled water and instead serve plain or filtered tap water in reusable jugs or bottles, to minimize transport and packaging waste.
7) Promote health and well-being by cooking with generous portions of vegetables, fruit and starchy staples like whole grains, cutting down on salt, fats, and oils, and cutting out artificial additives.

12.4 CHALLENGES OF SUSTAINABLE FOOD

12.4.1 FOOD SECURITY

The Food and Agricultural Organization (FAO) defines food security as a condition where all people at all times, have physical and economic access to sufficient, safe and nutritious food to meet their dietary needs and food preferences for an active and healthy life. Four broad dimensions of food security are usually identified: *availability*—the supply of food in an area, *access*—the physical and economic ability of people to obtain food, *utilization*—the proper consumption of food, and *stability*—the sustainability of food supplies. Food insecurity is the absence of food security implying that hunger exists as a result of problems with availability, access, and utilization or that there is susceptibility to hunger in the future (FAO, 2002).

Food safety has always been a concern of foodservice operators. However, the number of foodborne illness attributed to the foodservice industry is still significant. The food service industry is serving a public that is increasingly intolerant of any food safety risk. All food service operations have the potential to cause foodborne illness through errors in purchasing, receiving, storing, preparing and serving food. Foodservice operator cannot assume workers learned proper personal and food hygiene practices in their homes or that government health inspection will facilitate safe operations. The National Restaurant Association has reaffirmed its position that the responsibility for providing safe and wholesome food rests with the food service industry. It has called the establishment of and adherence to effective standards of safe food handling practices.

12.4.2 SANITATION

Sanitation is the creation and maintenance of healthful, or hygienic, conditions. It comes from the Latin word Sanitas meaning health. In a food service situation, the word sanitation means wholesome food, handled and prepared in a way that the food is not contaminated with disease-causing agents. In other words, sanitation is what helps food stay safe. But sanitary simply does not mean clean as which appears to be clean may not always be sanitary. Clean means free of visible soil. Sanitary means free of harmful levels of disease-causing microorganisms and other harmful contaminants. Clean refers to aesthetics and concerns outward appearance: a face without a smudge, a glass that sparkles, a shelf wiped clear of dust.

The following questions concerning food sustainability are central to this study: issues and potentialities of organic farming and food production and trends of implementation of management systems in the food sector. Where production sustainability is one of the most important factors influenced by development. There are two possibilities or directions for further sustainable food production development. The first direction for further food chain development is food production in medium and large companies (industrialization and cost-effective production) mainly for international and regional (EU, former USSR, etc.) markets.

12.4.3 ISSUE OF FOOD WASTE

Food waste is a significant sustainability issue for the world. Food waste prevention and consumer engagement are measures which can help to improve the sustainability of food. For example, the UK generates over 16 million tons of food waste every year, costing an estimated £22 billion a year. UK households throw away 8.3 million tons of food and drink waste, worth over £12 billion, every year. Most of this is avoidable; preventing it could save the average UK family £680 a year. The climate change benefits of tackling food waste would be equivalent to taking 1 in 4 cars off the UK's roads (WRAP, 2011).

The hotel industry can reduce the amount of waste produced by implementing and following a waste management system that is modeled around the concepts of reducing, reuse and recycle (Greenhotelier, 2004). Approximately 54 percent of a hotel's solid waste can either be recycled or reused (Alexander 2002). A study by Bohdanowicz (2005) identified that "a large proportion (50–60 percent) of the waste materials in an accommodation facility can be recycled or reused."

Increasing the primary production of food is only one of the many strategies that will be required to feed nine billion people sustainably and equitably by 2050. The challenge is to achieve optimal results right across the food system. This requires that substantial inefficiencies across the entire food supply chain – from the farm gate to the point of consumption – are addressed. Some estimates suggest that as much as half of all food grown is lost or wasted before and after it reaches the consumer (Lundqvist, de Fraiture, & Molden, 2008), although there are significant uncertainties and gaps in our understanding of losses in the food supply chain across the globe. In low-income countries, where infrastructure for storage and supply are often inadequate, food losses are greatest in early post-harvest stages, whereas in high-income countries, the greatest losses are usually incurred by the consumer (Lundqvist, de Fraiture, & Molden, 2011).

Food waste prevention can, Therefore, deliver significant environmental benefits, in terms of landfill avoidance, freshwater conservation and the mitigation of climate change. Raising awareness f food waste amongst consumers and providing practical advice to them can also help them to waste less and realize financial savings. Preventing food waste can also help to address other key strategic food issues, including supporting more healthy sustainable diets (through for example better control of portion sizes), food and water security.

12.4.4 FOOD SAFETY AND QUALITY

Food safety and food quality are critical to the bottom line. Food that has off flavors is dry, or appears stale is not likely to impress customers. Preserving food quality is another objective of sanitation. Sanitary handling is a major factor in both food safety and quality. Food that is stored, prepared and served properly is more likely to retain its quality. The standards of a food's quality include its safety, appearance, chemical properties, texture, consistency, nutritional value, and flavor. Anyone of these can be destroyed by unsanitary procedures from purchase to service. Proper food handling with is important to preserve food quality.

12.5 BENEFITS FROM HYGIENIC PRACTICES

In addition prevention from bad publicity that can follow an outbreak of foodborne illness catering unsafe food. Immediate returns can be gained through good food-protection practices. Food costs are lowered when waste is reduced and when portion control is improved. Quality control is improved when procedures and practices are standardized. Providing high-quality food is an important strategy in an increasingly competitive industry. Increased productivity is achieved when employees are trained on a continual basis and accountable for their actions. Thus, improved food protection results in improved operating efficiency, which ultimately results in improved profitability. Sanitary procedures will prevent outbreaks of foodborne illness, maintain goodwill, and keep the financial bottom line from bottoming out.

12.6 EMERGING LAWS AND PRACTICES

12.6.1 FOOD SAFETY AND STANDARDS ACT 2006 OF INDIA

The Much awaited Food Safety and Standards Act 2006, Rules 2011 and Regulations 2011 is implemented from 5th August 2011, all over the country by repealing the Prevention of Food Adulteration Act 1954 and other Food Laws/orders, in this regard the central govt. The Food Authority has notified the Rules and regulation on 5th May 2011 and 1st August 2011, respectively. The Central Govt. earlier indicated in the Supreme Court in response to a PIL that the Act would be implemented from 1st Jan 2010 but the machinery was not ready then. The FSS Act 2006 was enacted in the year 2006 by consolidating different food laws and to establish a single statutory regulator for the food sector namely "The Food Safety and Standards Authority of India." The main function of the Authority is for laying down science-based standards for articles of food and to regulate their manufacture, storage, distribution, sale, and import, to ensure the availability of safe and wholesome food for human consumption and for matters connected therewith or incidental thereto. The main features of the Act are: (a) Movement from multi-level and multi-departmental control to integrated line of command; (b) Integrated response to strategic issues like novel/genetically modified foods, international trade; (c) Licensing for manufacture of food products, which is presently granted by the Central Agencies under various Acts and Orders, would stand decentralized to the Commissioner of Food Safety and his officer; (d) Single reference point for all matters relating to Food Safety and Standards, regulations and enforcement; (e) Shift from mere regulatory regime to self-compliance through Food Safety Management Systems; (f) Responsibility on food business operators to ensure that food processed, manufactured, imported or distributed is in compliance with the domestic food laws; and (g) Provision for graded penalties depending on the gravity of offence and accordingly, civil penalties for minor offences and punishment for serious violations.

The Prevention of Food Adulteration Act 1954 that came into force on June 1, 1955, was aimed at preventing adulteration of food sold in hotels and packaged and raw food. The drawbacks PFA Act was that many of the cases filed so far have not reached a logical conclusion either because the witnesses were not traceable or other reasons like the enforcing machineries especially the health departments, municipalities/corporations, railways, Ports are overburdened with their own work. At least 10% of the cases pending were filed 20 years ago, and more than 40% were filed a decade ago. The PFA also did not address the core issue of Food safety, through Consumer education and was confined to only checking of adulteration in food items. Under PFA petty offenses also required a minimum of 6 months imprisonment together with a fine of Rs. 1000/- due to which a large number of unwanted cases mounted in courts. The PFA Act and other orders related with food do not have enough provisions to deal with recent developments in Food Science and Technology which has developed considerably in recent years and a whole new range of processed foods including functional food, nutraceuticals, proprietary foods, dietary supplements, genetically modified or engineered foods are hitting the market at a fast pace and remains unregulated.

Whereas the new Food Safety and Standards Act focuses on checking food quality and hygiene at outlets, there is scope for the food recall from the market by the manufacturer/wholesaler if the food is found to be substandard. Fines shall be imposed from 1 lakh to 10 lakh and will be dealt separately by the Adjudicating Officer (ADM) and cases which may require imprisonment will only go to court or to Food Safety appellate tribunal. This will speed up prosecution and will prove a deterrent for the defaulters.

In the new act, the petty food operators (shopkeepers/manufacturer) with less than 12 lakh annual rotation will not require a food license, and they will need a simple registration on payment of Rs.100/- fees. The State Food Licenses will be issued by the designated officer of each district appointed by the State Food Safety Commissioner on payment of the prescribed fees by the Govt. of India. All over India valid Food license will be issued by the Central Licensing Authority" appointed by the Chief Executive Officer of the

Food Safety and Standards Authority of India in his capacity of Food Safety Commissioner. In the State and UTs, the Act will be enforced through the Commissioner Food Safety, Designated Officers, Food Safety Officers, Food Analyst which will together constitute the State Food Authority.

12.6.2 CODEX INDIA

The Codex Alimentarius Commission (CAC) was created in 1961/62 by Food and Agriculture Organization of the United Nations (FAO) and the World Health Organization (WHO), to develop food standards, guidelines and related texts such as codes of practice under the Joint FAO/WHO Food Standards Programme. The main purpose of this Programme is to protect the health of consumers, ensure fair practices in the food trade, and promote coordination of all food standards work undertaken by international governmental and non-governmental organizations.

Codex India" the National Codex Contact Point (NCCP) for India, is located at Food Safety and Standards Authority of India (Ministry of Health and Family Welfare), FDA Bhawan, Kotla Road, New Delhi -110002, India. It coordinates and promotes Codex activities in India in association with the National Codex Committee and facilitates India's input to the work of Codex through an established consultation process.

12.6.3 CODEX ALIMENTARIUS

The Codex Alimentarius is the basic international standard of Food Safety and Hygiene that has been adopted by India and is being followed by all sector of the food production industry. This also affects hotels and restaurants being at tertiary level in the food chain. The codex principles uses the HACCP approach to identify various points of potential hazard to food safety, establishes critical limits and control points and systems of taking corrective actions and monitoring the effectiveness of these corrective actions in ensuring food safety. Hazard Analysis and Critical Control Point (HACCP) is a system of risk management developed to control food safety. It can be described as an operation-specific, internally managed system of preventative control that identifies, evaluates, and controls hazards of significance to food safety. While it has a relatively long history, originating as a means of assuring the safety of meals produced for the U.S. manned space program in the 1960s, it is only in the last 10 years that it has emerged as the primary approach to securing the safety of the food supply (Corlett, Jr., 1998). This is reflected in recommendation of HACCP by organizations such as the U.S. National Academy of Sciences, the International Commission on Microbiological Specifications for Foods and a Codex Alimentarius Commission decision to recommend its use by both the food industry and regulatory authorities. This latter development has had a direct effect on the countries of the European Union who have since built a requirement for risk management, based on HACCP principles, into the operational activities of all food businesses

12.6.4 THE CODEX ALIMENTARIUS COMMISSION AND THE FAO/WHO FOOD STANDARDS PROGRAMME

The Codex Alimentarius Commission implements the Joint FAO/WHO Food Standards Programme, the purpose of which is to protect the health of consumers and to ensure fair practices in the food trade. The Codex Alimentarius (Latin, meaning Food Law or Code) is a collection of internationally adopted food standards presented in a uniform manner. It also includes provisions of an advisory nature in the form of codes of practice, guidelines and other recommended measures to assist in achieving the purposes of the Codex Alimentarius. The Commission has expressed the view that codes of practice might provide useful checklists of requirements for national food control or enforcement authorities. The publication of the Codex Alimentarius is intended to guide and promote the elaboration and establishment of definitions and requirements for foods, to assist in their harmonization and, in doing so, to facilitate international trade.

12.6.5 BASIC TEXTS ON FOOD HYGIENE

In June 1997, the Codex Alimentarius Commission adopted three newly revised basic texts on food hygiene. These texts are published officially in Volume 1B of the Codex Alimentarius and have been republished in this compact format to allow their wide use and understanding by governments, regulatory authorities, food industries and all food handlers, and consumers.

12.6.6 ISO 22000:2005

ISO 22000 (Food Safety Management System) is an international standard, aligned with ISO 9001:2000 as well as HACCP, and is applicable to the entire food chain irrespective of the size and complexity, such as complete supply chain (production-processing-distribution); agricultural producers (animal based-plant based food);other suppliers of services and goods (packaging and veterinary products); processors and retailers

12.6.6.1 FEATURES OF ISO 22000:2005

a) Integrates the principles of Hazards Analysis and Critical Control Point (HACCP) system developed by the Codex Alimentarius Commission. It combines the HACCP plan with the prerequisite programme (PRPs) and operational PRPs.
b) Requires that all Hazards that may be reasonably expected to occur in the food chain are identified, assessed and controlled.
c) Can be applied independent of other management system standards or can be integrated with existing other management systems.
d) Intended for organizations seeking more focused, coherent and integrated food safety management systems.
e) Emphasis on preventions of food safety hazards of all types.
f) Ensures compliance with legislative and regulatory requirements.
g) Allows even small, tiny scale organizations to implement as an externally developed a combination of control measures.
h) Provides for management of potential emergency situations & accidents that can impact food safety.

12.7 CONCLUSION

Sustainable development is a holistic concept based on a simple principle- "development that meets the needs of the present without compromising the ability of future generations to meet their own needs." Sustainable food is holistic food, which is also a safe, quality, and tasty food creating a peaceful and healthy society. It has got a focus on hygiene and waste prevention. As per Indian Vedas; sustainable practices were part of day-to-day life and traditions of ancient India and food were prepared with the aim to generate a healthy mind and soul. The challenges of the present decade led to the concept of sustainability by Brundtland Commission in 1987. The food sector is facing the challenge of hygiene and sanitation, food safety and quality, food wastage, food security. The solution to these lies in the application of emerging laws & practices like HACCP-Codex, ISO 22000, India's FSSA-2006 for ensuring sustainable food.

KEYWORDS

- **challenges**
- **hygienic practices**
- **laws and practices**
- **sustainable development**

REFERENCES

Alexander, S. (2002). Green Hotels: Opportunities and Resources for Success. Portland: Zero Waste Alliance.

Bohdanowicz, P. (2005). 'European Hoteliers' Environmental Attitudes: Greening the Business, *Cornell Hotel, and Restaurant Administration Quarterly, 46* (2), 188–204.

Bramwell, B. (1998). Selecting policy instruments for sustainable tourism. *Global Tourism,* 2nd ed., edited by W. Theobald. Oxford: Butterworth-Heinemann.

Codex Alimentarius (1993). Guidelines for the application of the hazard analysis critical control point system. ALINORM 93/131, Appendix 11.

Codex Alimentarius (1997). Guidelines for the application of the hazard analysis critical control point system. Annex to CAC/RCP-1 (1969), Revision 3.

Corlett, Jr., D.A. (1998). Maryland Economics and Rural Development. *HACCP User's Manual,* Aspen Publication, Vol.3, No.2, 2007.

Council Directive on the Hygiene of Foodstuffs. Council Directive 93/43/EEC. OJL 175/1; 19.07.93.

FAO/WHO Codex Alimentarius Commission Annex to the Recommended International Code of Practice—General Principle of Food Hygiene, CAC/RCP 1–1969, Rev. 3, 1997.

Food and Agricultural Organization. (2002). The State of Food Insecurity in the World 2001; FAO: Rome, Italy.

Food Safety Using HACCP Quality Management System (ISO-22000:2005) Document, Food Safety and Standard Act, 2006.

Green Lodging News 2008, Retrieved January 31, 2008, from the World Wide Web: http://www.greenlodgingnews.com. Green Hotelier 2004, 'The know-how section: Better waste management,' Greenhotelier, vol. 32, pp. 1–4.

Lundqvist, J., de Fraiture, C., & Molden, D. (2011). Foresight Project on Global Food and Farming Futures – Synthesis Report C7: Reducing waste (2011).

Lundqvist, J., de Fraiture, C., & Molden, D. (2008). Saving Water: From Field to Fork – Curbing Losses and Wastage in the Food Chain (SIWI).

Manaktola K. & Jauhari V. (2007). Exploring consumer attitude and behavior towards green practices in the lodging industry in India, *International Journal of Contemporary Hospitality Management, 19*(5), 364–377.

Ramon Benedicto A. Alampay (2005). Sustainable Tourism: Challenges for the Philippines, Philippines.

UNWTO 2006, Retrieved August 16, 2007, from the World Wide Web: http://www.world-tourism.org/facts/menu.html – UNWTO 2008, Tourism Highlights, 2008 Edition.

Vereczi, G. (2002). The framework of the International Year of Ecotourism and conditions for the sustainable development and management of ecotourism. A paper presented at the

World Tourism Organization: Asian Productivity Organization Ecotourism Conference, 15–17 April 2002, Fiji.
World Commission on Environment and Development (1987). *Our Common Future.* Oxford: Oxford University Press.
World Tourism Organization. (2002). Voluntary Initiatives for Sustainable Tourism: Worldwide Inventory and Comparative Analysis of 104 Ecolabels, Awards and Self-commitments. Madrid: WTO.
World Travel and Tourism Council (WTTC), World Tourism Organization (WTO), & Earth Council (EC). (1995). Agenda 21 for the travel and tourism industry. London: WTTC.
World Travel and Tourism Council (1998). Agenda 21 for the travel and tourism industry—towards environmentally sustainable development. London: WTTC.
WRAP (2011). Website, http://www.wrap.org.uk/retailsupply_chain/voluntary_agreements/federation_house.html.

CHAPTER 13

INTRODUCTION OF COMMUNITY-BASED NATURE TOURISM (CBNT) PLANNING AND MANAGEMENT: WITH SPECIAL REFERENCE TO THE GARHWAL HIMALAYAN REGION OF UTTARAKHAND

SHIVAM PRAKASH BHARTIYA,[1] SHIVENDRA SINGH,[2] and VAIBHAV BHATT[1]

[1]*CMTHS, HNB Garhwal Central University, Srinagar (G), Uttarakhand, 246174, India*

[2]*Mahatma Gandhi Central University of Bihar, Mothihari, 845401, Bihar, India*

ABSTRACT

Tourism is always considered as a development strategy for sustainable development in rural areas, especially those areas which are rich in natural resources and having the abundant potential for tourism development. However, several past researchers have identified several problems allied with the development of tourism in these areas. Although tourism in rural areas utilizes inoperative local human resources, natural resources, in general, are more efficient with regard to generating income (Hjalager, 1996). Tourism is not a widespread solution since the economy of rural areas is best served when tourism is integrated into wider development plans in these areas (Briedenhann & Wickens, 2004).

13.1 COMMUNITY-BASED NATURE TOURISM (CBNT)

Community-based nature tourism considered as an amalgamation of those swaying activities that support objectives of socio-economic development and conservation of natural resources of the tourist destination. Growth in tourism development and an increase in the interest of travel to those destinations which have opulent natural resources and autochthonous culture, it is necessary to provide a broad definition of community-based nature tourism. The host-guest relationship is the crux of the community-based nature tourism and having a consequential participation by both sides for the natural resource conservation and the economic development of the local community.

Persisting changes in the natural environment of rural areas succor as a useful background to the emergence of interest in Community-based nature tourism. The natural heritage of rural areas of the developing nations is considered as an effectual source of economic, sociocultural, and environmental benefits.

Natural surroundings of the tourist destinations of developing nations considered as a primary attraction for foreign guests, and it also increases the number of domestic visitors.

13.2 RAISON D'KTRE FOR COMMUNITY-BASED NATURE TOURISM (CBNT)

1. CBNT constructs effective strategies for natural resource conservation and development. Strict enforcement, protection policies, and top-down, centralized decision-making process to conserve the natural resources have not been successfully implemented.
2. Involvement of host population in the management and decision-making process, considered as most accountable and preferable for long-term sustainable development.
3. Profit-orientation, monopolization, and amassing of benefits from a small number of beneficiaries from tourism support the development and promotion of CBNT. Another rationale for CBNT development is to find out the solutions related to the distribution of benefits of tourism among local communities because of the restrictions and loss of access to resources.

Community-based nature tourism not only focuses on the nature conservation but it also draws attention towards, the cultural conservation, community or gender empowerment and poverty alleviation linkages with economic development.

13.3 ASSUMPTIONS OF COMMUNITY-BASED NATURE TOURISM (CBNT)

- Community-based nature tourism must contribute to increasing and/or improve conservation of natural resources, including biological diversity, water, forests, landscapes, etc.;
- Community-based nature tourism must contribute to local economic development through increasing tourism revenues and other benefits to community participants, and ideally to an increasing number of participants;
- Community-based nature tourism must have a level of participation ideally progressing toward self-mobilization; and
- Providing a socially and environmentally responsible tourism products to the visitors is the main aim of the development of CBNT.

13.4 COMMUNITY-BASED NATURE TOURISM STRATEGIES

CBNT involves a wide range of activities, that collectively helps in tourism development and natural resources conservation. At one side of the range, lands used for tourism purposes may be owned and managed by the local community with their active involvement in decision-making and development process. The other side of the range the travel agency and tour operators make an agreement with the local tourism entrepreneur for their services and tourism products. Around the globe, CBNT may be considered as an alternative form of tourism and introduced in the rural areas and to the community, with an intention to propose intensive planning and to identify new market opportunities. CBNT emphasizes on the value-addition and constructing natural assets for the increase in revenue generation, providing income-generating opportunities to the local community and for the conservation of natural resources.

Since the main purpose of CBNT is to protect and marmalade natural resources for economic benefits, and for the profitability, CBNT development must be considered for the development and promotion of tourism products. In Table 13.1, four entrepreneur transforms are delineated in four squares consisting of different possible opportunities with business risk and potential of gaining profit with the development of CBNT.

TABLE 13.1 Tourism Entrepreneur Transform for Community-Based Nature Tourism (CBNT)

Existing Tourism products in the existing market	**Tourism products available in the new market**
This opportunity involves the selling of an existing product in the same market with the increase in the number of same customers	This preference involves the penetrating of new markets and new customers in addition to the existing customers for the same tourism products
New tourism products in the existing market	**New tourism products available in the new market**
In this option, suppliers seek to find new tourism products for the offered customers	This alternative includes a higher risk because the supplier intends to develop new products for new customers.

Source: Community-Based Tourism for Conservation and Development: A Resource Kit, The Mountain Institute, 2000.

13.5 COMMUNITY-BASED NATURE TOURISM ISSUES

With the potential of community-based nature tourism for the development of local community and environmental resources, there have been some key issues that can affect the outcome of CBNT. This section briefly explains the issues related to participation, policies, and markets.

13.5.1 PARTICIPATION IN NATURE CONSERVATION AND TOURISM DEVELOPMENT

Increase and improve participation by tourism stakeholder is an important objective of community-based nature tourism. Active participation of the local community in bottom-up and small-scale tourism development can diverge because of the socio-economic and political relationships. Participation of the local community can also vary because of the physiographic and ecological factors of a particular tourist destination.

There is a limitation to the availability and access to the suitable land and natural resources generally occurred by the ecological and physiographic resources. Type of ownership and stakeholding determines the economic benefits and an incentive to conserve natural resources. The participation of the local community will increase if they get reasonable returns of their participation. Poor and erratic source of income is the main cause of weak tourism linkages and natural resource conservation. Reasonable income to the owner of tourism services or employees will increase the active participation of local community in the decision-making process and management of natural resources.

Areas having affluent natural resources create a good foundation for achieving the objectives of the CBNT development. To support the livelihood and for the conservation of natural resources, local community willingly participates in the tourism development process. Besides natural conservation, the economic benefit is also considered as a rationale of community-based nature tourism. Since the development of CBNT is not linked with the income generating approach and majorly focus on the conservation of natural resources of the tourist destination (Table 13.2).

13.5.2 MEASURES AND POLICIES OF COMMUNITY-BASED NATURE TOURISM (CBNT)

The destinations in the development and promotion of CBNT are rural and monitored by the national level legislative organizations for regulating access by domestic and foreign tourists. In the initial assessment, rural areas show considerable potential for CBNT development. National policies impose several regulations to restrict tourist access by numbers and by seasons, and this regulation can affect the development of CBNT. Visitors need permits on some legal conditions like where, when fees, advance notice and maximum number by a government organization, travel, and tour agencies so on.

Visitors to protected natural areas and the local entrepreneur are linked with the international policies and actions. Visitation of tourists may affect the political instability, structural changes in economic, fiscal policy and subsequent changes in the currency rates. Demanding of payment by travel and tour agencies in the visitors' country of origin led to the economic benefits. In developing nations, ecotourism and community-based nature tourism activities to be subjugated by tour operators and/or local tourism entrepreneurs. This results from the different access to capital gain tendency with the development of necessary skills. Larger operators in the promotion of CBNT assist as an intermediary between consumer and local tourism entrepreneur by providing skills and covering costs.

A useful framework is necessary to explore the wider benefits of community-based nature tourism in the tourism market chain. The tourism market chain consists of the product or packages from the place of production

TABLE 13.2 National Level Strategy for Community-Based Nature Tourism (CBNT)

National tourism legislative bodies focus on the impact of tourism industry impacts on the competitiveness of the tourism market in the country. Sproule (1997) suggested three questions for the assessment of tourism industry impacts:

Q.1. Does the tourism industry expand the market capacity of the tourism sector?

Q.2. Is it necessary that CBNT will diversify the national tourism product by adding elements of eco-ethical, wilderness, or adventure?

Q.3. Does CBNT development increase the geographical spread of tourism potential of the nation around the globe?

For the purpose of gaining long-term competitiveness of the tourism sector and for strengthening the national tourism product and services, natural resource conservation and sustained natural resource are considered as key components. Several researchers in the past also noticed the economic and environmental benefits of the tourism industry by providing equity and welfare to the local community. They also suggested public and political support for the long run benefits of the tourism sector.

Consideration of several factors such as active community participation in decision-making and natural resource management, capacity building programs for increasing the number of tourism venture in the national tourism market, spreading awareness to the local community for their legal rights, proper distribution of revenue generated with the CBNT, and providing financial assistance to the local community for tourism development.

Five broad areas for the development and promotion of Community-based Nature Tourism:

1. Implementation of mechanism that directly benefits the communities with the development of CBNT
2. Development of financial and legal mechanism to facilitate the CBNT
3. Improved information and communication technology within and between CBNT and other principle suppliers of the industry
4. Improving promotional strategy for the continuous perusal of CBNT
5. Development of entrepreneurial skill or capacity building institutions for training and education for the local community because this facilitate their full participation in the tourism industry

Source: Sproule (1997).

to the ultimate customer. The concept and process of community-based nature tourism are critical until unless it places above than other factors, marketing context and follows the more effective method of assessment of its viability. Furthermore, a tourism market chain is the linkages between community-based nature tourism players, their function and opportunities for interventions.

13.6 CASE STUDY: GARHWAL HIMALAYAN REGION OF UTTARAKHAND

Garhwal lies in the center of the Indian Himalaya (Figure 13.1). In the Hindu scriptures spiritual and natural serenity is always associated with the snowy mountains of the region, and there are numerous popular natural destinations located in Garhwal. The region has some of the most majestic places, and also provides some outstanding natural beauty. It is a part of the Greater Himalayan ranges and is covered by beautiful Himalayan peaks and glaciers.

FIGURE 13.1 **(See color insert.)** Garhwal district.

The two very important and holy rivers of India, the Ganga and Yamuna start their journey from here. Garhwal considered truly as a paradise on earth, allures everyone from everywhere. The fresh air, the pure water, the chilling snow, the adverting mountains, the scenic beauty, the small villages, the simpler people and a tougher lifestyle is what that distinguishes Garhwal from rest of the world.

For those fascinated by the unexplored, the mountain ranges of India offer breathtaking, enthralling trekking routes. In the vast amphitheater of the Garhwal Himalayas, there are limitless possibilities for trekking, mountaineering, skiing, river rafting, and nature walking. The Garhwal region of Uttarakhand is not just for mountaineers, trekkers, and climbers. It provides the best way to enjoy natural beauty and camping in the foothills of Garhwal region of Uttarakhand conceivably the best way of luxuriating in the stupefaction of nature and sharing the love and the warmth of its people. Garhwal has such varied topography that it showcases different types of flora and fauna. It has some unique wildlife activities. Different destinations of the region have some unique features. So to explore the wildlife of Garhwal Himalayas jungle safari is the best option and it just not involves exploring the jungles of the region but also provides a chance to feel the natural serenity as well.

Garhwal Himalayan region of Uttarakhand is an ecstasy for nature lovers and adventure sports like river rafting, trekking, paragliding, Corbett jungle jeep safari, and enthusiasts. Rishikesh, Har-Ki-Doon, Mussoorie, Dhanaulti, Kanatal, Chamba, Landsdown, Khirsu, Chopta, Auli, Valley of flowers, Dayara Bugyal, Bedani Bugyal, Ghuttu, Panwali Kantha, Sahastra Tal are some of the prominent destinations for nature and adventure tourists.

13.7 SUMMARY

In tourism industry products based on natural heritage considered as an opportunity for the development of local community and rural areas. Nowadays tourism products are not only assessed as an environmentally responsible and ethical option for customers but also became an element of product development and marketing strategy of community-based nature tourism. The research article makes an effort to elucidate the concept of CBNT vis-à-vis its planning and management in the Garhwal Himalayan region of Uttarakhand. CBNT is a form of tourism that involves meaningful interaction and participation between the visitors and the host population to induce economic as well as nature conservation benefits for not only the local host population but also for the natural environment where the tourism activity is being performed. The rationale for the adoption of this form of tourism is to involve the local people in the planning and the decision-making process without whom the resources of the region would be blindly plundered leading to the destruction of the natural resources and disturbances in the ecological flora and fauna. Involvement of local people brings a local perspective and assures the sustainable utilization of indigenous resources. This also leads to a distribution of profits and benefits equal to the host population who were previously devoid of the tourism benefits and were only involved as an outsider being restricted to meager tourism jobs in their region. The assumptions of the given form of tourism include contributing to increasing and improving conservation of natural resources, local economic development through increased tourism revenues leading to progressing towards self-mobilization thus providing the tourists visiting the region with a socially and environmentally responsible product. The strategies for CBNT include involving the local people and encouraging them to get involved in this sustainable form of tourism in cooperation with the private entrepreneurs operating in the region. The Entrepreneur's transom can be applied for the better implementation of the designed strategies.

The research article also discusses the diverse issues related to CBNT. The ability of local people to actively participate is an important issue which needs to be addressed before venturing into the mentioned form of tourism. The sensitization of local people towards CBNT would be a key to ensure their active involvement in the economically viable tourism form. Private entrepreneurs should also be given some level of decision making freedom in the tourism system. The sustainable form of tourism should be seen as part of a long-term strategy to support conservation, and that would eventually result in greater involvement of key stakeholders in the natural resource management. The research article highlights the important role of strategic alliances in the successful operation of CBNT, i.e., alliances between communities, NGO's, Government agencies and commercial private sector and endorses for a symbiotic relationship between them.

The Garhwal Himalayan region of Uttarakhand is endowed with natural and cultural attractions. The region is an idyllic location for adventure, nature, spiritual, eco, and wildlife and pilgrimage tourism. The people of the region also have a unique culture and lifestyle to titillate the curiosity of city dwellers. CBNT devised with proper planning and management will not only act as a boon for the tourists visiting the Garhwal Himalayas by introducing them to the serene natural landscapes and fascinating local culture but will also do wonders for the ailing rural economy in the form of increasing employment opportunities for the local people thus opening doors for economic prosperity in the region.

KEYWORDS

- **community-based nature tourism**
- **community participation**
- **planning**

REFERENCES

Boo. E., (1990). Ecotourism: The Potentials and Pitfalls. Vol. 165. Washington. D.C. World Wildlife Fund.

Bornemeier, J., Victor, M., & Durst, P. B. (1997). Ecotourism for Forest Conservation and Community Development.Proceedings of an International Seminar, Chaing Mai, RECOFTC, Thailand.

Brandon, K. (1996). Ecotourism and Conservation: A Review of Key Issues. World Bank Environment Department. Paper No. 033. Washington. D.C. World Bank.

Briedenhann, J., & Wickens, E. (2004). Rural Tourism – Meeting the Challenges of the New South Africa. *International Journal of Tourism Research. 6*(3), 189–203.

Ceballos, L. (1996). Tourism, Ecotourism and Protected Areas. H.C. Gland. Switzerland: IUCN.

Cooperrider, D. L. (1990). Positive Image and Positive Action: The Affirmative Basis for Organizing, in Srivastava, S., Cooperrider, D. L., & Associates, (eds.). Appreciative Management and Leadership. Jossey-Bass. San Francisco.

Hammond, S., & Royal, C. (1998). *Lessons from the Field: Applying Appreciative Inquiry*. Thin Book Pub. Co.

Hjalager, A. M. (1996). Agricultural Diversification into Tourism: Evidence from a European Community Development Programme. *Tourism Manage. 17*(2), 103–11.

Ingram, C. D., & Durst, P. B. (1989). Nature-Oriented Tour Operators: Travel to Developing Countries. *Journal of Travel Research. 28*(2), 11–15.

International Resources Group Ltd. (1992). Ecotourism: A Viable Alternative for Sustainable Management of Natural Resources in Africa. U.S. Agency for International Development, Washington. D.C.

Jackson, W. J., & Ingles, A. W. (1998). Participatory Techniques for Community Forestry – A Field Manual.Australian Agency for International Development (AusAID), IUCN – The World Conservation Union. World Wide Fund for Nature.

Larson, P., & Svendsen, D. (1995). Participatory Monitoring and Evaluation: A Practical Guide for successful ICDPs. WWF-USA, Washington. D.C.

Lindberg, K., & Hawkins, D. (1993). Ecotourism. A Guide for Planners and Managers. North Bennington, Vermont, U.S.: The Ecotourism Society. Vols. I & II.

Pretty, J. N., Guijt, I., Scoones, I., & Thompson, J. (1995). A Trainer's Guide for Participatory Learning and Action.Sustainable Agriculture Programme. International Institute for Environment and Development. 3 Endsleigh Street. London WCIH ODD, UK.

Rietbergen-McCracken, J., & Narayan, D. (1997). Participatory Rural Appraisal.Module III. Social Policy and Resettlement Division, Environment Department. The World Bank.

Smith, V. (1989). Hosts and Guests: The Anthropology of Tourism. University of Pennsylvania Press.

Srivastava, S., & Cooperrider, D. L. (Revised Edition) (1999). Appreciative Management and Leadership: The Power of Positive Thought and Action in Organization. Williams Custom Publishing.

Welles, M., Brandon, K., & Hannah, L. (1992). People in Parks: Linking Protected Area Management with Local Communities. World Bank.WWF. USA1D. Washington. D.C.

Whelan, T., (ed.). (1991). Nature Tourism: Managing for the Environment. 223pp. Washington. D.C.: Island Press.

Worah, S., Svendsen, D., & Ongleo, C. (1999). Integrated Conservation and Development: A Trainer's Manual. WWF-UK, Godalming.

Ziffer, K. (1989). Ecotourism: The Uneasy Alliance. Washington. D.C.: Conservation International and Ernst & Young.

CHAPTER 14

COMMUNITY INTERVENTION IN DESTINATION DEVELOPMENT: A CROSS-DESTINATION ANALYSIS

A. VINODAN,[1] JAMES MANALEL,[2] and MANOJ EDWARD[3]

[1]*Assistant Professor Indian Institute of Tourism and Travel Management, Nellore, Andhra Pradesh, India*

[2]*Professor, School of Management Studies, Cochin University of Science and Technology, Kochi, Kerala, India*

[3]*Associate Professor, School of Management Studies, Cochin University of Science and Technology, Kochi, Kerala, India*

ABSTRACT

Purpose: The purpose of this chapter is to identify and evaluate the intervention of community in destination development of India based on cross-destination analysis.

Approach: The study adopted the exploratory sequential method. The employed focused group, interview, and structured questionnaire for primary data collection. One-way ANOVA was done for testing hypothesis. Data gathered from four ecotourism destinations of Kerala, India, viz Thenmala, Periyar, Parambikulam, and Wayanad.

Findings: The result indicates that the community intervention for destination development can be identified under three areas namely governance, ecodevelopment and commercial intervention with 16 variables. Except one factor pertaining to product development, all other aspects of community intervention found to be same across ecotourism destinations under study which underlines the diversity of resource profile and appropriation.

Research limitations: Study confined to community-based ecotourism destinations inside the protected areas (PAs) of India.

Practical implications: This study further extent scope for assessing the relationship between intervention strategies and destination sustainability, destination quality and quality of life of host communities. Further, a study on nature of intervention of community in tourism has significance in bringing more inclusive and sustainable resource management practices which can, in turn, support policy making at state or regional level and implementing the same at the destination level.

14.1 INTRODUCTION

The word community is commonly used to refer to a locality where people live, as well as to all the people who live there. It includes their shared experience, values, culture, religion or way of living. There are different ways of conceptualizing community such as geography, place, interest or identity.

Intervention comes from the Latin word *intervenire*, meaning "to come between, interrupt." Often an intervention is intended to make things better or improve the situation. When we examine the dictionary definitions of intervention, we can understand that intervention is the process of involvement for making some positive changes in the area involved. According to the Cambridge Advanced Learners Dictionary (2014, p. 311), intervention means to involve intentionally in a difficult situation in order to improve such situations. Macmillan English Dictionary for Advanced Learners (2007, p. 343) defines intervention as a situation in which someone involved in a particular issue or problem in order to influence the same. Oxford Advanced Learners Dictionaries (2015, p. 212) also defines intervention as a process of 'take part' in something so as to prevent or alter a result or course of events. Whereas Fritz (2014) defined intervention as a process of taking action in an existing situation and that situation may, or may not, be defined by one or all involved as something that is problematic. Parties may want to understand, improve or prevent something and still not think of the current situation as something that is a problem. There are seven levels of intervention such as global, international, national, local community, organization, small group, and individual and reiterated that each level has its own importance (Fritz, 2014).

Rothman (1995) examined the relationship between community and intervention, and he conceptualized grass-root community organizing as an intervention. The missions of most organizing groups explicitly involve instrumental goals such as local and societal change. Community intervention is considered as a process of involvement of community groups through various formal and informal mechanism to create positive changes in the respective actions or areas. However, the level of intervention coined by Graf (1995) is different. According to him, these are educational or other strategies that involve individuals, families, social networks, organizations, and public policy. Organizing models may be useful to build relationships between residents as part of efforts to increase capacity for collective action and collective decision making.

According to Merzel and D'Afflitti (2003), the term '*community-based*' often refers to the community as the *setting* for interventions. As a setting, the community is primarily defined geographically and is the location in which interventions are implemented. Intervention may be direct or indirect. If the intervention is practiced through media, it is an indirect intervention. Direct Intervention is being practiced through institutions such as neighborhoods, schools, churches, temples, work sites, voluntary agencies, or other organizations.

According to Responsible Ecological Social Tours Project (REST, 1997), CBT is tourism that takes all dimensions of sustainability into account. It is managed and owned by the community, for the community, with the purpose of enabling visitors to increase their awareness and learn about the community and local ways of life. The following model (See Figure 14.1) explains that relationship.

REST further clarified that CBT and community development are inherently connected, because they share the same natural and cultural resource. Culture and social norms determine not only resource use, but also structure internal and external relationships

In this juncture, Dassah (2013) content that participation is a human right, the execution of participation is essential to realize the other human rights and is also central in enabling the people to claim their rights. He has also clarified that declining civic participation, alienation of grass-root level citizens from resource management in a representative democracy and increased distance between the citizens and the governments are the reasons for the democratic deficits globally.

14.1.1 COMMUNITY INTERVENTION IN TOURISM

Community-based approaches allow each community to develop a management strategy which meets its own particular needs and conditions, enabling more flexibility. This approach also enhances recognition and respect for cultural differences on the local and regional levels and among nations. It

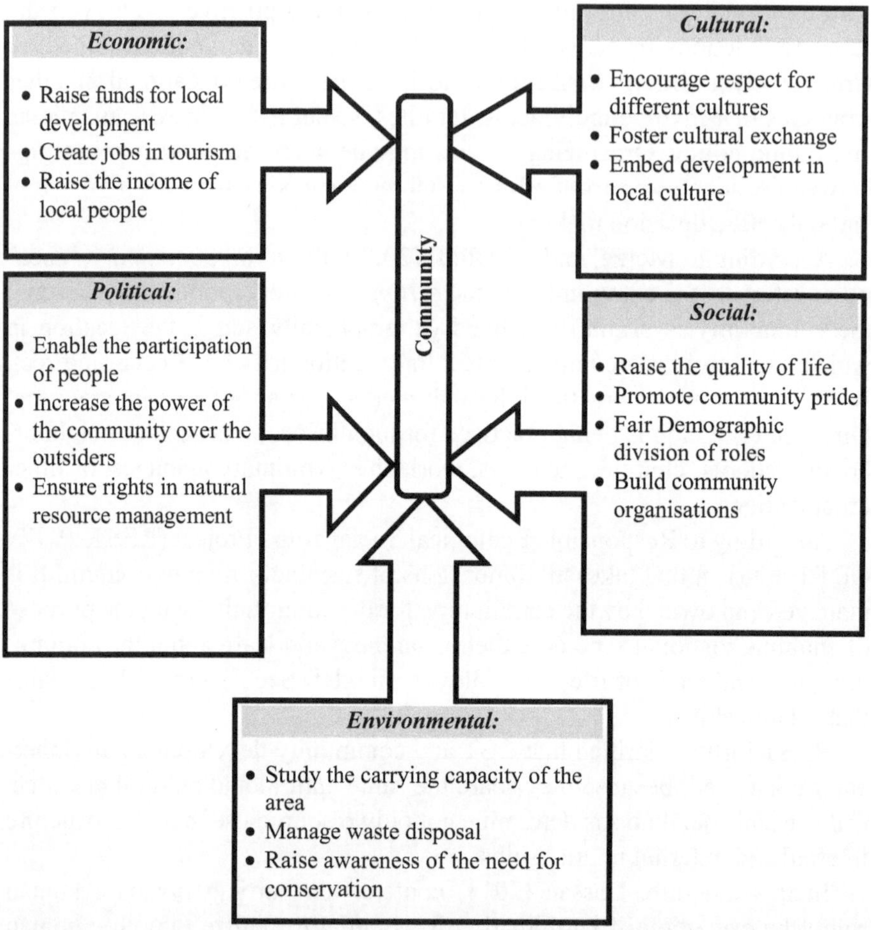

FIGURE 14.1 CBT model (Modified from REST, 1997).

strives to make maximum use of indigenous knowledge and experiences in developing management strategies. Community-based tourism (CBT) is more than a tourism business that aims at maximizing profits for investors; it is more concerned with the impact of tourism on the community and environmental resources. CBT emerges from a community development strategy, using tourism as a tool to strengthen the ability of community organizations that manage tourism resources with the participation of the local people. In order to connect the community with tourism, it is imperative to have a glance of such studies which connects community and tourism, community and sustainability and operational aspects of a community intervention in tourism (in general).

Community intervention in tourism can be classified into two: organized intervention and unorganized intervention. Unorganized intervention includes the voluntary involvement of the local community to provide various tourism and related services/commodities without proper guidelines or modus operandi in resource appropriation. Generally, these are practiced in two ways: (1) community member(s) involve in the appropriation of their/ his own resources for meeting visitors' needs and, (2) involvement of community member(s) in appropriating common property resources, mostly owned by the state with or without permission for creating and delivering tourism and related services. If this intervention is with permission, most often such appropriation is based on local specific norms or legislation [for example Provisions of Scheduled Tribes and Other Traditional Forest Dwellers Act (Recognition of Forest Rights), 2006] or traditions. In such intervention, the main focus was given to livelihood particularly in generating income and employment among identified communities. Experience shows that tourism and related activities were given less recognition as a means of livelihood in these kinds of intervention. However, the economic contribution of such intervention cannot be undermined.

In a nutshell, the unorganized intervention of communities in resource appropriation may bring economic sustainability whereas other dimensions of sustainability issues may not be addressed in a proactive manner. These other dimensions include the collective intervention of community for the holistic conservation of resources or more specifically environment management, democratic representation of all strata of the community concerned, strengthening neighborhood relations, etc.

Organized intervention, on the other hand, tries to spell out all dimensions of sustainability along with economic sustainability. These include: political, sociocultural and ecological in a holistic manner to address the multifaceted issues of society and ecology in an anthropocentric way. Organized intervention is most often initiated and institutionalized for the purpose of resource management based on well-documented rules, regulations, and guidelines based on the local specific norms or practices by considering the human being as the center of the development process. These institutionalized structures identify various dimensions of the sustainability and address them in a systematic manner. In the contemporary resource management practices particularly for common property management, community-based institutions are considered as the panacea for grass-root level, bottom-up approach of resource management.

From the above observation, we can conclude that CBT is a strategy to create the potential for empowerment of the community, enhancing their involvement in decision making, but it should also ensure that the will and incentive to participate comes from the community itself. In other words, this indicates that the control over an action should rest with the people who will bear its consequences. In ecotourism, the community is considered as an integral part of tourism activity owning many stocks of their credit including cultural and social properties, conservation and resource management strategies, etc., which are considered as attractions having good market demand. In this context, the present study tried to examine this organized or institutionalized intervention of communities in ecotourism destinations of Kerala.

14.1.2 REVIEW OF LITERATURE

Murphy and Murphy (1985) studied how to integrate community with tourism. Further, argued that the community approach to tourism development is an attempt to integrate the interests of all community stakeholders, including the local community as a primary group both for analyses and proposals for development. But Morais, Zhu, Dong, and Guihua (2006) tried to examine various factors constraining local residents from becoming involved in the tourism industry. They had outlined strategies and policies necessary to overcome those constraints, and also addressed the infrastructure improvements, training of villagers in basic hospitality skills, along with the introduction of key tourism policies. The study further stated that education and awareness programmes and local support are significant to preserve local culture and the natural environment. In this direction, UNWTO (2007) introduced the Visitor Industry Community Environment (VICE) model to showcase the role of community in destination management. It stated that destination management is the interaction between the visitors, the industry that serves them, the community that hosts them and the environment (built and natural) where this interaction takes place.

Michael (2007) also underlined the relationship between community and destinations. But he has further reminded that new tourism projects must take community choices and community welfare into greater consideration and optimize the consequences of economic growth caused by tourism to be relevant and to meet the expectations of the new tourists. In other words, there must be a match between the economic benefits created and the needs as well as the values of the local population.

However, various studies have also commented that through community-based models of tourism are widely advocated in these days as top-down models, even they have also been unsuccessful in meeting various dimensions of destination sustainability (Murphy, 1985; Scheyvens, 1999; Singh, Timothy & Dowling, 2003; Kiss, 2004; Jones, 2005).

When we examine operational aspects of CBT in general, the observation of Murphy (1997) with regard to the impact of tourism in connection with the individual entrepreneurs and community is found relevant. He has observed that tourism has managed to become a holy cow that can be let in anywhere. This indicated that tourism resources could be (mis)-appropriated. But Raik (2002) opined that capacity building of the community members should be considered as an important factor to handle tourism operations and to control and explore their environment, extracting, developing, and investing in local resources. Similarly, Mohamad and Hamzah (2013) suggested community cooperatives as a means of CBT to bring economic sustainability as well as sociocultural and environmental benefits. Further, they stated that shareholdership of the community enables higher involvement in the decision-making processes, increases tourism income distribution, solidifies the sense of ownership, and strengthens social cohesion.

14.1.3 OBJECTIVES

The objective of the study is to identify community intervention strategies in tourism particularly perceived community-based ecotourism destinations of India. In this study, the community's opinion on various intervention strategies has been examined to accelerate more inclusive resource appropriation of common property resources for tourism development leading to the overall development of the country. The study incorporates the following specific objectives:

- To identify the various community intervention strategies practiced in the tourism sector in India.
- To evaluate the community intervention strategies in tourism through a cross destination analysis.

14.1.4 RESEARCH QUESTIONS

The study tried to investigate answers for the following questions: What are the different intervention strategies of local/destination communities prevailing in Indian tourism destinations? And, how are these strategies different across destinations?

14.1.5 HYPOTHESIS

There is no significant difference in community intervention strategies' across tourism destinations in India.

14.1.6 SCOPE OF THE STUDY

In the context of inclusive and sustainable development of Indian economy, this study assumes significance as it attempts to examine various community intervention strategies practiced in Indian destinations; revealing the destination specific intervention strategies to strengthen the community-based tourism programmers. The study enables the policymakers, tourism service providers, and other intermediaries to gain a better understanding about the present strategies and design alternate or strengthen the existing practices to reduce the guest-host conflicts' as well as to ensure equity in resource sharing.

14.1.7 DESTINATION UNDERSTUDY

The present study has identified four Protected Area (PA)-based ecotourism destinations. These are Thenmala, Periyar, Parambikulam, and Wayanad. These four destinations have operationalized ecotourism activities through community support. Most of the destination community members were, one way or other, engaged in ecotourism and related activities of these PAs. Reports of the Department of Forest and Wildlife (DFW) showed that community-specific representation in ecotourism activities was ensured across destinations as shown in Table 14.1.

14.2 EXPLORATORY SEQUENTIAL APPROACH

The present study followed a sequential mixed method because one stage was conducted after the previous stage was completed (Creswell, Plano Clark, Gutmann, & Hanson, 2003). Furthermore, the results of stage one of this study were used to develop the measurement instrument for stage two. The rationale for adopting a sequential mixed method was justifiable on following grounds. The objective of the research was to identify certain factors which are unknown and its dimensional orientation that was not explored in prior studies. It is also observed that employing mixed methods could give more authenticity while identifying certain factors which are ever changing.

TABLE 14.1 Profile of the Study Area

Study Area	Area in Sq. Km	Year of Declaration as Wildlife sanctuary	Community profile (85% are Scheduled Tribes)	Community intervention framework	Community Members in Ecotourism	Active Members#
Parambikulam (Tiger Reserve)	265	1973	Kadar, Malasar, Muduvar, and Malamalasar,	EDC^	503	302
Periyar (Tiger Reserve)	777	1934	Mannan, Paliya, Urali, Mala-araya, Malampandaram.	EDC three types*	540	312
Thenmala (Senduruny Wildlife Sanctuary)	172	1984	Kanikkar, Malayarayar Malaipandaram, Malavedan Ulladan	EDC, SHG (for general tourism)	175	76
Wayanad (Wildlife Sanctuary)	344	1973	Paniyas, Adiyas, Kattunayakan, Kurichiyans, Urali Kurubas, Mulla Kurubas, and Jen Kurubas.	EDC	125	71

Source: Primary data.

^ Eco Development Committee.

Total No. of active members in ecotourism and related activities came to 761 in these 4 PAs as on December 2013.

*1. Village EDC: Consists of EDC in tribal settlement and hamlets;

2. User Group EDC: For Grazers, fuelwood, Thatching, Grass collection and assisting pilgrimage;

3. Professional Group EDC: Consists of Ex-Cinnamon Bark Collectors, Tribal Trekker-cum- Guides and Watchers.

Initial stages of the study adopted Focused group discussion (FDG), Expert interview (EI) to collect qualitative data and the questionnaire-based survey was adopted for quantitive analysis.

14.2.1 STAGES OF EXPLORATORY STUDY

Multiple data collection techniques were applied in stage one: focus groups and expert in-depth interviews. The output of the stage one study was analyzed, and the results, along with the existing literature, were used to develop the stage two instrument. Prior approval was sought from Chief Conservator of Forest (CCF), Government of Kerala (GOK). Stages of an exploratory study conducted for the present study are summarized in Table 14.2.

The stage of the exploratory study consisted of two phases. The primary stage consisted of Focused Group Discussions (FGD) with Tourism Eco-Development Committee (TEDC) office bearers, and secondary stage consists of Expert interviews with officials of Department of Forest and Wildlife (DFW) under Govt. of Kerala (GOK) and researchers educators of ecotourism.

14.2.2 RESPONDENTS' CRITERIA OF QUALITATIVE STUDY

For identifying the respondents in phase one of the study, the following conditions were stated:

a) Respondents with 4–7 years of experience as TEDC members as well as those holding the post of president/vice president.
b) Officials from the Department of Forest and Wildlife (DFW) associated with Eco and Tribal development, not below the rank of range officer or superintendent

TABLE 14.1 Stages of Exploratory Study

Stage	Action	Participant	Purpose
Primary	Focus groups	Tourism Ecodevelopment Committee (TEDC) office bearers	To Identify various CIS
Secondary	Expert interviews (Officials)	Officials of Department of Forest and Wildlife (DFW)	To discuss variables of CIS
	Expert interviews (Academic)	Researchers and educators in (eco) tourism	To discuss and finalize variables CIS

For phase two, only officials from DFW associated with Eco and Tribal development, not below the rank of range officer or superintendent, were interviewed for their expert views to finalize the variables related to community intervention as well as DS.

14.2.3 DEVELOPMENT OF QUALITATIVE QUESTIONS

A semi-structured interview outline was used in both the qualitative research methods: focus groups, and in-depth expert interviews. Each question was carefully and deliberately designed based on the methodology suggested by Thomas (2003).

14.2.4 FOCUS GROUP DISCUSSION

According to Kitzinger (1995), a focus group is beneficial to explore people's knowledge and experience and can be used to examine not only what people think but how they think and why they think that way. As a part of a mixed method, the present study used a focus group to increase the validity of findings. It tried to capture various intervention strategies of destination communities in ecotourism and tried to explore the depth and nuances of opinions regarding such strategies.

FGD was organized at Parambikulam, Periyar, Thenmala, Aralam, and Wayanad with TEDC office bearers present as well as past (see Table 14.3). Chain referral sampling method was followed for identifying past presidents and vice presidents. TEDC office bearers were considered as reliable sources of information with regard to various managerial aspects of community intervention. They were acting as an interlocutor between the community and various government departments. So it is understood that they can give more clarity about community participation and related aspects with regard to ecotourism.

According to Kitzinger (1995) and Krueger and Casey (2009), six participants in a group is considered adequate to gather information. As the number of office bearers (present as well as former) was less in number, focus groups had to be conducted even with less number than the stipulated limit in some cases. Only four each at Thenmala and Wayanad were available during the study period. So they were also considered as two groups for the study.

All the TEDC office bearers were first timers in FGD. It helped to get some spontaneous but genuine information to the investigator. They were seemingly least concerned about the outcome of their expressions. This might be on the basis of the assurance given by the investigator that the outcome of the discussions will be used only for academic purpose. Translators were used for FGD at Periyar and Parambikulam to get the information from Tamil speaking office bearers.

1) Face-to-face meeting where the investigator could assess the body language/gestures which support or oppose the response from fellow members also, to get participant's undivided attention was used.
2) Focus Group facilitation/content guide was prepared with open-ended questions which could provide in-depth responses.
3) The duration of each session varied, from 60 to 90 minutes. Twenty-three members participated across all four PAs under the study (see Table 14.3). After reviewing the focus group transcripts, the researcher assessed that there was potential unexplored information evident during the focus groups.

14.2.5 EXPERT INTERVIEW

The present study followed a semi-structured interview method. Semi-structured interviews are often used as the sole method of data collection for a qualitative research project and can be used to obtain data from individuals or groups (Bloom & Crabtree, 2006). Semi-structured questions allowed the respondents to express their own points of view and to describe situations, events and their experiences regarding CIS and their impacts on conservation and livelihood by ensuring tourist satisfaction in the destinations under study. As David and Sutton (2004) had pointed out, it also permits the researcher to obtain more details from the participants about their own views regarding the issue under study.

Expert interviews with Officials were organized during the same period of FGD with TEDC office bearer. This method was introduced especially to examine more deeply certain topics that remained unexplored in the focus groups and also where TEDC office bearers are not able to give a concrete view.

Experts consisted of 21 officials of DFW with the eligibility criteria as mentioned earlier, and they were interviewed. Employees of DFW in the respective forest ranges acting as the secretaries of TEDC were also included (see Table 14.3).

An in-depth interview guide was prepared with open-ended questions which could provide an in-depth response. Respondents were advised to

TABLE 14.3 Respondents for the Exploratory Stage

Expert interview respondents	No. of People Interviewed	Focused Group respondents (EDC Presidents/ Vice Presidents)	No. of People participated
Office of the Wildlife Warden **Parambikulam Tiger** Reserve (Wildlife warden and staff)	**8**	**TEDC Presidents/** Vice Presidents – Parambikulam Tiger Reserve	**8 (5)***
Office of Thenmala Ecotourism project (Officials) and Senduruny Wildlife Sanctuary Thenmala (Senior Forest Guards)	**7**	**TEDC Presidents/Vice** Presidents – Periyar Tiger Reserve	**7 (4)***
Office of the Tribal and Eco-development, Periyar Tiger Reserve (Tribal and Eco-development Officer and staff)	**2**	**TEDC President/Vice** President- Senduruny Wildlife Sanctuary (Thenmala)	**4 (1)***
Office of the Wayanad Wildlife Sanctuary (Assistant Wildlife Warden, Range officer, Superintendent, and staff)	**4**	**TEDC President/** Vice President-Wayanad Wildlife Sanctuary (Muthanga Range)	**4 (1)***
Total	**21**		**23**

Source: Primary data, * in parenthesis denotes no. of TEDC.

put a tick mark on the variables pertinent to their CIS. The interviews ran from 45 to 90 minutes for each interviewee. Interviews were held in their chambers only. Eighteen variables relating to CIS were presented before officials from four PAs (Thenmala, Periyar, Parambikulam, Wayanad). It may be noted that the variables identified by the investigator with regard to CIS had been informally discussed with the officials during the interviews. They had identified a few redundant variables and suggested to remove them for a comprehensive view of community intervention across the states. A few items were reworded to get a holistic view, and thus 16 variables were finalized for CIS.

After the verification from the officials of DFW, the details of the research design were presented before academic experts in (eco) tourism from reputed organizations and universities for review. Eight experts were consulted in person and/ or by mail. They confirmed the variables with minor modification in the wordings.

14.2.6 RESULTS OF THE EXPLORATORY STUDY

Based on the literature review related to Community-Based Natural Resource Management (CBNRM) (Agrawal & Chhatre, 2006; Danielsen et al., 2009), focus group and the recommendations of the experts, all those 16 variables of various Community Intervention Strategies (CIS) were broadly classified into three distinct activities for the purpose of the present study. They are as follows:

1) commercial intervention;
2) eco-development intervention; and
3) governance intervention.

14.2.6.1 COMMERCIAL INTERVENTION

According to Buckley (2009) commercial opportunities in ecotourism go beyond direct employment, which includes production and distribution of goods and services. In the context of a community intervention in ecotourism in PAs, Commercial Intervention (CI) has rarely been acknowledged, because the purpose of Community-Based Resource Management (CBRM) framework is to bridge the conservation and livelihood, mostly through subsistence activities like farming, fishing, collection of minor forest produce, etc. These activities are to meet communities' basic or bare minimum needs, by exploiting the available natural resources. Subsistence activities comprise all diverse sets of the endemic local system of production and distribution of various goods and services representing economic, ecological and sociocultural factors. The creators and beneficiaries of this subsistence operation of an area are mainly the members of the community concerned (Robert, 2001).

As a market-led programme, tourism has its own facets in transactions. In PAs, tourism is considered as one of the means of addressing livelihood issues; where communities are not direct consumers of the resources, they organize community's resources and natural resources for satisfying the visitor's needs. The earning from these transactions are utilized for meeting the conservation and livelihood issues of local communities. In a nutshell, such activities of communities are not a direct subsistence activity like other communities of PAs do; rather they meet the livelihood requirements through the sale of destination resources *per se* both cultural as well as natural. The present study, Therefore, considers such activities as commercial activities. Most often community activities are destination specific, so a holistic definition of commercial activities is appropriate. The following section will discuss the major commercial activities of destination communities in the context of PA-based ecotourism in India.

- *Development and Organization of tourism activities/products:* Identification of destination resources, and organizing them in a manner which meets tourist needs and sustainability of the destinations is considered as one of the major intervening areas of communities in ecotourism. Most of these resources are intangible as well as endemic to the regions, characterized by the type of destinations like hill stations, river basin, plantation, etc. Widely observed such products are trekking, bird watching, nature walk, etc. Presentation and/or reintroduction of various cultural properties are also major components of ecotourism products. Indigenous art forms are organized and presented for visitors on a regular basis, in most of the PAs coming under the present study. Moreover, in order to provide employment throughout the year, the community often introduces special products like monsoon tour, spice tour, and other specific tour packages or even diversifying existing products to attract tourists. Further, community-based tourism programmes often showcase their skills by producing various tangible products to the tourists. Most of such products are not available in the open market, because, the raw material or the skill or the process of these products carry certain kind of endemism. Accordingly, these products have been considered as authentic products of the region, to attract visitors. The present study identified community products like souvenirs, food and food products, traditional medicines, art, and craft works, etc., in the tourism market due to these reasons. In practice, these products were sold directly or through agencies to the tourists.

- *Enterprise development*: In order to avoid middlemen in selling the products of the destination communities, various shops and establishments with the support of FDA have been created by the community members in each destination. Such establishments are instrumental by ensuring sale through exhibitions, besides serving as outlets for sale of various endemic products of the region including souvenirs, food items, etc. Community operated accommodation facilities like huts and camping sites, are also major attractions of these destinations. Majority of these enterprises are organized as microenterprises or operated as ventures owned or operated by the institutions formed as a registered society like EDC or VSS in PAs.
- *Guiding, Interpretation, and Escorting*: One of the major wage employment available to most of the destination communities under the study is community assisted guiding, interpretation and escorting. Community members are trained to take-up these jobs and provide authentic information about the destination. Guiding and escorting by the community members are unique as the same by outsiders may undermine the importance of destination-specific features. Guiding is mostly arranged for day trips and escorting for trekking, and other adventurous activities. Nature walk and other special interest tours require interpretation services. The group consisting of guides, interpreters, and escorts are the major educators of ecotourism, and their institutional form is called as naturalist EDCs in the study area.
- *Product diversification*: Community members were considered as drivers of product diversification in ecotourism destinations. Most often diversification was made on the basis of linkages with other resources and practices in and around PAs. Major identified diversifications were usage of indigenous medicines, organic farming, exploring unexplored areas, value addition on existing tourism services and local products, etc.
- *Promotional activities*: As a community-driven initiative, TEDCs do often promote the natural and cultural heritage of the community.

Mostly they participate in various trade shows related to tourism in nearby areas, present indigenous art forms, and practices and also exhibit traditional medicines and rituals.

14.2.6.2 ECODEVELOPMENT INTERVENTION

As the name indicates, ecodevelopment is the major objective of community-based ecotourism where the destination community directly takes part in eco maintenance and restoration programme. According to United Nations Glossary of Environmental Statistics, eco-development can be defined as development at regional and local levels, consistent with the potentials of the area involved, with attention given to the adequate and rational use of natural resources, technological styles and organizational forms that respect the natural ecosystems and local social and cultural patterns (United Nations, 1997). Within the framework of ecotourism operation, the community concerned assumes responsibility for ecodevelopment and also encourages other stakeholders including tourists to involve in it.

Following are the major community initiated ecodevelopment activities found in PA-based ecotourism programme:

- *Conservation of natural resources*: Destination communities engage in various conservation activities to protect their endemism as well as to ensure prudent use of their community resources. In order to reduce the ecosystem degradation conservation activities like terracing, afforestation, etc., and the application of 3Rs: reuse, reduce and recycle are practiced. The provision for renewable and nonrenewable sources of resources are identified, and initiatives are made to conserve it to reduce the rate of ecosystem destruction or degradation.
- *Monitoring of resources*: As a custodian of destination resources, communities are entrusted with the task of monitoring the destination under ecotourism programmes of PAs of Kerala. They are deployed as watchers, guards, etc., to safeguard the resources. Poaching and other illegal activities are invigilated by members and reported to the DFW. Most often measures suggested by the communities are operationalized to check such menace. Besides, community members are also functioning as surveyors for environmental reporting programmes conducted by the DFW.
- *Environmental education and awareness programmes*: The common understanding is that conservation is possible only through proper

awareness among stakeholders. As a custodian of destination resources, awareness among destination communities is *sine qua non*. At the same time awareness among tourists is also considered essential for resources conservation as they are the ultimate consumers of the destination resource. Community members through Tourism EDCs (TEDCs) frequently organize various environmental awareness programmes for destination communities and practice it at their surroundings. These community members are further engaged in educating tourists about the environmental importance of the destination as well as the conservation of natural resources in general.

- *Financing for Conservation*: Beyond community volunteerism, financial support is important for executing various conservation programmes. The amount generated from various tourism-related activities including entrance fees, fees from tourism activities, revenue from other services rendered by the community at the destinations are the major sources of financial support for conservation of natural resources. As per the agreement between the community and DFW, certain percentages of such earnings are earmarked for conservation.

14.2.6.3 GOVERNANCE INTERVENTION

Governance is about the use of power and authority and how a country manages its affairs (DFID, 2007). This can be interpreted at many different levels, from the state down to the local community or the household. As categorized by OECD (2012), governance has four dimensions with various indicators. These are: political system (democracy, human rights, rule of law and decentralization), public administration (corruption, public management, public financial management, transparency and fiscal policy), social governance (efficient public service delivery, citizen empowerment, community development) and market governance (creation of favorable business environment). When we examine these variables, we find that most of them are pertinent to CBE as community intervention is aimed to improve and maintain conservation and livelihood objectives of ecotourism by exercising community power and authority. As mentioned earlier, local community members are mobilized within PAs, to make them involved in various tourism and related activities on a regular basis to meet the stated objectives. But there are a large number of other areas where the destination community is involved beyond ecodevelopment and commercial intervention at the destination. These relate to the formulation and implementation of policy decisions, which include representation, resource sharing, skill development, action for social cause, etc. These interventions are collectively referred to as Governance Intervention for this study. In other words, these activities are administrative in nature and are organized within the community for the betterment of community intervention. Governance Intervention under CBE encompasses the following:

- *Democratic representation*: Democratic representation of community members by following the stipulated procedure, i.e., one man one vote and also ensuring equality of opportunity for marginalized sections like women, tribals, scheduled communities, rehabilitated people and other vulnerable groups are the essence of community participation framework of ecotourism. As mentioned earlier, Tourism Eco Development Committees (TEDCs) have an Executive Committee which comprises of 11 persons elected by members, and 2–4 ex-officio members (non-voting). At least five elected members must be women, with a woman Chairperson or Vice-Chairperson. Further, there must be at least one member from each family in TEDC for better representation of destination communities in ecotourism and related activities.
- *Decision making and benefit sharing*: In order to regularize the pattern of intervention, community members are regularly holding meetings under the aegis of TEDC to discuss issues or new policies and programmes. Most of the Executive committee members and the General body meet frequently, the former at least twice a month and the general body members on a monthly basis. A minimum number of quorum is also ensured in all meetings as per the guidelines prepared by the DFW. Often new norms and rules are also evolved while resolving earlier decisions. In principle, they follow a stipulated method for sharing the benefit to ensure equity in resource and benefit sharing. It has been noted that such intervention has also helped to ensure standardization of wage structure across the destination, by improving the bargaining power of the community through negotiation.
- *Intermediary*: Most often community members act as an intermediary between destination communities and other stakeholders of tourism of the respective region including government agencies. In the study area, it has been observed that the representatives or the leaders of the local community help their members to connect with FDA, or the

next immediate (Range level) arm of DFW and other agencies like Directorate of Ecotourism (DOE), Department of Tourism (DOT) of Government of Kerala (GOK). These representatives also connect destination communities with local self-government institutions and other government agencies for the overall development of the region.

- *Linkage with other sectors*: Activities emanating from the integration of tourism with other sectors include agriculture, floriculture, aquaculture, etc. For example, visit to farmhouses, and spice villages are widely practiced in ecotourism destinations under study. The purpose of such integration includes better market access and extension of stay of visitors at the destinations. The communities are extending their services to such areas to attract visitors and thereby improve their livelihood options.
- *Consultant*: As the opinion of destination communities are of utmost importance for managing destination issues, the authorities like FDA of the respective regions seek the suggestions and recommendations of these community members in important matters. It has been observed that in matters like tackling of poaching, and other anti-social elements in and around PAs, community members play a significant role and their advice is useful for handling such issues.
- *Programme for Social wellbeing*: Community members themselves organize large numbers of social awareness programmes pertaining to health issues, i.e., alcoholism, drugs, etc., ethical issues, i.e., illegal transactions, poaching, smuggling, etc., and social issues like child marriage, dowry, etc., among its members. These programmes one way or the other affect the health and hygienic environment in and around the destination.
- *Capacity building*: Community members organize and participate various capacity building programmes with the support of FDA as part of the skill building process to equip its members to explore various entrepreneurial and wage employment opportunities in the ecotourism sector. Capacity building programmes include training on hospitality services, customer handling and various self-employment programmes, like the production of fancy items from waste, extraction of minor forest products, handicrafts, etc., for sale at the destinations.

Based on the above analysis it can be concluded that there are significant community intervention strategies are existing across ecotourism destinations. List of variables pertaining to various community intervention strategies and its latent dimensions are summarized in Table 14.4.

TABLE 14.4 Lists of Intervention Areas and Its Corresponding Variables

Intervention areas	Indicator variables
Governance Intervention	Democratic Representation, Capacity building, Linkage with other sectors, Consultant, Programme for Social wellbeing, Decision making and benefit sharing, and Intermediary
Commercial Intervention	Development and Organization of tourism activities/ products, Promotional activities, Enterprise development, Guiding, Interpretation and Escorting and Product diversification
Eco-development Intervention	Conservation of natural resources, Monitoring of resources, Financing for Conservation, and Environmental education and awareness programmes

14.2.7 RELIABILITY AND VALIDITY OF QUALITATIVE MEASUREMENT

In order to assess the reliability and validity of the qualitative part of the present study, the Trochim (2006) criteria have been adopted. These are credibility (on the basis of criteria and informal conversations), transferability (replicability in similar or identical community-based tourism cases), dependability (methodological consistency) and conformability (since all respondents are from similar settings conformity can be established). Credibility and transferability correspond to the internal and external validity of the quantitive research. Creditability seeks to ensure that their study measures or tests what is actually intended. Lincoln and Guba (1985) also stated that ensuring credibility is one of the most important factors in establishing trustworthiness.

14.2.8 DESCRIPTIVE STAGE

The second stage of the present study is descriptive research. According to Ethridge (2004), in descriptive research issues and problems are analyzed through data collection to describe the situations more effectively. In this study, a descriptive study was planned on the basis of primary knowledge of the subject matter obtained from the exploratory study. For the purpose of this study, a cross-sectional study was considered as the appropriate technique. As Jick (1983) has suggested, the survey is also one of the major methods of data collection used in descriptive research that contribute to a greater confidence in the generalizability of the results of the study.

14.2.9 SCALE DEVELOPMENT FOR CIS

Developing a scale to measure CIS, was difficult because every case is unique due to locations, situations, operational diversity and diversified perception of the individual, culture and visitors characteristics as revealed from the literature review. Hence, an attempt was made to develop a scale for CIS. More importantly, all the scale developed for the study were destination specific. The scale development developed was based on Churchill's (1979) guidelines, and content validity was ensured on the basis of C-OAR-SE procedure (Rossiter, 2002). A 5-point Likert agreement scale was designed to understand CIS. A short and simplified questionnaire consisting of items which had been pretested with the advice of expert officials of DFW has been used for the final study.

14.2.10 SAMPLING DESIGN

The sampling design explains the definite plan for obtaining a sample from the population, i.e., the entire group of people whom the researcher is interested to know about (Gupta & Gupta, 2013). In this study, the local community members of ecotourism destinations in south India were considered as the population for the study. As mentioned above, the investigator identified four destinations of south India, i.e., Thenmala, Periyar, Parambikulam, and Wayanad. These selections were made on the basis of purposive sampling. Purposive sampling is justified on the following grounds:

- It helps the researcher to use his discretion to select the respondents, so as to get the best samples to meet the purpose of the study.
- Purposive sampling is widely used in mixed method research (Maxwell & Loomis, 2002).
- In terms of the sampling frame, it was basically judgmental in nature as the researcher's judgment has been combined with the expert opinion which had been checked through the exploratory study.
- In this research, the sampling unit is the community members of the protected area (PA) based ecotourism destinations of south India; Therefore, purposeful sampling is appropriate.

14.2.11 SAMPLING TECHNIQUE

In this study, convenience sampling was used to select the sample units based on the following respondent criteria:

- All respondents have membership in Tourism ecodevelopment committees (TEDCs);
- All respondents have experience in involving (eco)-tourism activities.

Convenience sampling refers to sampling by obtaining units or people who are most conveniently available. For example, it may be convenient and economical to select samples of local community members who are approachable and reachable to get feedback on the subject matter. Moreover, convenience sampling is the only feasible way to proceed while attempting to learn about groups whose spatial representation is wider.

The samples were selected on the basis of availability or presence of community members under study without any prejudice for considering or rejecting a particular respondent. Selection of the respondents at the time of visit was purely by chance.

14.2.12 DATA COLLECTION

As indicated in Table 14.5 destination wise data collection was done personally by meeting the respondents. The structured questionnaires were distributed to respondents, and the purpose of the study was explained. The questionnaires were administered using a direct face-to-face survey methodology because of the strength of this method in achieving high response rates.

TABLE 14.5 Destination-Wise Data Collected and Analyzed for Final Study

Unit of analysis	Thenmala	Periyar	Parambikulam	Wayanad	No. of Data Collected	No. of data Analyzed	%
Community members	52 (43)	152 (132)	154 (134)	47 (41)	405	350	86

Source: Primary data.

14.2.13 ANALYSIS OF DESCRIPTIVE STUDY

Table 14.6 shows the categorical variables sourced during the study indicating the income, age, gender, education, and ecotourism experience.

TABLE 14.6 Demographic and Ecotourism related Variables of Community Members

Variables	Status in average	Percentage
Income	4800 (Average)	
Number of family members	6 (Average)	
Age of the respondents	Below 25	12%
	25–40	40%
	40–55	30%
	Above 55	18%
Gender in ecotourism	Male	72.4%
	Female	27.6%
Education	Below matriculation	58%
	Higher secondary	33%
	Graduates/Diploma	8%
No. of family members engaged in ecotourism	2 (average)	
Experience in ecotourism	6 years (average)	

Source: Primary data.

14.2.14. RESULTS

The results are shown in Tables 14.7 and 14.8.

The cross destination analysis of various intervention strategies of destination communities in (eco)tourism destination based on one way ANOVA shows that there no significant difference in 15 indicators of CIS across tourism destinations in India.

Levene's test for homogeneity was administered to check whether the assumption about homogeneity is violated or not for these 15 indicators of CIS. The test result revealed that it was not significant ($p > 0.05$) as shown in Table 14.7. Hence, it is concluded that the population variance of 15 indicators is more or less equal. The resultant F statistics shows that there existed no significant difference in 15 indicators of CIS at 0.05 levels (Table 14.8). However, one indicator pertaining to development and organization of tourism activities/ products indicates significant difference across destinations, i.e., Thenmala, Periyar, Parambikulam, and Wayanad at 5% significance level.

14.3 DISCUSSION

Communities play a very important role in the management of ecotourism destinations in India, particularly in PA-based ecotourism destinations. A

TABLE 14.7 Test of Homogeneity of Variances

Sl. No.	Test of Homogeneity of Variances				
	CIS	**Levene Statistic**	**df1**	**df2**	**Sig.**
1.	Democratic	1.146	3	346	0.331
2.	Capacity	2.594	3	346	0.053
3.	Linkage	3.028	3	346	0.030
4.	Intermediary	3.093	3	346	0.027
5.	Awareness	1.159	3	346	0.325
6.	Benefit	2.269	3	346	0.080
7.	Social wellbeing	0.950	3	346	0.416
8.	Watchers	1.064	3	346	0.364
9.	Reporting	0.790	3	346	0.500
10.	Consultant	1.725	3	346	0.162
11.	Finance	3.162	3	346	0.025
12.	Guiding	0.236	3	346	0.871
13.	Tourism service	5.258	3	346	0.001
14.	Promotion	1.663	3	346	0.175
15.	Diversification	2.862	3	346	0.037
16.	Enterprise	1.850	3	346	0.138

number of initiatives have been taken, both at the national and state level, to improve community intervention in PA-based ecotourism destinations. The Government of India had issued ecotourism guidelines in 1998, which emphasized the role of the community at all levels of operation in meeting the twin objectives of PA management, i.e., Conservation and livelihood, and instructed state governments to issue ecotourism policy guidelines in their respective state to improve community-based resource management practices in ecotourism destinations. The purpose of this study is to identify various community intervention strategies in tourism destinations and is there any difference in these strategies across (eco)tourism destinations of India. Accordingly, the study identified 16 intervention areas through an exploratory study by adopting a focus group discussion and expert interview.

The results of the exploratory study shows that there are sixteen interventions strategies (Democratic Representation, Capacity building, Linkage with other sectors, Consultant, Programme for Social wellbeing, Decision making and benefit sharing, and Intermediary, Development and Organization of tourism activities/products, Promotional activities, Enterprise development, and Product diversification, Conservation of natural resources, Monitoring of resources, Financing for Conservation, Environmental

TABLE 14.8 F-Test for Significance Among Destinations on CIS Variance

CIS	Periyar		Parambikulam		Thenmala		Waynad		F Value	P Value
	Mean	SD	Mean	SD	Mean	SD	Mean	SD		
Democratic	3.2197	1.1346	3.3209	1.0157	3.3023	1.1027	3.3902	1.0459	0.345	0.793
Capacity	3.4470	0.95960	3.4776	0.82002	3.3488	1.1523	3.5122	0.86954	0.271	0.846
Linkage	3.6894	0.77294	3.6940	0.73809	3.5581	.90770	3.4146	0.94804	10.104	0.188
Intermediary	3.4394	0.98273	3.5746	0.82605	3.6279	.90035	3.6829	0.81973	10.104	0.347
Awareness	3.2045	1.0964	3.2313	0.97267	3.1628	1.1112	3.3659	0.99388	0.319	0.812
Benefit	3.3485	1.0410	3.4701	0.86442	3.5581	1.0072	3.3659	0.99388	0.695	0.555
Social wellbeing	3.3561	0.97396	3.2985	0.95799	3.2791	1.0982	3.3902	0.86250	0.170	0.916
Watchers	3.2500	1.0512	3.2090	0.97390	3.2093	.96506	3.3171	0.90662	0.142	0.934
Reporting	3.2197	1.0655	3.2239	0.98593	3.3488	.97306	3.1707	1.0465	0.246	0.864
Consultant	2.9924	1.0665	3.1567	0.94872	3.1628	1.0675	3.2195	1.0877	0.879	0.452
Finance	3.6136	0.84391	3.4478	0.89751	3.3488	1.0665	3.3659	1.0667	1.471	.222
Guiding	3.0000	1.1722	2.8806	1.1309	3.0000	1.1270	3.0488	1.1694	0.369	0.775
Tourism service	2.9545	1.1713	3.0597	1.1088	3.3256	0.99333	3.5366	0.89715	3.593	0.014*
Promotion	3.2803	1.0138	3.1940	1.1009	3.4651	0.93475	3.1707	1.0465	.850	0.468
Production	3.3485	1.0113	3.1642	1.1118	3.4651	1.0027	3.3902	0.89101	1.326	0.266
Enterprise	3.4167	0.96537	3.4627	0.96290	3.6047	0.76031	3.3902	0.99695	0.498	0.684

*indicates items significant at 5% significance level.

education and awareness programmes and Guiding, Interpretation and Escorting) prevailing in (eco) tourism destinations, through three major modes of intervention, i.e., governance, ecodevelopment and commercial. Accordingly, the first objective of the study has been fulfilled.

The descriptive analysis of the study used to test the hypothesis to check whether any significant difference exists among various intervention strategies across destinations. ANOVA results show that there existed no significant difference in CIS across (eco)tourism destinations of India. However, one indicator pertaining to development and organization of tourism activities/products indicates significant difference across destinations, i.e., Thenmala, Periyar, Parambikulam, and Wayanad at 5% significance level.

The result reiterates the concept that sustainability is always destination specific as the resources are endemic to different regions (Orams, 1995; Diamantis, 1997). Accordingly, the development and organization of various tourism activities and products become destination specific. For instance, the development and organization of bird watching, nature trail, cultural elements like handicraft, food, etc., purely depend on local level resources. So destination level variations found in the study is in tune with the existing literature (Mowforth & Munt, 1998; Parkins, Stedman & Varghese, 2001; UNEP & UNWTO, 2005; Young, 2008). Other interventions of local communities are found to be similar across destinations as most of these strategies are developed and operationalized through Forest development agency (FDA), a range level arm of DFW of the state government. As the ecotourism operations are organized as a part of ecodevelopment and livelihood programmes in India, the results of this study found significance in all PA-based tourism programmes in the country. Accordingly, study further, strengthen the theory that sustainability is always destination specific.

This study further extent scope for assessing the relationship between intervention strategies and destination sustainability, destination quality and quality of life of host communities. Further, a study on nature of intervention of community in tourism has significance in bringing more inclusive and sustainable resource management practices which can, in turn, support policy making at state or regional level and implementing the same at the destination level.

14.4 CONCLUSION

The objective of the study was to understand various intervention strategies of destination communities. Through the exploratory method, the study could identify 16 widely used intervention strategies across detonations. The study further tried to examine the variations in the operational aspects of these intervention strategies through hypothesis testing. The statistical analysis shows that there was no significant difference in CIS across destinations except one strategy of organization and development of tourism activities and products as different the destinations are endemically different.

KEYWORDS

- **commercial intervention**
- **community intervention**
- **destination analysis**
- **destination development**
- **ecodevelopment intervention**
- **exploratory sequential method**
- **governance intervention**

REFERENCES

Agrawal, A., & Chhatre, A. (2006). Explaining success on the commons: community forest governance in the Indian Himalaya. *World Development, 34*(1), 149–166.

Bloom, D. B., & Crabtree, B. (2006). The qualitative research interview, *Medical Education, 40*(4), 314–321.

Buckley, R. C. (2009). Evaluating the net effects of ecotourism on the environment: A framework, first assessment, and future research. *Journal of Sustainable Tourism, 17*(6), 643–672.

Cambridge Advanced Learners Dictionary, (2014). 3rd edn. Cambridge: Cambridge University Press.

Churchill, G.A. (1979). A paradigm for developing better measures of marketing constructs. *Journal of Marketing Research, 16* (2), 64–73.

Creswell, J. W., Plano Clark, V. L., Gutmann, M. L., & Hanson, W. E. (2003). Advanced mixed methods research designs. In: Tashakkori, A., & Teddlie, C. (Eds.), *Handbook of Mixed Methods in Social and Behavioral Research* (209–240). Thousand Oaks, CA: Sage.

Danielsen, F., Burgess, N. D., Balmford, A., Donald, P. F., Funder, M., …Yonten, D. (2009). Local participation in natural resource monitoring: a characterization of approaches. *Conservation Biology, 23*(1), 31–42. doi: 10.1111/j.1523-1739.2008.01063.

Dassah, M. O. (2013). Longitudinal analysis of South Africa's performance on the Mo Ibrahim Index of African Governance. *Journal of Public Administration: Governance and Leadership: Supplement 1, 48*(4), 730–750.

David, M., & Sutton, C. D. (2004). *Social Research the Basics*. London: Sage.

DFID. (2007). Governance, Development and Democratic Politics: DFID's work in building more effective states. London: DFID. Retrieved May 12, 2011, from http:/www.dfid.gov.uk/pubs/files/governance.pdf .

Diamantis, D. (1997). The development of ecotourism and the necessity of using environmental auditing in its planning agenda. *General Technical Report*, Northeastern Forest Experiment Station, USDA Forest Service, 19–23.

Ethridge, D. E. (2004). *Research Methodology in Applied Economics.* Chichester: John Wiley & Sons.

Fritz, J. M. (2014). Essentials of Community Intervention. In: Fritz, J. M. & Rhéaume, J. (Eds.). Community Intervention: *Clinical Sociology Perspectives*, (15–29). New York: Springer.

Graf, A. (1995). Economic development or social development? A strategy for rebuilding inner cities. *Review of Black Political Economy, 24*(2/3), 251–255.

Gupta, M., & Gupta, D. (2013) *Research Methodology*. New Delhi, PHI.

Jick, T. D. (1983). Mixing Qualitative and Quantitative Methods: Triangulation in Action. In Van Maanen, J. (Ed.), *Qualitative Methodology* (135–148), Beverly Hills, CA: Sage.

Jones, S. (2005). Community-based ecotourism the significance of social capital. *Annals of Tourism Research, 32* (3), 303–324.

Kiss, A. (2004). Is community-based ecotourism a good use of biodiversity conservation funds? *Trends in Ecology and Evolution, 19* (5), 232–237.

Kitzinger, J. (1995). Qualitative research: Introducing focus groups. *British Medical Journal, 311* (4), 299–302.

Krueger, R., & Casey, M. (2009). *Focus groups* (4thed.). London: Sage.

Lincoln Y. S., & Guba, E. G. (1985). *Naturalistic Inquiry*, Beverly Hills: Sage.

Macmillan English Dictionary for Advanced Learners, (2007). (2nded.). London: Macmillan Education.

Maxwell, J., & Loomis, D. (2002). Mixed methods design: An alternative approach. In: A. Tashakkori & C. Teddlie (Eds.), *Handbook of Mixed Methods in Social and Behavioral Research*, (241–271). Thousand Oaks, CA: Sage.

Merzel, C., & D'Afflitti. J. (2003). Reconsidering community-based health promotion: Promise, performance, and potential. *American Journal of Public Health, 93*(4), 557–574.

Michael, E. J. (2007). *Micro-Clusters and Network, the Growth of Tourism*, London: Elsevier Science.

Mohamad, N. H., & Hamzah, A. (2013). Tourism cooperative for scaling up community-based tourism. *Worldwide Hospitality and Tourism Themes*, *5*(4), 315–328.

Morais, D. B., Zhu, C., Dong, E., & Guihua, Y. (2006). Promoting sustainability through increased community involvement: The Sangri La Ecotourism demonstration projects. *Tourism Review International, 10* (3), 131–140.

Mowforth, M., & Munt, I. (1998). *Tourism and Sustainability: A New Tourism in the Third World.* London: Routledge.

Murphy, P. E. (1985). *Tourism: A Community Approach*. New York & London: Methuen.

Murphy, P. E. (1997). *Quality Management in Urban Tourism*. Chichester: Wiley.

OECD. (2012). *Public Governance Reviews. Strategic Agility for Strong Societies and Economies*. OECD Publishing.

Orams, M. B. (1995). Using interpretation to manage nature-based tourism. *Journal of Sustainable Tourism*, *4*(2), 81–94.

Oxford Advanced Learners Dictionaries (9th edn.). (2015). Oxford: Oxford University Press.

Parkins, J. R., Stedman, R. C., & Varghese, J. (2001). Moving towards local-level indicators of sustainability in forest-based communities: a mixed-method approach. *Social Indicator Research. 56* (1), 43–72.

Raik B. (2002). *Capacity Building for Co-management of Wildlife in North America.* Retrieved May 12, 2016 from http://www.dnr.cornell.edu/hdru/PUBS/HDRUReport02-2.pdf .

REST. (1997). *Community-based Tourism: Principles and Meaning, Community-based Tourism Handbook, Thailand.* Retrieved May 12, 2011 from http://mekongtourism.org/wp-content/downloads/2011/02 /CBT-Handbook-by-CBT-I-Thailand.pdf .

Robert, J.W. (2001) Subsistence Food production and distribution in Alaska: Social Organization, Management, and Development issues. *INUSSUK- Arctic Research Journal, 1* (2) 213–231.

Rossiter, J. R. (2002). The C-OAR-SE procedure for scale development in marketing. *International Journal of Research in Marketing,* 19 (2), 305 – 335.

Rothman, J. (1995). Approaches to Community Intervention. In: Rothman, J., Erlich, J. L. & Tropman, J. E. (Eds.). *Strategies of Community Intervention,* Fifth Edition. (26–63) Itasca, IL: Peacock Publishers, Inc.

Scheyvens. R. (1999). Ecotourism and the Empowerment of Local Communities. *Tourism Management, 20* (3), 245–249.

Singh, S., Timothy, D. J., & Dowling, R. K. (Eds.). (2003). *Tourism in Destination Communities*. Cambridge, USA: CABI.

Thomas, R. M. (2003). *Blending Qualitative and Quantitative Research Methods in Theses and Dissertations*. Thousand Oaks, California: Corwin Press, Inc, Sage.

Trochim, W. M. K. (2006). *Introduction to Validity. Social Research Methods.* Retrieved May 12, 2011 from www.socialresearchmethods.net/kb/introval.php.

UNEP & UNWTO. (2005). *Making Tourism More Sustainable – A Guide for Policy Makers*. Retrieved May 12, 2011 from http://www.unep.fr/shared/publications/pdf/DTIx0592xPA-TourismPolicyEN.pdf.

United Nations. (1997). *Glossary of Environment Statistics, Studies in Methods*. Series F, No. 67. New York: United Nations.

UNWTO. (2007). *A Practical Guide to Destination Management*. Madrid Spain.

Young, M. (2008). *Development of Criteria and Indicators for Evaluating Forest-Based Ecotourism Destination: A Delphi Study*. Master's thesis. Morgantown: West Virginia University. Retrieved June 12, 2015, from https://www.researchgate.net/publication/228826578.

CHAPTER 15

DESTINATION MANAGEMENT STRATEGY: WITH SPECIFIC REFERENCE TO ECOTOURISM DESTINATION

S. MEERA[1] and R. ANNAPOORANI[2]

[1]*Assistant Professor, Indian Institute of Tourism and Travel Management, Nellore, Andhra Pradesh, India, E-mail: mirasmadhav@yahoo.com*

[2]*Professor, Department of Economics, Avinashilingam University, Coimbatore, Tamilnadu, India*

ABSTRACT

Ecotourism as a product and approach has gained popularity over the years, due to its contribution towards sustainability. Prudent maintenance and effective management strategies at ecotourism destinations can proffer substantially towards economic contribution and foster towards a semblance of sustainability.

This chapter has attempted to identify problems faced by tourists in ecotourism destination. The Periyar Tiger Reserve in Kerala, India, and examine if any difference persists in problems faced by tourists in peak and lean season. The sequential exploratory design has been adopted. Qualitative approach through Expert Interviews and quantitative analysis using Garrett's Ranking, Kruskal-Wallis H test has been deployed to examine the statistical significance of the difference in problems faced in the peak and lean seasons.

The study has highlighted the need to evolve appropriate destination management strategies to address the requirements of tourists visiting ecotourism destinations in the case of protected areas, exclusively, in peak and lean seasons, to provide better holiday experience along with paradigm shift towards sustainability, thus, contributing both from short and long run perspective.

15.1 INTRODUCTION

Ecotourism is relatively a new concept in tourism, kindled by the idea of creating harmonious co-existence with nature. It pivots around activities focusing on combining conservation, community, and sustainable travel. Any tourism program which is nature-based, ecologically sustainable, where education and interpretation is a major concept and where local people are benefitted can be called ecotourism. The International Ecotourism Society (TIES, 1990), in the US, defines ecotourism as responsible travel to natural areas that conserves the environment and improves the well-being of the local people. Ecotourism principles are intended to minimize the impact, build environmental and cultural awareness and respect by providing positive experiences for both visitors and hosts. It provides direct financial benefits for conservation, helps in the empowerment of the local people, and raise sensitivity to political, environmental and the social system of the host country. As a whole ecotourism is intended to conserve biological and cultural diversity by strengthening protected area management systems at public or private places to increase the value of stable ecosystems, promote the sustainable use of biodiversity by generating income, jobs and business opportunities in ecotourism and related business network and share the benefits of development equitably with local communities and indigenous people by seeking their consent and full participation in planning and management of the ecotourism business (Rajasenan et al., 2012).

Ecotourism destinations witness temporal trends and repeated cycles with upward and downward trends in visitor footfalls indicating seasonality in ecotourism (Nzengy'a, 2004). Seasonality poses challenges through negative impact and provides opportunities through positive impact on Tourism (Petrevska, 2012).

Apart from analyzing socio, economic, and environmental changes created by seasonality on tourism (supply perspective), an examination into the specific problems faced by tourists (demand perspective) during the peak and lean season is required to offset or minimize them; to ensure a continuous flow of tourists, irrespective of variation in season. Supply perspective of seasonality is well documented in tourism literature, while demand perspective, especially in the context of tourism, is not explored.

15.2 REVIEW OF LITERATURE

The term ecotourism has been traced back to the work of Hetzer (1965), who used it to explain the intricate relationship between tourists and the environments and cultures in which they interact. Ceballos-Lascuráin was the first to coin the phrase in the early 1980s.

Laarman and Durst (1987) in their early reference to ecotourism, defined it as a nature tourism in which the traveler is drawn to a destination because of his or her interest in one or more features of that destination's natural history. The visit combines education, recreation, and often an adventure. According to Boo (1990), Eagles (1992), and Wight (1993) ecotourism is called nature tourism, alternative tourism, cultural tourism, soft tourism, adventure tourism, responsible tourism or green tourism.

According to the International Ecotourism Society (1990) "Ecotourism is responsible travel to natural areas which conserves the environment and sustains the wellbeing of the local people." Ryel and Grasse (1991) defined ecotourism as purposeful travel that creates an understanding of cultural and natural history, while safeguarding the integrity of the ecosystem and producing economic benefits that encourage conservation. According to Wheller (1991), the actors in an ecotourism system are visitors, natural areas and their managers including both public and private areas, communities, businesses, including various combinations of local businesses, inbound operators, outbound operators, hotel and other accommodation providers, restaurants and other food providers, Government and Non Governmental organizations.

The Australian National Ecotourism Association (1992) defined ecotourism as a nature-based tourism that involves education and interpretation of the natural environment and is managed to be ecologically sustainable

In the view of Eagles (1992) and Wight (1993), Ayala (1995), ecotourism is nature-based, advancing conservation and sustainable development. A subsequent definition by Laarmanand Durst (1993) identifies a conceptual difference between ecotourism and nature tourism. In recognizing the difficulties in defining nature tourism, they establish both a narrow and broad scope to its definition. In a narrow sense, it refers to operators running nature-oriented tours; however, broadly it applies to tourism's use of natural resources including beaches and country landscapes.

According to Wallace and Pierce (1996), ecotourism is travel to relatively undisturbed natural areas for study, enjoyment, or volunteer assistance. It is travel that concerns itself with the flora, fauna, geology, and ecosystems of an area, as well as the people (caretakers) who live nearby, their needs, their culture and their relationship to the land. He suggested that tourism may be said to be true ecotourism if it addresses six principles:

- It entails a type of use that minimizes negative impacts on the environment and to local people.
- It increases the awareness and understanding of an area's natural and cultural systems and the subsequent involvement of visitors in issues affecting those systems.
- It contributes to the conservation and management of legally protected and other natural areas.
- It maximizes the early and long-term participation of local people in the decision-making process that determines the kind and amount of tourism that should occur.
- It directs economic and other benefits to local people that complement rather than overwhelm or replace traditional practices (farming, fishing, social systems, etc., and
- It provides special opportunities for local people and nature tourism employees to utilize and visit natural areas and learn about wonders that other visitors come to see.

The most conceptual definition of ecotourism can be reduced to the following: ecotourism is both nature-based and sustainable. The definition clarifies the descriptive and the prescriptive components of ecotourism concept. The nature component is descriptive or positive because it simply describes the activity location and associated consumer motivations. The sustainable component is prescriptive or normative because it neglects what people want the activity to be. Sustainability incorporates environmental, experiential, sociocultural and economic dimensions.

Ecotourism has assumed a lot of significance in the contemporary context, and it is one of the fastest growing segments of the tourism industry. Fillion et al. (1992) identified, through an analysis of inbound tourist motivations to different worldwide destinations, that 40–60% of all international tourists are nature tourists and that 20–40% are wildlife-related tourists. According to a study conducted by UNWTO in 2010 ecotourism is the fastest growing sector of the tourism industry, expanding globally between 20% to 34% per year. It predicts that ecotourism will grow rapidly over the next two decades and global spending on ecotourism is expected to increase at a higher rate than the tourism industry as a whole. According to the International Ecotourism Society (2012), ecotourism would grow to 25% of the global travel market within 6 years and account for US$ 470 billion revenue per year.

Seasonality has been well documented in tourism literature as seasonality as a phenomenon had been empirically found to affect destination performance irrespective of forms of tourism, across nations, from a holistic perspective. Seasonality is defined as the cyclical variation in a number of tourist visits which is repeated every year (Ahas et al., 2007). The impact of seasonality on destination performance assumed a lot of significance. The implication of seasonality can have both positive and negative impacts on tourist destinations. According to Boniface and Cooper (2005), seasonality affects the profitability and overall well-being of a destination. According to Lee et al. (2008), seasonality threatens the viability of investing in tourism. Seasonality has an implication on the utilization of facilities, especially in the context of fixed costs (Chung, 2009); which leads to lesser investment in tourism. Thus seasonality largely has a negative implication on the destination performance as it tends to reduce revenues available for finance operations hampering the possibilities of earning attractive profits. Increase in visitor footfalls due to seasonality can also have a negative impact on destination performance. According to Roberto and Giuseppe (2012), congestion during the peak season may have negative impacts on the local communities residing in and around the destination. Crowding due to seasonality has a negative bearing on the quality of service as service delivery may be poor, for suppliers may not be able to deliver according to customers' expectations which can affect the destination image and reduce the repeat visitation (Boardman, 2008). Some of the benefits assured by seasonality is also found in tourism literature. Offseason, a period when visitor footfalls are less provides management of small enterprises with recuperation time (Commons & Page, 2001). According to Page and Connell (2006), aspects influencing destination performance can be addressed during the off-season period like staff training and development to enhance productivity.

15.3 CONTEXTUAL BACKGROUND

15.3.1 PERIYAR TIGER RESERVE

Periyar Tiger Reserve, Thekkady is one of the primary ecotourism destinations in Kerala. It is known as Periyar National Park since 1981, and its ecotourism development project is well known since 1996. Periyar Tiger Reserve is often cited as one of the best-managed tourism destinations in South India and is known for its local community-forest department mutual collaboration (Narain et al., 2005). According to Saji and Narayan (2012) from the ecological perspective, Periyar Tiger Reserve is viewed by many as a critical test of the State's ability to preserve a fragile ecosystem in which nature and human live together so intensely. Periyar Tiger Reserve has received the award for Excellence in Environmental Concern in 2000, and Periyar Tiger Reserve had also been accepted as the Conservation Model by Tiger Task Foundation in 2005.

As an ecotourism destination with immense scope for growth, Thekaddy is the most sought after destination due to its locational advantage, situated in the border areas of Tamil Nadu and Kerala, offering easy assessability to both the States and is en route to other destinations like Munnar, Gavi. Thekkady is a unique destination on the tourism map due to its climate, landscape, and possibility of watching wildlife closely on the lakeshore while taking a boat cruise. The prime attraction of tourists visiting Thekkady is a boat cruise in the Periyar lake. This lake is artificially formed, due to the submergence of low-lying forest areas, following the construction of the Mullapperiyar Dam in 1895. It offers a variety of opportunities to cater to the needs of various classes of tourists. Tourism has become an important management issue ever since the area was declared a Tiger Reserve in 1978 (tenth of its kind in India) and there has been a tremendous increase in the number of tourists visiting the park year after year. The day visitors from 2005–2010 in Periyar Tiger Reserve had increased from 425.22 thousand in 2005 to 480.4 thousand in 2010 (Official Records, District Tourism Promotion Council, Department of Tourism, Government of Kerala, Trivandrum). The revenue earnings have increased from Rs. 4,247,723 lakhs to Rs. 11,565,257 crores in 2010–11. (www.periyarfoundation.com).

The Project at Periyar has two basic objectives: to manage protected areas and to get local people involved in it. To implement the project initially, local communities living off the forest were organized into eco-development committees (EDCs). The basic objectives for the EDCs were to reduce the negative impact of local people on the sanctuary and to involve encroachers in conservation, instead of exploitation. This was done by addressing the economic needs of those living in and around the park by finding viable and innovative livelihood alternatives for them. The project has so far benefited about 40,000 people of 5,540 families. Some of the ecotourism activities at Periyar include a nature walk, clouds walk, green walk, border trekking, bamboo rafting, bamboo groove, jungle inn, bullock cart discoveries, jungle scout, Periyar Tiger Trial, etc.

15.4 NEED FOR THE STUDY

Tourism is a vibrant industry known for its considerable contribution to the economy and evaluation of challenges affecting the tourism business, and destination performance require to be examined. Seasonality in tourism is well documented in terms of causes, types, and the impact of seasonality on tourism. As ecotourism is nature-based, the impact of seasonality needs to be analyzed to improve destination performance and tourist footfalls. Evaluation of seasonality both from the tourism businesses and tourists perspective is required to minimize the demand and supply mismatch for effective destination management during peak and lean seasons.

Hence an analysis of the problems faced by tourists visiting ecotourism destination; protected area namely Periyar Tiger Reserve, is done both in peak and lean season, as this is not explored in earlier studies.

15.5 RESEARCH GAP

A critical analysis of existing literature on ecotourism revealed that there were only limited studies on seasonality in ecotourism destinations especially in India and problems faced by tourists in peak and lean season. Further, in India, the existing studies on Periyar Tiger Reserve have tried to analyze ecotourism activities, visitor satisfaction and community participation in Periyar Tiger Reserve. But there had been lack of research attempt focusing on seasonality aspect, destination performance due to seasonality, destination management strategies, problems faced by tourists (demand perspective) in peak and lean seasons of Periyar Tiger Reserve.

In this context, the current study assumes significance.

15.6 OBJECTIVES

- To identify the peak and lean season at Periyar Tiger Reserve.
- To identify the problems of ecotourists in peak and lean season in Periyar Tiger Reserve.
- Destination management strategies to manage seasonality in the context of protected area parks.

15.7 SCOPE OF THE STUDY

The economic contribution of tourism may be enhanced through better destination planning and development strategies which could be evolved on the basis of effective feedback mechanism through evaluation of problems of tourists who had visited the destination during peak and lean seasons and attempts may be made to evaluate the possibilities of revisitation in a particular season. Thus this study has contributed to the field of knowledge pertaining to seasonality in ecotourism. One of the significant contributions made in this study was the identification of the problems faced by tourists at Periyar Tiger Reserve. On the basis of the arrival of tourists at the Periyar Tiger Reserve, the problems have been analyzed in peak (footfalls are high) and lean season (footfalls are less), to evaluate whether there are significant differences in problems faced in peak and lean seasons. This may be of immense help to destination managers and policy makers working in the domain of destination planning and development, considering the aspects of seasonality at destinations.

The research further would provide necessary inputs to managers for examining the method of measuring an abstract concept—ecotourism seasonality paradigm. This knowledge will help both the destination planners and marketers in a substantial way as it will empower them to evaluate the requirements of the ecotourists effectively through ecotourism seasonality paradigm which could be used for making strategic, operational and innovative decisions for ensuring a quality experience to ecotourists throughout the year.

15.8 RESEARCH METHODOLOGY AND DATA COLLECTION

15.8.1 SELECTION OF THE SAMPLE

The study adopted a proportionate random sampling method for selecting the sample respondents from the chosen destination. The study covered both foreign and domestic tourists in peak and lean seasons. The peak seasons in the context of Periyar Tiger Reserve ranges from September to January and the lean seasons are June, July, and August.

Table 15.1 represents the number of domestic and foreign tourists visiting the Periyar Tiger Reserve during peak and lean seasons (during the study period).

By adopting proportionate random sampling, (1% of the total number of tourists), the number of sample respondents in peak and lean season were finalized.

Table 15.2 represents a number of sample respondents selected in peak and lean season.

The study covered 477 domestic tourists and 46 foreign tourists. Hence the total size of the sample was 523.

TABLE 15.1 Number of Domestic and Foreign Tourists Visiting Periyar Tiger Reserve on An Average in Peak and Lean Seasons (During the Study Period)

Type of tourists	Peak season	Lean season
Domestic tourists	41067	6712
Foreign tourists	3916	74

Source: Author compilation.

TABLE 15.2 Number of Sample Respondents Selected in Peak and Lean Season

Type of tourists	Peak season	Lean season	Total
Domestic tourists	410	67	477
Foreign tourists	39	7	46

Source: Calculated values from author compilation.

15.8.2 COLLECTION OF DATA

The data has been collected from secondary and primary sources.

The secondary sources include:

- Reports and Statistics on Tourism- Ministry of Tourism, Government of India, Government of Kerala.
- Reports on International Tourism Trends, United Nations World Tourism Organization, World Tourism, and Travel Council.
- Websites of Government and private Organizations – Ministry of Tourism, Government of Kerala, United Nations World Tourism Organization, World Tourism, and Travel Council. Eco-Club, Periyar Foundation, Periyar Tiger Reserve, International Ecotourism Society, etc.
- Working papers on Ecotourism, Periyar Tiger Reserve.

Whereas, the primary data were collected in two phases:

A. Phase I – Exploratory Research Through Interviews and Focus Group Discussion

Exploratory research forms the foundation of a good study (Churchill and Iacobuci, 2004). Exploratory research has been adopted in this study, and it has helped to determine the research design as such and methods of data collection and in the selection of relevant subjects for the study.

The study conducted interviews and focus group discussions on identifying and understanding the relevant dimensions and variables to be considered or modified appropriately to suit the destination considered for the study. In the current study, interviews with ten guides operating in Periyar Tiger Reserve, as they interact directly with the tourists, was conducted. Group discussions were held with 12 watchers, 5 guards, and 5 wildlife officials for identification and evaluation of the seasonality component in visitor footfalls. Five experts from the academia were also interviewed to provide their contribution to the relevance of the research problem identified.

B. Phase II – Descriptive Research Through Survey Method

Survey method was adopted in collecting responses from the respondents both in the peak and lean season of Periyar Tiger Reserve. This approach was adopted as it gives an opportunity to generalize conclusions from a given context on the basis of penetrating and relevant questions for capturing the richness of the concept – differences in identified variables in peak and lean season.

15.8.3 PERIOD OF STUDY

The pilot study was conducted during October–November, 2014 for 63 respondents. It was undertaken to check the reliability of items. Testing the reliability of the various constructs is the pre-requisite for data analysis and inference. Reliability analysis tests whether a scale consistently reflects the subset it measures (Churchill, 1979; Nunnally and Bernstein, 1994). Cronbach's alpha coefficient is used to measure reliability. The closer Cronbach's alpha coefficient is to 1.0 the greater the internal consistency of the items in the scale. George and Mallery (2003 p. 231) provide the following rules of thumb: "Cronbach's alpha coefficient ≥0.9, Excellent if Cronbach's alpha coefficient ≥0.8, Good if Cronbach's alpha coefficient ≥0.7, Acceptable if value is ≥0.6, Questionable if the value is ≥0.5, Poor if value is ≥0.5, lesser than 0.5 is Unacceptable." So a high value for Cronbach's alpha indicates good internal consistency of the items in the scale. In the present study the indicators were having Cronbach's alpha coefficient (calculated using SPSS 16) between 0.7 to 0.9; thus indicating good reliability. The final data collection was done during the period December 2015 to April 2016 (for peak season) and from June 2017 to August 2017 (for lean season).

15.9 HYPOTHESIS

There is no significant difference in constraints faced by tourists visiting Periyar Tiger Reserve in peak and lean seasons.

15.10 RESULTS AND DISCUSSIONS

15.10.1 SOCIO-ECONOMIC PROFILE OF THE RESPONDENTS

With the intention of developing a proper perspective analysis, the current study tried to analyze the socio-economic profile of the respondents. This helps in understanding the influence of these factors on their interests and activities. Table 15.3 represents the socio-economic profile of the selected respondents.

TABLE 15.3 Socio-Economic Profile of the Selected Respondents

Items	Classification	Domestic Tourists	Foreign tourists	Total
Age (in years)	Below 30	63 (13.20)	3 (6.50)	66 (12.62)
	30–40	187 (39.20)	25 (54.34)	212 (40.53)
	40–50	123 (25.78)	12 (26.08)	135 (25.81)
	50–60	101 (21.17)	4 (8.69)	105 (20.07)
	Above 60	3 (0.63)	2 (4.34)	5 (0.96)
Sex	Males	379 (79.45)	30 (65.21)	409 (78.20)
	Females	98 (20.55)	16 (34.78)	114 (21.79)
Education	Secondary	Nil (0.00)	Nil (0.00)	Nil (0.00)
	Higher secondary	3 (.63.00)	Nil (0.00)	3 (0.57)
	Graduation	323 (67.71)	34 (73.91)	357 (68.26)
	Post-graduation and above	151 (31.65)	12 (26.08)	163 (31.16)
Employment	Unemployed	54 (11.32)	2 (4.34)	56 (10.70)
	Employed	423 (88.67)	44 (95.65)	467 (89.29)
Annual Income (in rupees)	Below 5 lakhs	73 (15.30)	Nil	73 (13.95)
	5–10 lakhs	367 (76.93)	1(2.17)	368 (70.37)
	Above 10 lakhs	37 (7.75)	45(97.82)	82(15.68)
Marital status	Married	355 (74.42)	42 (91.30)	397 (75.90)
	Unmarried	122 (25.57)	4 (8.69)	126 (24.09)
Total		477	46	523

Source: Field survey.

Figures in parentheses indicate percentage to total.

Majority of the domestic tourist's respondents, (39.20%), foreign tourist respondents (54.34%) and total respondents (40.53%) belonged to the age group of 30 to 40 years. This finding is in line with that of Ballantine and Eagles (1994) and Diamantis (1999) that, soft ecotourists are below 45 years of age.

Majority of the respondents (89.29%) are employed with substantial high income between 5–10 lakhs in the case of domestic tourist respondents and 97.82% were having an annual income greater than 10 lakhs in the case of foreign tourist respondents.

15.10.2 ANALYSIS OF SEASONALITY IN ECOTOURISM

Seasonality has assumed a lot of importance in tourism and tourism activities are associated with 'seasons' as the inflow of tourists, visitation patterns and tourism resources change depending on seasons. Butler (2001) has described seasonality in tourism as a temporal imbalance in the phenomenon of tourism, which may be expressed in terms of dimensions of such elements as a number of visitors, expenditure, traffic on highways, and other forms of transportation, employment, and admissions to attractions.

There are two causes of tourism seasonality – natural and institutionalized (Baron, 1975). Natural seasonality is the result of nature and its forces, related to the elements of weather and climate. According to Koenig-Lewis and Bischoff (2005), these factors are predictable and recur with only small changes. On the other hand, institutionalized seasonality remains in the domain of people and is associated with legislation, religion, custom, historic conventions, etc.

The aspect of the seasonality is bound to have an influence in the case of ecotourism as ecotourism is associated with nature. Periyar Tiger Reserve is also affected by 'seasonality.'

Table 15.4 represents the number of tourists visiting the Periyar Tiger Reserve in peak and lean seasons.

Table 15.4 reveals that 85.95% of domestic sample respondents and 84.78% of sample foreign respondents visited Periyar Tiger Reserve in peak season. During the peak seasons, from September to April, there had been long and short vacations, and this encouraged the domestic tourists to visit this destination. As for the foreign tourists, generally, the summer season of Kerala and festival seasons like Onam, Christmas and New Year encourage them to visit Kerala, including ecotourism destination of Periyar. Most of them visit during Summer season, not severe summer, as forest fires destroy

TABLE 15.4 Number of Sample Respondents Visiting Periyar Tiger Reserve in Peak and Lean Seasons

Type of season	Domestic respondents	Foreign respondents	Total respondents
Peak	410 (85.95)	39 (84.78)	449(85.85)
Lean	67 (14.04)	7 (15.21)	74(14.15)

Source: Field Survey.
Figures in parenthesis indicate percentage to the total.

the flora and animals (fauna) move into deeper areas of the jungle to quench thirst, whereas, in mild summer, wild animals gather near rivers and ponds to drink water and so it is easy to have a glimpse of them.

Lean season in the context of Periyar Tiger Reserve is during the monsoon season – June, July and August. Probably, as most of the schools and colleges in India reopen around this time, families tend to visit less in this time and also due to heavy rains, nature walk paths become slippery, and these inconveniences cause the reduction in a number of tourists.

On the basis of the expert interviews, the peak and lean seasons were identified. The visitor footfalls on the basis of secondary data were also considered for this purpose. On the basis of group discussions with the tourist guides on the visitor feedback, the general constraints which are destination specific were identified.

15.10.3 CONSTRAINTS FACED BY TOURISTS VISITING PERIYAR TIGER RESERVE

On the basis of primary data collected from tourists, it was evident that tourists had encountered certain problems at the destination both in peak and lean seasons, which was authenticating the opinion of tourist guides on the problems faced by tourists visiting Periyar Tiger Reserve. The constraints include overcrowding, lack of coordination among staff at the entrance/ticket counter and boat jetty, serpentine queue leading to long waiting time, restrictions on dress when the tourists opt for nature walk, lack of cleanliness and hygiene, technological constraints pertaining to communication inside the protected area and restrictions on non renewable source of energy, lack of amenities like toilets and canteens inside the protected area, etc.

The current study tried to calculate the mean score of the constraints faced and Garrett ranking for the constraint faced in peak and lean season. Table 15.5 represents the mean scores of the constraints faced and Garret

TABLE 15.5 The Mean Scores of the Constraints Faced and Garrett Ranking for the Constraint Faced in Peak and Lean Season

Constraints	Peak Season		Lean season	
	Mean scores	Rank	Mean scores	Rank
Overcrowding	92.01	I	79.55	VIII
Lack of coordination among staff at entrance and boat jetty	89.77	III	83.58	V
Long waiting time	90.43	II	81.12	VII
Dress restrictions for a nature walk	85.54	V	82.86	VI
Entry and exit restrictions	76.83	X	80.25	IX
Age restrictions for a nature walk	77.33	IX	85.09	IV
Lack of amenities like toilets, canteens	87.21	IV	86.18	II
Cleanliness and hygiene	80.30	VII	90.55	I
Nonavailability of accommodation facilities inside the protected area	83.21	VI	77.13	X
Lack of orientation about the insect bites.	79.26	VIII	87.24	III

Source: Calculated values based on a field survey.

ranking for the constraint faced in peak and lean season at Periyar Tiger Reserve.

In the peak season, the problem of overcrowding had been assigned the first rank while in the lean season lack of cleanliness had been assigned the first rank.

The study tried to apply the Kruskal Wallis test to find out whether there is a significant difference in the rank assigned for the problems faced by the tourists in the peak and lean season.

The calculated Kruskal Wallis test indicates that there is a significant difference in ranking of the problems faced by tourists in the peak and lean season since the calculated Kruskal Wallis value (19.20) is higher than $X^2_{0.05} = 3.84$.

15.10.4 DISCUSSIONS

Seasonality management strategy, development, and implementation must be in an integral part of destination management strategies, as seasonality management strategy varies across destinations because of the nature of the seasonality phenomenon which is quite complex.

The recommended strategies for managing seasonality in the context of Periyar Tiger Reserve can include infrastructure development at the Reserve, introduction of new products, differential pricing, innovative marketing involving local participation, carrying capacity assessment and establishing linkages leading to sustainability, developing inventories and facilities which can be used to promote events in offseasons, government initiatives, effective training of employees, developing eco-friendly technologies to reduce the negative impact. Thus a coordinated effort is warranted to combat seasonality.

Destination management strategies must be formulated on the basis of a continuous study of the progress of a destination. Destination management must be construed as coordinated management and include the different aspects including destination, attractions, shows, amenities, access and infrastructure, activities at the destination, food and tour services. Therefore, an ecosystem needs to be developed encompassing all these elements with a focus on visitor experience, footfalls, repeat visitation, continuous product improvement, destination quality and competitiveness, local knowledge and local economic development. Higher growth in terms of seasonal trends poses threats to sustainability (Martin, 2014). Therefore, impact on sustainability parameters as a result of seasonality needs to be examined.

On the basis of problems faced by tourists in peak and lean season, innovative and sustainable approaches must be adapted to convert the challenges faced by seasonality into opportunities. Appropriate strategies must be followed on the basis of SWOT analysis, collaborations to raise investments through a partnership with players of different segments of the tourism industry like accommodation travel, etc. Destination-specific policies may have to be formulated to convert a day's visit to many days visit by tourists, irrespective of seasons. Seasonality may also bestow some specific benefits like in the context of Periyar, monsoon seasons, which are construed as the lean season can be promoted for biodiversity resources in terms of flora at the destination. Destination competitiveness can be assessed periodically. Destination management strategies considering the aspect of seasonality in the context of protected areas must consider the impact of seasonality. The 'threshold of sustainability' (Drumm, 2008) must be assessed. Destination management strategies in the context of protected areas must consider the primary goals of protected areas: promotion of sustainability and Therefore, offsetting seasonality can lead to increase in cost, which needs to be reviewed from an economic perspective. Thus destination management must consider the specific problems faced by tourists in different seasons to enhance visitation throughout the year.

15.11 LIMITATIONS OF THE STUDY

- Seasonality phenomenon is studied only from a demand perspective.
- Limitations of primary data collection can apply in this study as well.
- The constraints considered for the study are limited.

15.12 SCOPE FOR FURTHER RESEARCH

New destinations can be explored, and comparison of constraints across regions or destinations can be evaluated. Problems faced by stakeholders including tourism businesses can be made, and the impact of seasonality can be analyzed across destinations.

15.13 CONCLUSION

Ecotourism in protected areas is prone to destination specific seasonal problems. Identification of destination-specific problems and its intensity in different seasons will help the destination managers to chalk out strategies to maximize income generation from the destination, irrespective of seasonality component. Destination management strategies must be chalked out based on strategic tourism planning efforts considering sustainability from political, social, economic dimensions, thus promoting quality tourism without neglecting the problems faced by tourists visiting the destination.

KEYWORDS

- **destination management**
- **economic contribution**
- **ecotourism**
- **ecotourist**
- **lean season**
- **peak season**
- **Periyar Tiger Reserve**
- **protected area**
- **seasonality**
- **sustainability**

REFERENCES

Activities at Periyar Tiger Reserve. http://periyarfoundation.com/ (3rd August 2015).

Activities at Periyar Tiger Reserve. http://periyartigerreserve.org/ (3rd August 2016).

Ayala, H. (1995). From Quality Product to Ecoproduct: Will Fiji Set a Precedent? *Tourism Management, 16*(1), 39–47.

Ballantine, J. L., & Eagles, P. F. J. (1994). Defining Canadian ecotourists. *Journal of Sustainable Tourism, 2*(1), 1–6.

Baron, R.; Seasonality in Tourism. (1975). *A Guide to the Analysis of Seasonality and Trends for Policy Making.* The Economist Intelligence Unit Ltd., Technical Series N. 2.: London.

Boardman, K. (2008). Addressing Seasonal Visitor Flows in Australian Resort Destinations. *EDA Economic Development* Australia, *49*(3), 26–28.

Boniface, B. & Cooper, C. (2005). *Worldwide Destinations: Geography of Travel and Tourism* (4th ed.). Oxford, Butterworth-Heinemann.

Boo, E. (1990). Ecotourism: The Potentials and Pitfalls. *World Wildlife Fund*: Washington, D.C.

Butler, R. (2001). Seasonality in tourism: issues and implications. In: *Seasonality in Tourism.* Wallingford, UK, pp. 5–22.

Ceballos-Lascurain, H., (1987). The future of ecotourism, *Mexico Journal*.

Chung, J. Y. (**2009**). Seasonality in Tourism: A Review. *e-Review of Research (eRTR), 7*(5), 35–44.

Churchill, G. A., & Iacobucci, D. (2004). *Marketing research: Methodological Foundations.* 9th ed., Thomson South-Western, Ohio, p. 133.

Commons, J. & Page, S. (2008). Managing Seasonality in Peripheral Tourism Regions: The Case of Northland. In: Lee, C., Bergin-Seers, S., Galloway, G., O'Mahony, B., & McMurray, A. (Eds.), *Seasonality in the Tourism Industry: Impacts and Strategies, Queensland,* CRC for Sustainable Tourism.

Diamantis, D. (1999). The characteristics of UK's eco-tourists. *Tourism Recreation Research, 24*(2), 99–102.

Drumm Andy (2008). The threshold of sustainability for protected areas. *Bioscience 58,* 782–783.

Eagles, P. F. J. (1992). The travel motivations of Canadian ecotourists, *Journal of Travel Research, 31*(2), 3–7.

Ecotourism Association of Australia (1992), Newsletter 1, 2, *The Ecotourism Society of Australia* (1991 a, b).

Filion, F. L., DuWors, E., Boxall, P., Reid, R., Hobby, E., Bouchard, P., Gray, P., & Jaquemot, A. (1992). *The Importance of Wildlife to Canadians in (1987) 'Trends in participation in wildlife-related activities' (1981 to 2006),* Environ. Can., Ottawa, p. 36.

Hetzer, N. (1965). Environment, tourism, culture. Links (July): reprinted in *Ecosphere, 1*(2), 1–3.

Jan G. Laarman, & Hans M. Gregersen, (1996). Pricing policy in nature-based tourism, *Tourism Management, 17*(1), 247–254.

Koenig-Lewis, N., & Bischoff, E. E., (2005). Seasonality research: The state of the art, *International Journal of Tourism Research, 7*, 201–219.

Lee, C., Bergin-Seers, S., Galloway, G., O'Mahony, B. & McMurray, A. (2008). *Seasonality in the Tourism Industry: Impacts and Strategies.* Queensland, CRC for Sustainable Tourism, 12–24.

Narain, S., Panwar, H. S., Gadgil, M., & Thapar V. (2005). "*Joining the Dots: The Report of the Tiger Task Force"* New Delhi. The Ministry of Environment and Forests, Government of India.

Nzengy'a, D. M., (2004). Temporal trends in ecotourism in the eastern highlands of Zimbabwe. *Journal of Ecotourism, 3*, 129–146.

Page, J. S. & Connell, J. (2006). *Tourism: A Modern Synthesis (2nd ed.).* Thomson Learning: London.

Petrevska, Biljana. (2012). Forecasting international tourism demand: The evidence of Macedonia. *UTMS Journal of Economics*, *3*(1), 45–55.

Rajasenan. D., & Bijith George Abraham, (2012). "Livelihood Security and Socioeconomics of Community-based Ecotourism (CBET) in Kerala," *South Asian Journal of Socio-Political Studies*, January–June, *XII* (2), 23–27.

Roberto, C., & Giuseppe, R. (2012). Private and Public Incentive to Reduce Seasonality: A Theoretical Model. *University of Catania the Open* Access, *Open Assessment E-Journal*, Discussion Paper No. 2012–16.

Wallace, G. N., & Pierce, S. M. (**1996**). An evaluation of ecotourism in Amazonas, Brazil. *Annals of Tourism Res*earch, *23*(1), 843–873.

Wight, P. (1993). Sustainable Ecotourism: Balancing Economic, Environmental and Social Goals Within an Ethical Framework. *Journal of Tourism Studies*, *4*(2), 54–66.

CHAPTER 16

AN INVESTIGATION OF TOURISM INVESTMENT ON ACCOMMODATION SECTOR OF TOURISM INDUSTRY OF SRI LANKA

HARINI K. AMARASINGHE[1] and D. A. C. SURANGA SILVA[2]

[1]*University of Sri Jayewardenepura, Sri Soratha Mawatha, Nugegoda, 10250, Sri Lanka, E-mail: harinik@sjp.ac.lk*

[2]*University of Colombo, Kumaratunga Munidasa Mawatha, Colombo, 00700, Sri Lanka, E-mail: drsuranga3@gmail.com*

ABSTRACT

The study is aimed at investigating the growth potentials and the challenges that are confronted with Sri Lanka Tourism of the present day. The study is focused on to investigate the literature on the tourism accommodation sector investment as it has been identified as the main challenge of today's industry within the country. The problem statement was identified as what are the major determinant variables of the tourist accommodation sector of the tourism industry in Sri Lanka? The article intends to examine the major implications of the existing investment shortage in the tourism accommodation sector of Sri Lanka tourism industry. The main objective of the study was to identify the key determinants of the level of investment in Sri Lanka tourism accommodation sector with specific objectives being to understand the current situation of the investment in tourist accommodation sector of Sri Lanka and to stipulate policy reforms in the tourism investment promotions within the country for the next 5 years.

The methodology adopted by the study is basically a desk review of the present literature on the hotel sector investments around the world. The objective is to identify the key determinants of hotel sector investments around the world, specifically within the Asian region.

The study concluded with the identification of five major variables that affect the level of investment into the accommodation sector for tourism namely; the tourist attraction for a particular destination, the national output of the destination, the rate of inflation of the destination, exchange rates of the destination and the availability and the quality of human capital of the destination. This finding is useful for quantitative model building, testing casualty or even to conduct further exploration.

16.1 TOURISM GROWTH AND CHALLENGES IN SRI LANKA

16.1.1 INTRODUCTION

This chapter introduces the research study to the reader. The first section will describe the nature of the country as a tourist destination its attraction and outlook. The second section discusses the research issue and the research gap which the researcher intends to investigate. This section will further elaborate on the problem statement, research questions, and research objectives. It will be followed by the significance of the study and the limitations of the study. It is the researcher's attempt to make it clear and understandable to anyone who wishes to use the study for knowledge or as inputs for their work.

Sri Lanka is an island located in the Indian Ocean to the southwest of Bay of Bengal. It lies between latitudes 5° and 10°N, and longitudes 79° and 82°E separated from India by the Gulf of Mannar and Palk Strait. In extent it is 65,610 km^2, 62,705 km^2 of land and 2,905 km^2 of inland waters in area with the island classified in to dry and wet zones depending on the annual average rainfall mainly by monsoons ranging from 1,000 mm to 5,000 mm with an average temperatures ranging from 9.4°C to 34.7°C, having landscape of tropical coasts, plains and central hills running beyond 8,000 ft from the sea level. The island is rich with fresh waters, diversified flora and fauna and also natural minerals. Availability of petroleum by the Gulf of Mannar has been confirmed where extraction efforts are underway (Department of Census and Statistics, 2013; www.wikipedia.com, 2013).

The social and cultural construct of Sri Lanka is mostly influenced by Buddhism and Hinduism where the majority of the people are Sinhalese (74.88%) and Sri Lankan Tamils (11.2%). However there is a growing population of Muslims (9.2%) and the rest of the other minority groups such as Malays, Burgher and native Vedda people constitute the balance population (Ministry of Finance and Planning, 2014). Official languages by the constitution are Sinhalese and Tamil where English is identified as a linking language. Few other languages are also spoken by minority nationalities with unique features to the island (www.wikipedia.com).

Sri Lanka is a lower-middle income developing nation with the GDP being US$ 74.9 billion by 2014 with a per capita GDP of US$ 3,625 at the average annual exchange rates of 2014. The economy of Sri Lanka is driven by its service sector which contributes roughly to 56% of the GDP while industrial and agricultural sectors contribute 34% and 10%, respectively (Central Bank of Sri Lanka, 2014). Sectorial growth rates over the last year for agricultural, industry and service sectors were 0.3%, 11.4%, and 6.5%, respectively for 2014. Some important indices could be identified as inflation rate on GDP Implicit Price Deflator (%) 2006/07 = 100 was at 5.1%, unemployment rate at 4.3%, GDP growth at 7.4%, and the annual average exchange rate of Rs./US$. 130.56 (Ministry of Finance and Planning, 2014).

The tourism industry has emerged to be the fourth highest foreign exchange earner with all the efforts made to capitalize on the post-conflict peaceful environment in boosting tourism in Sri Lanka. Tourism sector contributes by Rs. Million 221,720 a proportion of 7.9 to the total foreign exchange income which is impressive compared to other established foreign exchange sources such as remittances and core export products by 2014 (Central Bank of Sri Lanka, 2014). It has been continuously identified as one of the best tourist destinations due to its authentic and exotic nature with diversified and compacted elements, placed in a strategic location of the Indian Ocean (Lonelyplanet, 2012). The growth rate has well exceeded 25% over last year's arrivals leaving hope to prosper with potentials by 2012. The last reported was around 19.8% in 2014 (Sri Lanka Tourism Development Authority, 2014).

Western Europe had been identified as the main source of arrivals to Sri Lanka for a long time. Other significant contributors to the total are East Asia and Pacific, North America, Eastern Europe, Australasia, and Middle East in 2014 (Figure 16.1 and which is supported by Figure 16.2). There had been a positive upward movement in per capita daily spending of US$ 160.8 by 2014 mostly accounted for by the escalated tariffs and hotel charges (Sri Lanka Tourism Development Authority, 2013).

The tourism industry of Sri Lanka accounts for 8.4% of the GDP by 2012 (World Travel and Tourism Council, 2012) while the hotel and restaurants as a sub-sector contribute 0.5 to the GDP of 2012 and 0.9 to the total services sector which grew by 18% during the first nine months of 2013. The sector grew its inflows by 41% during 2014 (Ministry of Finance and Planning, 2014). Due to the established perception as an unsecured Therefore, cheap

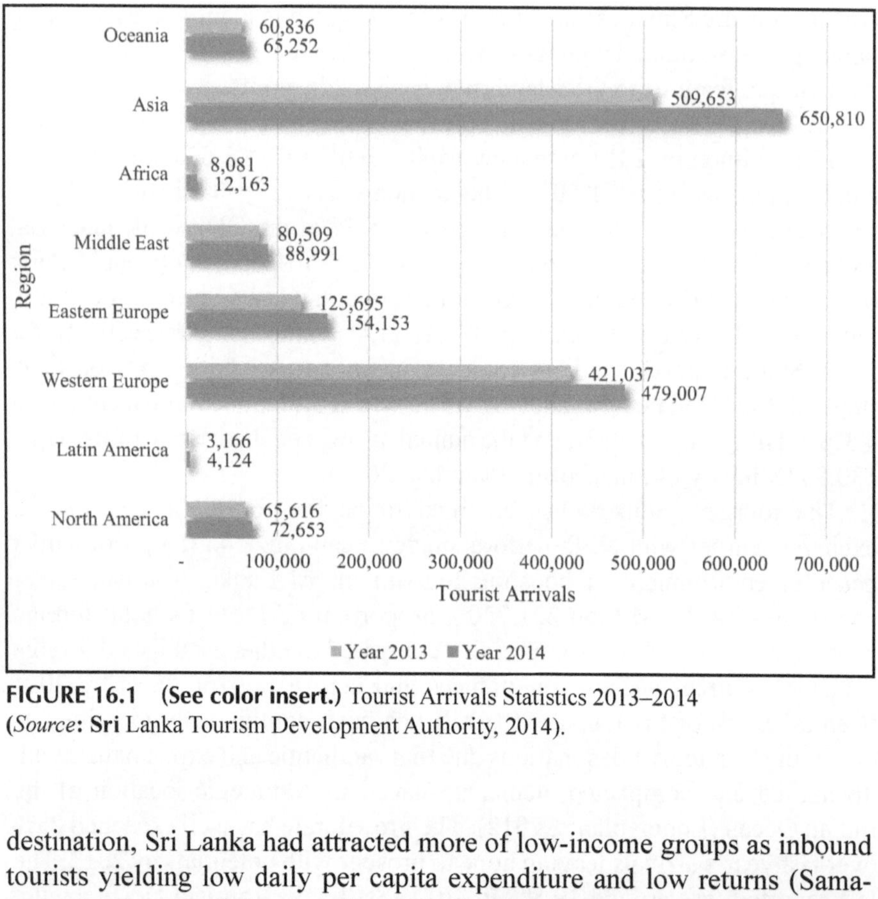

FIGURE 16.1 **(See color insert.)** Tourist Arrivals Statistics 2013–2014 (*Source*: **Sri** Lanka Tourism Development Authority, 2014).

destination, Sri Lanka had attracted more of low-income groups as inbound tourists yielding low daily per capita expenditure and low returns (Samaranayake, Lantra, & Jayewardene, 2013). This needs to be addressed at the strategic planning level.

The country has attracted attention more recently in the international arena in many aspects due to the ending of decade's long terrorism in May 2009, which lead to an overnight boost in the tourism industry potential. As per the policy document of the previous regime, Mahinda Chinthana – Vision for Future 2010, it was expected to harvest a total of US $2.5 billion by the end of 2016 with an inbound traffic of 2.5 million tourists. It was well presented that an integrated plan covering the controlling and liaison bodies, major tourist sites and capacity developments and agro and ecotourism will be well weighted on in achieving this goal (Department of National Planning, Ministry of Finance and Planning, 2010).

The administration of the tourism as a sector is currently a subject of the Ministry of Economic Development which is controlled under four responsible

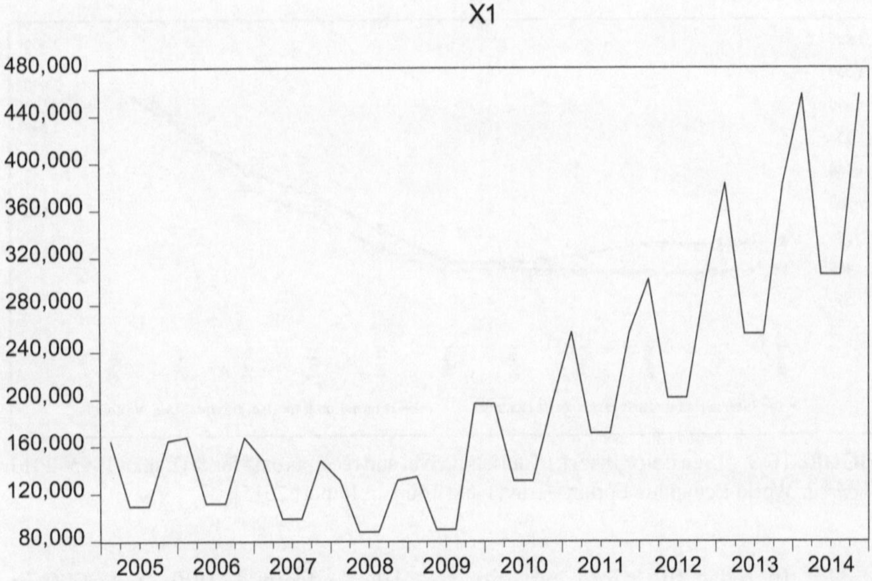

FIGURE 16.2 Growth of Tourist Arrivals from 2005 Q1 to 2014 Q4
(*Source*: Developed by Author based on the collected data).

bodies; Sri Lanka Conventions Bureau, Sri Lanka Tourism Development Authority, Sri Lanka Tourism Promotion Bureau and Sri Lanka Institute of Tourism and Hotel Management (Ministry of Economic Development, 2010). The industry establishments are classified into nine categories as tourist hotels, guesthouses, bed and breakfast units, restaurants, travel agencies, boutique villas, tourism shops, heritage homes, and water sports (Figure 16.3) (Sri Lanka Tourism Development Authority, 2013).

The Ministry of Economic Development had developed a five-year master plan (2011–2016) to accelerate growth in the sector emphasizing on the sustainable development of tourism recognizing the multiplier effects it can bring in to the national economy, predominantly to the small and medium sectors along with the emphasis on the private sector of the economy.

The key objectives of the master plan were (Ministry of Economic Development, 2010);

1. Increase inbound traffic (by 285%) from 650,000 to 2,500,000.
2. Increase sectoral job opportunity (by 300%) from 125,000 to 500,000.
3. Attract US$ 3,000 million as FDI (Foreign Direct Investment) to the country.

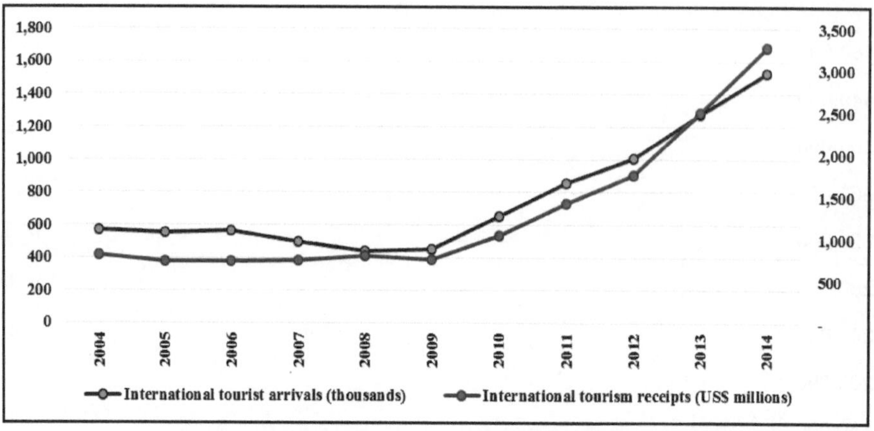

FIGURE 16.3 (See color insert.) Tourists arrival and receipts data for Sri Lanka 1995–2013 (*Source*: World Economic Forum –Travel and Tourism Report 2015).

4. Increase the room capacity (by 104%) from 22,000 to 45,000 in hotels and the informal accommodation sector (guest-houses, rest-houses, holiday bungalows, homestays and youth hostels).
5. Increase tourism foreign exchange earnings (by 500%) from US$ 500 million to US$ 3 billion (making tourism, Sri Lanka's third largest foreign exchange earner, after remittance by Sri Lankans working overseas and the export of garments).

Though there is an undoubted growth potential of Sri Lanka tourism as a foreign exchange earner, there are some criticisms from experts and researchers that Sri Lanka is a destination which has overpriced itself (Fernando & Jayawardena, 2013). It has been identified that for the opportunity to be optimized further commitment towards the growth of the four Travel and Tourism Competitiveness Index (TTCI) pillars; tourism infrastructure, human resources, Information communication technology, and environmental sustainability is vital (Jayawardena, 2013).

Tourism as a sector has potentials in emerging as a major foreign exchange earner for Sri Lankan Economy. It is expected to have positive impacts on (1) economic stability, (2) infrastructure and related services, (3) foreign exchange earnings, (4) employment, and (5) primary and secondary sectors of the economy.

It is emphasized in the literature that Sri Lanka must consider its potential with incentives and policy reforms in the light of eco-friendly, pro-poor and community-based tourism. Further, the six challenges that Sri Lanka is confronted with are highlighted: (1) infrastructure development (logistics and room capacity), (2) attracting investment, (3) developing industry-specific human resources, (4) environment conservation, (5) building strong destination positioning, and (6) prevention of all types of tourist harassment (Lokuhetti, Jayewardene, & Mudadeniya, 2013).

16.1.2 PROBLEM STATEMENT

As per Table 16.1: Doing Business Index (DBI) has ranked Sri Lanka as 107 in 2016 which is clear drop by the 2013 rank which was at 85 out of 189. Thus doing business within the country is identified to be tougher than two years back. The Travel and Tourism Competitiveness Index 2015 (Table 16.2) have placed Sri Lanka at 63 out of 140 countries. This is a clear uplifting of overall ranking 74 in 2013 (Table 16.3). This indicates that the country's competitiveness has improved over the last two decades amidst of the hardships in doing business.

As per the statistical appendix of Annual Report 2014, with a provisional room capacity of 18,510 in graded accommodation and 9,916 rooms in supplementary establishments totaling to 28,4426 rooms (Table 16.4), an average annual room occupancy ratio of approximately 75–80% was enjoyed during the year (Ministry of Economic Development, 2013).

To accommodate the tourist arrivals of 2.5 million by 2016, the room capacity must have increased up to 55,650 rooms. Even if the average stay drops to 8 nights, still the room capacity should increase up to 44,520 rooms to accommodate the inbound tourist arrivals as predicated by Jayewardene et al. (2013) which are more than 50% increase in within three years. But when the patterns of average annual growth of tourist traffic are studied the planned targets seems to be too optimistic.

Though the investment climate is made attractive in recent years with policy reforms the capital formation in the tourist accommodation sector is not noticeably picking up the relative to the growth. Even though there is an enormous amount of literature on multiple aspects of tourism such as tourism product, promotion, sustainability, tourism planning, trends and management of foreign exchange; literature available in the field of hotel sector investment of Sri Lanka is limited.

16.1.2.1 PROBLEM

There is an increasing need to critically examine the gap between the existing level of investment and the expected level of investment with

TABLE 16.1 Doing Business in Sri Lanka: Economy Overview 2016

Region	**South Asia**	
Income Category	Lower middle income	
Population	20,639,000	
GNI Per Capita (US$)	3,400.00	
City covered	Colombo	
Topics	**DB 2016 Rank**	**BD 2015 Rank**
Starting a Business	98	104
Dealing with Construction Permits	77	106
Getting Electricity	81	101
Registering Property	153	151
Getting Credit	97	90
Protecting Minority Investors	49	46
Paying Taxes	158	157
Trading Across Borders	90	90
Enforcing Contracts	161	161
Resolving Insolvency	78	84
Topics	**DB 2016 DTF (% points)**	**DB 2015 DTF (% points)**
Starting a Business	84.98	83.01
Dealing with Construction Permits	70.39	66.54
Getting Electricity	70.82	63.91
Registering Property	46.76	46.75
Getting Credit	45	45
Protecting Minority Investors	60	60
Paying Taxes	55.23	55.08
Trading Across Borders	70.7	70.7
Enforcing Contracts	39.31	39.31
Resolving Insolvency	46.4	44.69

Source: World Bank Group – Extract from Doing Business Index – Sri Lanka.

special reference to hotel accommodation sector of the tourism industry in Sri Lanka. Furthermore, a distinction of the key determinants of investment decision in the sector is to be identified in illuminating the key challenges faced by the tourism growth of Sri Lanka.

Thus the study is focused on to investigate the literature on the tourism accommodation sector investment in Sri Lanka as it has been identified as the main challenge of today's industry within the country. The problem statement was identified as what are the major determinant variables of the tourist accommodation sector of the tourism industry in Sri Lanka?

TABLE 16.2 Travel and Tourism Competitiveness Index 2015 – Asia – Pacific

	TTCI Index		**Enabling Environment Pillars and Values**				
	Regional Rank	Global Rank	Business Environment	Safety and Security	Health and Hygiene	Human Resource and Labor Market	ICT Readiness
South East and Southern Asia							
Singapore	3	11	6.13	6.40	5.44	5.49	5.98
Malaysia	7	25	5.60	5.79	5.18	4.98	4.52
Thailand	10	35	4.78	3.75	4.87	4.98	4.34
Indonesia	11	50	4.48	5.16	4.24	4.70	3.73
India	12	52	4.02	3.82	4.32	4.03	2.83
Sri Lanka	**13**	**63**	**4.59**	**5.58**	**5.24**	**4.33**	**3.49**
Philippines	14	74	4.54	3.84	4.79	4.59	3.71
Vietnam	15	75	4.45	5.31	4.99	4.68	3.37
Bhutan	16	87	4.56	6.02	4.57	4.26	3.51
Lao PDR	17	96	4.58	5.51	4.25	4.50	3.09
Nepal	19	102	3.98	4.52	4.90	4.22	2.36
Cambodia	20	105	3.56	4.96	3.92	4.48	3.19
Pakistan	21	125	4.19	3.04	4.39	3.07	2.54
Bangladesh	22	127	4.06	4.43	4.29	3.85	2.70
Myanmar	23	134	3.20	4.04	4.41	3.85	1.60
South East and Southern Asia average			**4.45**	**4.81**	**4.65**	**4.40**	**3.40**
South East and Southern Asia standard deviation			0.72	0.98	0.44	0.58	1.04
Asia Pacific average			**4.65**	**5.20**	**5.14**	**4.62**	**4.06**
Best performer (global)			**6.13**	**6.70**	**6.97**	**5.64**	**6.37**

Source: Extract from Travel & Tourism Competitiveness Report 2013: World Economic Forum, 2013.

TABLE 16.3 Travel and Tourism Competitiveness Index 2013 – Overall Ranking

	2013		2011
Country/Economy	**Rank/140**	**Score**	**Rank/139**
Morocco	71	4.03	78
Brunei Darussalam	72	4.01	67
Peru	73	4.00	69
Sri Lanka	**74**	**3.99**	**81**
Macedonia, FYR	75	3.98	76
Ukraine	76	3.98	85
Albania	77	3.97	71
Azerbaijan	78	3.97	83
Armenia	79	3.96	90
Vietnam	80	3.95	80
Ecuador	81	3.93	87
Philippines	82	3.93	94
Trinidad and Tobago	83	3.93	79

Source: Extract from Travel & Tourism Competitiveness Report 2013: World Economic Forum, 2013.

16.1.2.2 OBJECTIVES OF THE STUDY

With the intention to examine the major implications of the existing investment shortage in the tourism accommodation sector of Sri Lanka tourism industry this study was conducted. The main objective of the study was to examine the factors that may determine the level of investment in Sri Lankan tourism accommodation sector with specifically focusing to understand the current situation of the investment in tourist accommodation sector of Sri Lanka and to stipulate policy reforms in the tourism investment promotions within the country for the next 5 years.

16.1.3 ORGANIZATION OF THE STUDY

This study is organized as desk research on the literature available on the subject. The terminology is introduced first, and then the empirical literature on several countries is lined up in the following section. The study concludes with identifying the most significant and commonly applied determinants in determining the level of investment to Sri Lankan tourist accommodation sector.

16.1.4 CONCLUSION

This section has introduced the background of the research project, along with the establishment of the research problem. The problem statement, research questions, and objectives are presented along with the major limitations of the study. The next chapter may present the findings on literature applicable for the study with a thorough investigation which may be used for further studies and model building on investment decision evaluation.

16.2 LEARNINGS FROM THE WORLD

16.2.1 INTRODUCTION

The first section provides a detailed examination of tourism terminology related to the study; where the term *tourism* is elaborated. The next section is dedicated for identification of *tourism expenditure*, *fixed capital formation in Tourism* and establishment of the concept of *tourist accommodation* and related classification. Preceding the establishment of the operating definitions on tourism terminology for this particular research the third section is a compilation of theories and concepts related to *investment* in light of economic growth along with different perspectives presented by a variety of scholars. The next section will present the relevance of the study to the contemporary economy of Sri Lanka followed by an explanation of both economic and financial evaluations relevant to the investment decision of the interested field.

16.2.2 TOURISM TERMINOLOGY

16.2.2.1. TOURISM

Temporary movement of people which could simply be defined as tourism is not just an individual activity but a social, cultural and economic phenomenon having multiple effects on an economy. The magnitude and the impact of tourism in the global economy today have nurtured a whole discipline to be established for measurement and control of the array of activities that it encounters. Consequently, a formal body of knowledge has evolved comprising of several organizations on world tourism, passing several stages of development. They contribute by introducing concepts, frameworks, tools, and techniques

TABLE 16.4 Room Capacity vs. Occupancy 2013–2014

Class of accommodation	No. of Units		No. of Rooms		No. of Beds		Total Guest Nights		Foreign Guest Nights		Local Guest Nights		Room Occupancy Rate	
	2013	2014	2013	2014	2013	2014	2013	2014*	2013	2014*	2013	2014*	2013	2014*
(A) Hotel**	304	334	16,655	18,510	34,840	36,883*	8,408,722	10,054,996	6,696,239	7,982,110	7,439,483	2,072,886	71.70	74.30
5 Star	14**	14	3,152	3,152	6,265	6,265*	1,926,189	1,979,280	1,740,306	1,773,396	185,883	205,884	71.90	74.80
4 Star	18	18	2,070	2,084	3,863	4,064	1,369,709	1,416,389	1,170,496	1,232,036	199,213	184,353	70.20	74.90
3 Star	14	14	1,061	1,061	2,142	2,142*	934,192	894,836	853,392	808,912	80,800	85,924	70.30	73.90
2 Star	33	33	1,717	1,725	3,381	3,381*	930,592	1,046,087	830,174	882,389	100,418	163,698	69.90	73.40
1 Star	34	34	1,325	1,333	2,627	2,657*	830,224	927,537	700,958	769,420	129,266	158,117	73.10	72.90
Small luxury hotel	24	24	432	432	890*	907*	227,376	249,670*	176,217	195,991	51,159	53,679	72.10	75.40
Unclassified	167	197	6,898	8,723	15,672	17,467*	2,190,440	3,541,197	1,497,696	2,319,966	692,744	1,221,231	74.90	75.10
(B)Supplementary Establishments**	1,021	1,265	8,513	9,916	16105*	18,340*	3,367,882	4,108,565	2,728,607	3,596,548	639,275	512,017	77.1*	77.8*
***Estimated**														
****Revised**														

Source: Sri Lanka Tourism Development Authority, 2014.

and conducting scholarly discussions in the prompt measurement of tourism contribution in the world economy as a multi-disciplinary economic activity.

In the academia tourism is defined as the sum of processes, relationships and activities arising from travel and stay of non-residents; with the assumption that they do not lead to permanent residency and are of no interest what so ever of earning at the destination (Hunziker & Krapf, 1941). The term tourism is further elaborated as any particular set of selected activities of choice that are undertaken at a location outside home location (International Association of Scientific Experts in Tourism, 1981). Additionally, another explanation was put forward for tourism as a temporary motion of people to a substantial location outside their normal environment of living or employment and related activities during the stay, by the Tourism Society of England (Beaver, 2002).

The demand side definition was presented above while a supply-side definition could be established as an aggregate of productive activities that facilitates in delivering the needs of a visitor is tourism. An industry is a collection of individual business enterprises engaged in a common economic activity. The tourism industry, in particular, is the whole value chain that delivers the tourist with the tourism product offer that was purchased (Technical manual No. 02: Collection of Tourism Expenditure Statistics, 1995).

Travel and tourism are defined as the related activities of a traveler on a trip outside his usual environment for a period of less than one year by the World Travel and Tourism Council. As per the World Tourism Organization, tourism definition is presented as activities of a visitor taking a trip to a main destination outside his/her usual environment for less than a year and for any main purpose (business, leisure or other personal purpose) other than to be employed by a resident entity in the country or place visited. Tourism is classified into three segments in a broader sense as; (1) domestic tourism, the movement of local tourists within the local boundaries (2) inbound tourism, the inward movement of foreign tourists into local boundaries and (3) outbound tourism, the movement of local tourists out of the borders. Further classification is available as (1) internal tourism; comprises of domestic and inbound tourism (2) national tourism; comprises of domestic and inbound tourism and (3) international tourism; comprises of inbound and outbound tourism (World Tourism Organization, 2008).

16.2.2.2 TOURISM EXPENDITURE

The tourism expenditure has been classified by the United Nations World Tourism Organization. It is mentioned that tourism expenditure includes the amounts paid to acquire consumption goods and services and valuables for the purpose of consumption or gifts. The expenditure includes the sums paid by the visitors, paid for on visitor's behalf or those that are reimbursed. There are certain kinds of expenditure such as social transfers, owned accommodation services and financial intermediation that are excluded from the tourist expenditure at the demand side. There are two types of tourism-related products as identified by UNTWO in accordance with the Systems National Accounts 2008; (1) tourism characteristic product and (2) tourism connected products. It is accepted that the production process, capital input, intermediate consumption and labor requirement of tourism characteristic products should only be accounted for in relation to tourism supply. Further classification is provided for the purpose of statistics of tourism characteristic products as (1) package travel, (2) accommodation, (3) food and beverage, (4) local transport, (5) international transport, (6) recreation, (7) shopping, and (8) others (World Tourism Organization, 2008).

16.2.2.3 FIXED CAPITAL FORMATION IN TOURISM

It is vital to study the stock and the flow of physical investment as a certain minimal level of infrastructure is required to operate tourism as an industry and benefits to be harvested with a considerable impact in a country (World Tourism Organization, 2008).

Tourism-driven investment is categorized in to three: (1) tourism specific fixed assets which are used exclusively or almost exclusively in production of tourism characteristic products, (2) investment by tourism industries in non-tourism specific industries which includes those fixed assets that are used to facilitate tourism and (3) tourism-related infrastructure which are commonly developed by the state or government authorities. While there is no such defined proposal as to aggregate tourism fixed capital formation, tourism specific fixed assets and investments by tourism industries in tourism specific products are encouraged to recognize in much particular terms and detail (Tourism Satellite Account: Recommended Methodological Framework, 2008).

16.2.2.4 TOURIST ACCOMMODATION

Tourist accommodation falls under tourism specific fixed assets along with four other categories of fixed capital formation as non-residential buildings and structures, passenger transportation equipment and infrastructure, machinery and equipment especially for the production of tourism characteristic products and improvements of land for the purpose of tourism (Tourism Satellite Account: Recommended Methodological Framework, 2008).

Accommodation, in particular, is comprised of the specific places of stay with or without services for tourists either in the form of paid commercial establishments or unpaid non-commercial owner-occupied vacation homes. Depending on the facilities and amenities available, accommodation could be classified from luxury self-contained units with regular housekeeping along with other services (such as parking, gym, and swimming pool, laundry, and recreation) to youth hostels only with the bed and shared sanitary facilities. They are further classified based on class/grade categories, geographical characteristics of location, facilities provided or proximity to places of interest (World Tourism Organization, 2008).

Sri Lanka has identified two types of broadly defined accommodations as graded and supplementary establishments with respective capacities being 15,510 rooms in 269 units (Table 5) and 6,577 rooms in 620 units. A further decomposition of accommodation sector is made as boutique hotels and villas, (holiday) bungalows, guest houses, heritage bungalows, heritage homes, homestays, hotels, rented apartments and houses and Sri Lanka tourism resorts (Sri Lanka Tourism Development Authority, 2014).

The study will follow the United Nations terminology with relevant local adaptations. The term tourism shall mean the inbound tourism into Sri Lanka, and accommodation shall mean paid commercial accommodation hereafter for this specific study report.

16.2.3 INVESTMENTS

Capital or investment is a term used interchangeably by economics and financial management, to refer the forgone present consumption of funds or property with the expectations of returns in future with different applications. In an economic stance, investment and capital may mean two different things. Capital is the stock of productive assets accumulated over time where investment is the fresh addition made on to the same available stock (Dwiwedi, 2010).

Tourism had long been identified as a multi-disciplinary field of which the involvement of many industries and natural setting would bring out the compound impact. Planning in detail of every element and discipline involved is a pre-requisite in realizing intended results. Failure to do so would bring about undesirable consequences such as degradation of natural resources and environment, destruction of cultural and social value system and lost potential economic impacts. Literature supports that proper planning destinations ensure the success of the trade while unplanned will suffer from negative consequences of the same. Planning in particular for tourism should include a detailed assessment of the stock of resources at the destination and their potential in the attraction (Gum, 1994; Inskeep, 1994; Blank, 1989; Formica, 2000; cited by Kim, 2002).

WTO has classified the stakeholders of sustainable tourism development into three clusters as (1) tourism industry, (2) environmental support, and (3) local community and Government. The tourism industry has a role to play in an economy to create opportunities for business and employment, generate income and foreign currency exchange via provision of tourism services which include but are not limited to transportation, accommodation, food, and beverage. On the other hand, it is the environment's capacity and role to attract the tourist with culture, nature, and architecture based experience and exposure. The local community and the Government play the role of making decisions and facilitating/controlling the industry (Anuar et al., 2012).

Many Government establishments and private researchers have studied the areas of tourism performance measurement and development of tourism plans preceding assessment of resources. Long-term planning is essential for both developed and developing tourist destinations in revitalizing and sustaining the attractiveness of the destination to unleash the potentials of the industry. Though early planning had been centralized around the tourist without much concern of the social or environmental impacts of the destination of today's planning is more oriented towards the welfare of the destination population and the sustainability of the industry in the long term (Dragicevic, 1991; Witt, 1991; McIntosh, Goeldner & Ritchie, 1995, Formica, 2000; cited by Kim, 2002).

This study attempts to address issues related to the planning gap of in fixed physical capital formation, specifically in the tourist accommodation sector in Sri Lanka which falls under properties.

State-supported planned tourism development is always affirmed in literature. Being the leading tourist attraction in Latin America today, Mexican tourism growth today is highly accredited to its State role to polarize sectoral growth with high infrastructure facilitation and active investment during 1970s. Subsidized lending for hotel investment apart from the active participation in hotel operating by the Government have supported immensely in developing the sector up to the standard today. The government agency backed by the Central Bank of Mexico being the bearer of the risk of hotel erections funding, the country accumulated a stock of hundred thousand rooms by 1974. The role of the entrepreneur and the risk taker undertaken by the Mexican Government and active policy change set examples for most of the developing nations nurturing the tourism industry (Clancy, 2001).

FDI is considered to be a promising tool in financing the investment gap in developing countries. The World Bank reports continuously have shown an increasing trend of FDI inflows to developing nation during 1990s and beyond. Commercial banks are constrained in their capacity of lending in most of the developing countries due to the nature of the vicious cycle that they experience. Thus the greater openness brought in to developing nations investment climate, and the level of reinforcement applicable to FDI have accelerated domestic growth in host countries. In 2000 more than 50% of private capital inflows to developing nations had been in the form of FDI (World Bank, 2000).

A study had been conducted in Canada to investigate trends in investment in hotels and managing finance (Willie et al., 2013). This study is a descriptive research of the Canadian hotel investment climate. This study provides insights as to how the government has supported the tourism accommodation sector as a real estate investment and the business models that were suitable at different stages of the business cycle faced by a developed country.

The nature and the fluctuations of debt and equity markets influenced the volume of transactions in hotel investment in any country. It is identified that the change in the fundamental structure of the hotel investment environment in Canada over the last three decades has clear relationships with these fluctuations. In the point of this research, hotels are much risky investment in real estate than other classes of property due to the volatility of returns embedded into the asset. The rates and occupancy ratios both are capricious leading the returns to have the same effect in the short and long run (Willie et al., 2013).

The hotel industry in Canada particularly had been undoubtedly affected by the economic downturn owing to the Great Recession during 2007. Yet as per the research, the Canadian hotel industry was least affected compared to its counterparts of other regions. In fact, the impact would have resulted adversely on hotel investment and hotel property valuation. This was shown by the volume of global hotel transaction decline of 80%, from CDN $ 113 billion in 2007 down to CDN $ 23 billion in 2008. Hence the hoteliers including investors, owners, operators and financial managers were pushed into challenging decisions making situations. Thus the hotel investor support services were provided by the Government under Canada's Federal Tourism

Strategy in the form of advocatory and facilitator services including communication, collaboration, promotion along with development and improvement of the output (Willie et al., 2013).

The amount of hotel properties was over 8,500 with a labor force of 285,000 having direct jobs from the sector of Canada in 2013. The sector has accounted for CDN $16.5 billion to national income in addition to the contribution of CDN $7 billion made to local authorities including municipal, provincial and federal governments. Average revenue per available room (RevPAR which is the combined measure of occupancy rate and the average daily room rate) had been CDN $83 which dropped to CDN $73 due to the recession. It was expected to increase up to CDN $82 (forecast) for 2013 (Hotel Association of Canada, 2013).

The Canadian hotel investments have passed through a few cycles during the past few decades. The private ownership, institutional ownership, lenders as well as real estate investment trusts (REITs) and credit unions have played vital roles time to time in hotel assets investment in Canada as depicted in Table 16.5.

Another study had been conducted with respect to Antarctica for the purpose of surfacing the elements of national tourism. They have investigated how the important key factors have been supportive in defining Antarctic tourism development and the role of those in the shaping the future of the industry.

TABLE 16.5 ATOS Developed by Lamrs, Haase, &Amelung, 2008

Period	Equity type	Debt type
1985 – 1989	Tax efficient Limited Partnerships, foreign capital	Canadian Schedule A Banks, trust companies, insurance companies
1990 – 1993	Private Capital	Vendor take back (VTB) mortgages (seller financing)
1994 – 1996	Opportunity funds	US and International lenders
1997 – 1999	Opportunity funds and REITs	US and International lenders
2000 – 2004	Private capital and REITs	US and International lenders, commercial mortgages backed securities (CMBS), private institutional
2005 – 2007	Private equity, pension funds and REITs	CMBS and all cash
2008 – Current	Private capital, hotel investment companies, private equity	Credit unions, regional banks, private institutional

Source: Willie et al. (2013).

The paper used the Recreational Opportunity Spectrum (ROS) which is a combined tool of the analytical framework with an action-oriented management perspective proposed by Clark and Stankey in 1979. This model had later been applied by many scholars such as Butler and Waldbrook, 1991; Boyd and Butler, 1996 in tourism-related studies (Lamers, Haase, & Amelung, 2008). To be precise, the Ecotourism Opportunity Spectrum adopted from the original ROS by Boyd and Butler (1996) was modified to develop the Antarctic Tourism Opportunity Spectrum (ATOS presented in Figure 16.4) (Lamers, Haase, & Amelung, 2008).

With the advancement of means of transportation and recreational activities, tourism in Antarctica has boomed rapidly during the past two decades. Thus proactive tourism management in developing and regulating the industry rather than reactive is encouraged by the scholars due to the increasing importance that it has gained today. The regulatory framework of Antarctica consists of Antarctic Treaty System (ATS) complimented by other regulatory frameworks of International Maritime Organization (IMO), International Association of Antarctica Tour Operators (IAATO) and Treaty Consultative Meetings (ATCMs), etc. These are demanded to be further strengthened to refer site-specific regulations in later times for sustainability concerns (Lamers, Haase, & Amelung, 2008).

ATOS adopted six sub-factors out of the original eight sub-factors as (1) access, (2) other users, (3) attractions, (4) tourism infrastructure, (5) required human resources, (6) acceptability of visitor impacts, (7) level of social interaction, and (8) the appropriate regulatory frameworks. Under the six selected sub-factors, 2 to 9 variables were captured for in-depth analysis as shown in Table 16.6.

All in all an increasing and positive trend have been established by the researchers for Antarctic tourism development. Under these six tangible factors, access factor for a remote region like Antarctica was proven to be significant. Activities related to science were identified to be a hindering factor and a promoting factor in certain cases as it limits ordinary tourism activities. Contextual factors were often referred to as the factors that are uncontrolled by the stakeholders such as global energy prices, climate changes, inflationary situations, etc. The opportunity windows opened with this framework have been useful in planning the industry in Antarctica as a luxurious and adventurous destination with cautions by both the industry and self-regulatory ATS of environmental issues that may arise.

TABLE 16.6 Variable Selection and Measurement Criteria

Sub factor	Measurement criteria
1. Access	Remoteness and extremeness
	Availability of suitable ships
	Ship building
	Availability of aircraft
	Global environmental change
	Infrastructure development
	Compatibility with global itineraries
	Information channels/media
	Marketing resources
2. Other users: science	Logistical operation
	Infrastructure sharing
	Facility sharing
3. Attractions/activities	Available landing sites
	Wildlife
	Dramatic scenery
	Heritage sites
	Adventure activities
4. Operational factors	Interaction in the field
	Qualified staff and crew
5. Acceptance of impacts	Environmental impacts
	Human safety
	Intrinsic value
6. Regulations/management	Antarctic treaty system
	IAATO Self-regulation

Source: ATOS presented above in Figure 16.4.

16.2.4 EVALUATION OF DETERMINANTS OF PROPERTY INVESTMENT DECISION

In any industry investment either in the form of liquid capital or physical capital needs to undergo rigorous appraisal process before a decision is made whether or not to invest. This decision is not a laboratory decision or a clinical decision; rather an effort undertaken by imperfect people of imperfect industry conditions with imperfect information. It is well known that the exploration into the process of decision making in investment in property is limited and inadequate. It is further bounded by the rules, tools, and techniques that are widely accepted to be the common practice though far deviated from the representation of the true market scenario. Therefore, the area needs further scrutiny in improving the understanding of the patterns of property investments (Roberts & Henneberry, 2007).

The academic literature on investment decision making and related topics of property investment is poorly integrated and sparsely organized. The rationalist perspective is the dominant model in the field though it is heavily reliant on formally structured investment behavior by investors which is far from the truth. Thus the decision makers are assumed to be rational and fully informed in their decision making. Hence rely largely on techniques such as Modified Internal Rate of Return (MIRR), Payback period, Net Present Value (NPV), Modern Portfolio Theory (MPT) and Capital Asset Pricing Model (CAPM) (Dubben & Sayce, 1991; Gallimore, Hansz, & Gray, 2000; Sangster, 1993). Though the use of rational models sounds perfect in an economic sense, the field work and literature does not necessarily support the use of these methods in the realistic behavior of investors (Roberts & Henneberry, 2007).

Furthermore, the sustainability concerns over the preservation of social and environmental credentials with property investment decision making are also of increasing importance. The property investment and assets which are in compliance with principles of sustainable and ethical development which is termed as Socially Responsible Investment (SRI) are in great demand more recently (US Social Investment Forum, 2006). Hence the property investment as a commodity is changing its shape to ensure the sustainable design elements and product attributes due to the prominence of those in the valuation of the property. Far more the negligence of the same is taken as an appreciation of investment risk factor where sustainable corporate practices are increasingly seen as a precondition for higher returns. It is thus a nasality to adopt principles of sustainable development in the property sector and to introduce the use of relevant consultations and valuations to safeguard and increase competitiveness. Therefore, the sustainable business is no longer seen costly but ensuring long-term profits (Lorenz & Lutzkendorf, 2008).

It is well presented that the decision making on property investment is largely influenced by the close integration between property valuation, risk assessment and management, financial analysis and the social and environmental concerns. The players in the property industry comprise of several stakeholder categories for example; constructors, designers, engineers, and other related professionals, researchers and academics and other institutions, along with Government officials and authorities. Though the industry has been sluggish in the development of sustainable property development philosophy and strategies, there had been noticeable achievements in the fields of design, technology and material elements along with assessment tools and techniques. The attention had been growing faster than the practice in this field. This serious concern raises a question as to how reinforcement of sustainable property investment should be framed (Lorenz & Lutzkendorf, 2008).

Advantages of sustainable property investment had been compiled by many scholars. The outcomes can be summarized into basically two categories and direct and indirect. In addition, it was established that an additional 2% upfront investment on sustainable design of property would add a saving of 20% on the total construction cost along the lifecycle of the property (Kats, Alevantis, Berman, Mills, & Perlman, 2003; Katz, 2003). Indeed with a higher level of cost than traditional, the benefits are overwhelming as higher efficiency rates in energy consumption, the effectiveness of functional elements, adaptability and serviceability, ability to market the investment, higher levels of profitability and, higher levels of comfort and well-being for occupants. At the same time, it offers loss preventions benefits, reduced risk potentials and conservation of natural environment (Wilson et al., 1998; Heerwagen, 2002; Mills, 2003; Kats et al., 2003; Royal Institution of Chartered Surveyors, 2004, 2005 cited in Lorenz & Lutzkendorf, 2008).

Generally accepted and compiled property valuation methods are presented in Table 16.7. Hotels as in particular accommodation sector properties of the tourism industry are quite open and vulnerable to internal and external economic and situational factors that determined the demand for tourism (Newell & Seabrook, 2005). Therefore, hotel sector assets investment had

TABLE 16.7 Property Valuation Techniques

Traditional valuation methods	Advanced valuation/ data analysis methods
Sales comparison method	Hedonic pricing method
Investment method/discounted cash flow method	Artificial neural networks
Cost method	Spatial analysis method
Profits method	Fuzzy logic
Residual method	Autoregressive integrated moving average
	Real options
	Rough set method

Source: Lorenz & Lutzkendorf (2008).

been recognized as a hybrid property which incorporates both business risk and property risk. Apparently, it had been controversial whether to recognize hotel properties as a part of core property portfolio or to treat as an asset class with additional risk profile. Following sector-specific characteristics have been identified as differentiating features of hotels investments from other office, retail or industrial property investments as (1) higher volatility in cash flows, (2) risk-adjusted returns being low, (3) attributes of an operating business rather than a property, (4) investor support of institutions being low, (5) dependence on other sectors, (6) Lack of know-how and (7) industry-specific features (Property Council of Australia, 2003). It had been noted that in Australia hotel sector property was the worst performing sector on a risk-adjusted basis entailing higher levels of risk (Newell & Seabrook, 2004).

Like in any sector of business hotel investment also demands of prosperous, strong and optimistic business climate assessed by constantly growing GDP, strong demand conditions, availability of specialized human resources, stability of the relevant financial institutions, low indebtedness, federal Governmental and other Government institutional support and attractive cost structures inclusive of fair tax structure and interest rates. The decision to invest in a hotel ultimately is dependent upon the fact that the market condition is accountable to be strong. Higher levels of employment, high confidence level of consumers, low level of inflation, etc. will motivate the parties involved in making the hotel investment decision (Willie et al., 2013).

There are three distinctive parties involved in a hotel investment as (1) investor, (2) lender and, (3) operating team. Each of these three parties has their own motives in making the hotel investment. The investor (an individual, a group or a fund/trust) is making the hotel property investment to acquire an attractive asset. He should particularly be interested in earning a healthy return on investment (ROI) and an appreciation of the real estate value of the asset over time. The lender is the party who finances the investment with the expectation of interest income with a low-risk income source with a substantial borrower. The operator is interested in gaining a hotel management opportunity as a professional service provider in exchange of a service fee. The service fee, in general, is set at 3–5% of gross operating profits and an incentive of 5–10% of gross operating profit. Mutual understanding between the parties, respect of everyone's motives and requirements and coming into win-win solutions after effective negotiation will make the investment more attractive (Willie, Pirani, Jayawardena, Sovani, & Davoodi, 2013).

A study on influential factors of hotel investment decision had been undertaken (Newell & Seabrook, 2005) in Australia which concentrates on five core areas of interest as (1) location factor (2) economic factors (3) financial factors (4) diversification factors, and (5) relationship factors. Under each factor, the research identifies a few relevant sub-factors as shown in Table 16.8.

Specifically, distinctive multi-criteria decision-making models are available as the Analytic Hierarchy Process (AHP), Equal Weight Averaging Model (EWAM), and Simple Multi-Attitude Rating Technique (SMART).

TABLE 16.8 Sub Factors Under Each Variable

Factor	Sub-factors
1. Location	Volatility of demand
	Number of international visitors
	Number of domestic visitors
	Site attributes, i.e.; profile, visibility, transport, convention facilities
	Age of target hotel
	Current hotel supply (hotel cycle)
2. Economics	Trends in interest rates
	Tourist spending patterns
	Business spending patterns
	Employment growth (office, professional)
	Extent location market is mature.
	Extent location market is emerging
3. Financial	Medium-term forecast return on investment (five years)
	Historical rates of return
	Unsystematic risk, i.e., liquidity, business, hotel market risk
	RevPAR as a measure of return
	Gross Operating Profit (GOP) as a measure of return
	Economies of scale advantages
4. Diversification	Geographical diversification
	Brand diversification
	Segment diversification
	Relationship of target property to current portfolio
5. Relationships	Regulatory influence
	Independent hotel asset management
	Alignment with stakeholders

Source: Newell & Seabrook, 2005.

AHP had been continuously recognized as the superior technique allowing for both weighting and scoring of factors. A record of extensive use of AHP is compiled by the researchers in the fields of residential property, urban environmental quality and planning, foreign investment in direct property, segment assessment in valuation and facility location selection. Hence AHP had been used in this study to weight the factors and sub-factors along with the respective scorings (Newell & Seabrook, 2005).

The methodology adopted had been a two-stage survey where respondents were interviewed preliminarily to identify the most important factors and then to gain 4.5 average rating out of 7 (1 being equal importance, and 7 being absolute importance) for each sub-factor to be established as critical. Initially, a paired comparison of the relative importance of the factors was required to be made leading to identifying the degree of importance. Any pair was allowed to have equal importance among the 64 pairs studied. Hotel investors and owner/operators and public listed hotel investors and private investors in hotels were separately analyzed for overall AHP analysis. The consistency of the respondents was also measured using AHP consistency ratio measures where only > 0.20 consistency ratio was accepted to be suitable. The Expert Choice Software was the decision-making support system tool used for AHP analysis in this study which is widely accepted to be the best in use (Newell & Seabrook, 2005).

It was identified that financial factor and its sub-factors were the most influential factors in hotel investment decisions where location factor, economic factor, diversification factor, and relationship factors along with relevant sub-factors respectively were identified to be important. AHP output had identified three levels of importance within which each level had interdependencies as (1) financial and location, (2) economic and diversification, and (3) relationships.

Hotel owners have prioritized the location factor as the most influential factor whereas investor has identified financial factor to be most influential. Hence it was concluded that hotel owners were more knowledgeable of the fact that hotel investments were vulnerable to drivers of operational performance of the property. On the other hand, the public investors have identified location factor to be of highest weigh while the private investor identified the economic factor to be of the highest importance. This leads to the conclusion that private owners have regarded hotel property investment to be more of a financial investment rather than an operating business (Newell & Seabrook, 2005).

In the Sri Lankan context, a formal body does not seem to exist in assessing, valuing or archiving the trade and the indices on property transactions. Yet general data can be gathered from deferent sources which can be amalgamated to produce useful information. There had been an increasing interest and prospects especially in coastal areas of the island for hotel sector investment as per a series of recent reports published and of researcher's own observations as well.

16.2.5 CHAPTER SUMMARY

The chapter intended to identify the most significant determinants of the Sri Lankan tourism sector investment. In materializing the intention, the researcher conducted a desk review of the available literature on tourism investment and investment attractiveness tourist destinations. This chapter is based on the findings and the conclusions of such an exercise aimed at localizing the findings.

The findings include that many countries have gone through cycles of ups and downs in hotel sector investments. Some countries had provided government support during downturns for private investments and have even established public-private partnerships to maintain the level of investment and the industry. The attractiveness of the investments is highly dependent upon a multitude of factors including the tourist destination attractiveness, macroeconomic indicators of the destination state and nature of human capital within the target destination.

The chapter concludes with the identification of five major variables that affects the level of investment in accommodation for Sri Lankan tourism as the tourist attraction for a particular destination, the national output of the destination, the rate of inflation of the destination, exchange rates of the destination and the availability and the quality of human capital of the destination. This finding may be used for investment determinants model building and further examination at a macro and micro level.

KEYWORDS

- **investment**
- **literature**
- **Sri Lanka**
- **tourist accommodation sector**

REFERENCES

Anuar, A. N., Ahamad, H., Jusoh, H., & Hussain, M. Y. (2012). Understanding the Role of Stakeholder in the Formation of Tourist Friendly Destination Concept. *Journal of Management and Sustainability*, 69–74.

Beaver, A. (2002). *A Dictionary of Travel and Tourism Terminology.*

Central Bank of Sri Lanka. (2014). *Economic and Social Statistics of Sri Lanka.* Colombo: Central Bank of Sri Lanka.

Clancy, M. (2001). Mexican Tourism: Export Growth and Structural Change since 1970. *Latin American Research Review*.

Department of Census and Statistics. (2013). *Annual Statistical Report.*

Department of National Planning, Ministry of Finance and Planning. (2010). *Mahinda Chinthana – Vision for Future, Development Policy Framework.* Government of Sri Lanka.

Dubben, N., & Sayce, S. (1991). *Property Portfolio Management: An Introduction.* London.

Dwiwedi, D. N. (2010). *Macroeconomics: Theory and Policy* (3rd ed.). New Delhi: McGraw Hill Education Private Limited.

Fernando, P., & Jayawardena, C. (2013). Focusing on realistic tourism targets for post-war Sri Lanka. *Worldwide Hospitality and Tourism Themes, 5*(5), 495–504.

Gallimore, P., Hansz, J., & Gray, A. (2000). Decision making in small property companies. *Journal of Property Investment and Finance*, 602–12.

Hotel Association of Canada. (2013). *Hotel Industry Fact Sheet.* Ottawa: Hotel Association of Canada.

Hunziker, W., & Krapf, K. (1941). *Grundriß der allgemeinen Fremdenverkehrslehre.*

International Association of Scientific Experts in Tourism. (1981). *The AIEST, its character, and aims.*

Jayawardena, C. (2013). Innovative Solutions for Future (2013–2026). *Worldwide Hospitality and Tourism Themes*.

Kats, G., Alevantis, L., Berman, A., Mills, E., & Perlman, J. (2003). *The Costs and Financial Benefits of Green Buildings – A Report to California's Sustainable Building Task Force*. Retrieved January 9, 2014, from www.usgbc.org: www.usgbc.org/Docs/News/News477.pdf.

Katz, G. (2003). *Green Building Costs and Financial Benefits*. Retrieved from www.cap-e.com.

Lamers, M., Haase, D., & Amelung, B. (2008). Facing the elements: analyzing trends in Antarctic tourism. *Tourism Review*.

Lokuhetti, A., Jayewardene, C., & Mudadeniya, D. (2013). Developing a shared vision for tourism in post-war Sri Lanka. *Worldwide Hospitality and Tourism*, 486–494.

Lonelyplanet. (2012, October 23). *Best in Travel 2013 – Top 10 countries*. Retrieved November 20, 2013, from www.lonelyplanet.com.

Lorenz, D., & Lutzkendorf, T. (2008). Sustainability in Property Valuation: Theory and Practice. *Journal of Property Investment & Finance*, 482–521.

Ministry of Economic Development. (2010). *Tourism*. Retrieved December 2, 2013, from www.med.gov.lk.

Ministry of Economic Development. (2013). *Tourism*. Retrieved December 2, 2013, from www.med.gov.lk.

Ministry of Finance and Planning. (2014). *Fiscal Management Report 2014.* Colombo: Ministry of Finance and Planning.

Ministry of Finance and Planning. (2014). *Annual Report 2014.*

Newell, G., & Seabrook, R. (2004). The role and performance of hotels in investment portfolios. *Proceedings of 10th Press Conference.* Bangkok.

Newell, G., & Seabrook, R. (2005). Factors influencing hotel investment decision making. *Journal of Property Investment & Finance.*

Property Council of Australia. (2003). *New Investment Frontiers: An Industry Action Plan For Reshaping Hotel Investments.* Sydney: PCA.

Roberts, C., & Henneberry, J. (2007). Exploring office investment decision-making in different European contexts. *Journal of Property Investment & Finance*, 289–305.

Samaranayake, H. M., Lantra, N., & Jayewardene, C. (2013). "Forty-six years of organized tourism in Sri Lanka (1966–2012). *Worldwide Hospitality and Tourism Themes*, 5.

Sangster, A. (1993). Capital Investment Appraisal Techniques: A Survey of Current Usage. *Journal of Business Finance & Account.*

Sri Lanka Tourism Development Authority. (2013). *Statistics.*

Sri Lanka Tourism Development Authority. (2014). *Annual Statistical Report* . Colombo, Sri Lanka: Sri Lanka Tourism Development Authority. see, http://www.sltda.lk/statistics.

Technical Manual No. 02: Collection of Tourism Expenditure Statistics. (1995). World Tourism Organization.

Tourism Satellite Account: Recommended Methodological Framework. (2008). OECD, WTO, UNSD, EUROSTAT.

US Social Investment Forum. (2006). *Report on socially responsible investing trends in the.* US Social Investment Forum.

Willie, P., Pirani, A., Jayawardena, C., Sovani, A., & Davoodi, R. (2013). Trends in hotel investment and financial management in Canada. *Worldwide Hospitality and Tourism Themes.*

World Bank. (2000). *World Development Report 2000.* The World Bank Group.

World Tourism Organization. (2008). *International Recommendations for Tourism Statistics.* United Nations.

World Tourism Organization. (2008). *The Conceptual Framework for Tourism Statistics – International Recommendations for Tourism Statistics.* United Nations.

World Travel and Tourism Council. (2012). *Travel and Tourism, Economic Impact 2012 Sri Lanka.* London: WTTC.

www.wikipedia.com. (2013, November 29). *Sri Lanka.* Retrieved December 1, 2015, from Wikipedia.

CHAPTER 17

IMPACT OF INFORMATION TECHNOLOGY ON MARKETING STRATEGIES OF ONLINE TRAVEL AGENCIES (OTAs) IN INDIA: A COMPARATIVE STUDY OF GOIBIBO.COM AND MAKEMYTRIP.COM

SUJOOD[1] and SHEEBA HAMID[2]

[1]*Research Scholar (Junior Research Fellow-Tourism), Department of Commerce, Aligarh Muslim University, Aligarh, India, E-mail: sjdkhancool@gmail.com*

[2]*Professor (Tourism), Department of Commerce, Aligarh Muslim University, Aligarh, India, E-mail: sheebaranyah@gmail.com*

ABSTRACT

This chapter examines the impact of IT on marketing strategies of online travel agencies in India. In the previous 5 years, travel and tourism industry of India has selected technologies, modern trends, and marketing strategies and given an advanced dimension to this sector. In recent time, due to the advancement of the Internet technology and favorable conditions in India, online travel portal has become very alluring and suitable. IT has a subtle impact on the industry of travel and tourism. This sector of travel and tourism has always been one of the most critical ones for the financial system as it is not only contribute to GDP and the Forex Reserve of the country, but also produce the employment. Marketing strategies of Online Travel Agencies include all the activities in the field of marketing which directly deal with the analysis of the initial strategic situation of a company and the formation, evaluation, and selection of market-centered strategies and contribute to the aims of the company and its marketing objectives. This paper dwells on the present marketing strategies scenario of OTAs in India. The authors have given some recommendations for better integration of Marketing Strategies and use of Information Technology to flourish Online Travel business.

17.1 INTRODUCTION

In the last 5 years, the Indian travel and tourism industry has selected technologies, innovative trends, and marketing strategies and given an advanced dimension to this sector. In recent time, due to the advancement of the Internet technology and favorable conditions in India, online travel portal has become very alluring and suitable. The sector has constantly been one of the most critical ones of the financial system as it is now not only a giant contributor to GDP and the Forex reserve of the country, but also creates tremendous employment. It is estimated that the Indian travel and tourism industry will make billions through 2018, so hoping for a very sanguine future. India is a nation with many varieties, cultural inheritances and ratings of languages and dialects, presents very extraordinary challenges and possibilities for Online Travel Portals (OTAs) (Hajra, 2013).

A few years ago when the scheme of booking a journey online has taken birth, and it was the revolutionary attempt of Captain Gopinath of Air Deccan that pave the way of this revolution in the travel and tourism industry. India has the benefit of always being forward of the curve and has quickly learned from practices in developed nations and adapted well. The travel explosion, development of the internet, increasing recognition of e-commerce and the appearance of budget carriers, have been key factors that have contributed to the advancement of this industry. Today, clients are aware and understand the significance and various profits of using information technology to book any travel product. Consumers have motivated from traditional means counter booking to online booking. And as the Internet usage rate increasing in the cities, this industry will definitely be blooming immeasurably in the upcoming years. The marketing strategies of an organization are those strategies which integrate all of its marketing objectives into one inclusive plan. It has the primary objective of rising sales and attains sustainable development. Marketing strategies of Online Travel Agencies include all the activities in the field of marketing which directly deal with the analysis of the initial strategic situation of a company and the formation, evaluation, and selection of market-centered strategies and contribute to the aims of the company and its marketing objectives.

17.2 CHOICE OF COMPANIES

This chapter is focused on the information technology and marketing strategies of online travel agents. In recent times, due to the development of the information technology and favorable condition in India, online travel and tourism industry has become very fascinating and suitable.

IT has a subtle impact on the industry of travel and tourism. To compare online travel industry with offline is important. Even though many research workers think that the online travel industry will replace offline in the near future, but the results do not validate it, and Therefore, we need to analyze the causes behind it. Combing the aspects outlined and the purpose of the study, Indian online travel agents, are taken as samples.

The authors choose www.Goibibo.com and www.MakeMyTrip.com as two Indian online travel companies to study further. Both of the companies are very famous in India and even abroad, representing new and old business form as well as the progress of OTAs in India.

17.2.1 GOIBIBO.COM

Www.Goibibo.com is an online travel portal, formed as part of the Ibibo group in 2009. Its headquarter situated in Gurgaon, India. India's top online Bus ticketing platform, redBus.in owned by Ibibo Group and newly launched car-pooling booking app, "ibibo Ryde." Under the travel category, Goibibo is first ranked mobile app. Fastest delivery is Goibibo's core value differentiator and the most trusted consumer experiences, in terms of quickest booking procedure and search, fastest settlement, payments, and the quickest process of refund. Over the previous year, Hotel booking volumes have grown 5 times in 2015 by Goibibo. 70% of hotel reservations take place on the mobile application of Goibibo ("Goibibo," n.d.).

Goibibo.com, which takes place among the leading Online Travel Agencies in India, enables consumers to buy train, air, bus tickets, hotel booking, car rentals, etc. In a very short time, Goibibo has created a prestigious image for itself. With the fastest process for refunds, cancellation, and rescheduling, Goibibo has given an ultimate tourist experience in last years. It won the 'Best Website' award by Website of the Year India (WOTY) in the leisure and travel category. It is an annual online public award of its kind in India.

17.2.2 MAKEMYTRIP.COM

Www.MakeMyTrip.com is an Indian OTA, founded in April 28, 2000, by Deep Kalra, Headquartered in Gurgaon, Haryana. MakeMyTrip's core products and services include train, air, bus tickets, hotel booking, car rentals, and related travel requirements, as make easy access to 3rd-party travel insurance and also visa processing. The air ticketing reservation system gives the facility to book domestic as well as international air tickets via an IT-based distribution channel. The hotels and holidays sector provide holiday packages and hotel bookings via an IT-based channel, BPO and branch offices. Its other functions mainly comprise ads revenue from advertisements on its Internet websites, revenue by sale of bus and train tickets and revenue from assisting website access to a travel and tourism insurance firm ("MakeMyTrip," n.d.).

The organization provides travel and tourism-related products and services which cater to the needs of people in India and NRIs and others coming to India from the US, Europe, and other such countries. On its website, www. MakeMyTrip.com, and mobile apps, explorers can research, plan and book any of the travel services and items in India, and in addition abroad. The Company provides travel items and arrangements in India, the US, Australia, Mauritius, Malaysia, Japan, and the UAE. Till March 31, 2016, it had 16 company-possessed travel stores in 14 urban areas, incorporating one of its offices in Gurugram, 49 franchisee-claimed travel stores, which basically offer packages in 43 urban areas, and counters in four air terminals in India.

From Figure 17.1, it can be analyzed that the Indian online travel industry was growing with 49% Compound Annual Growth Rate (CAGR) during 2007

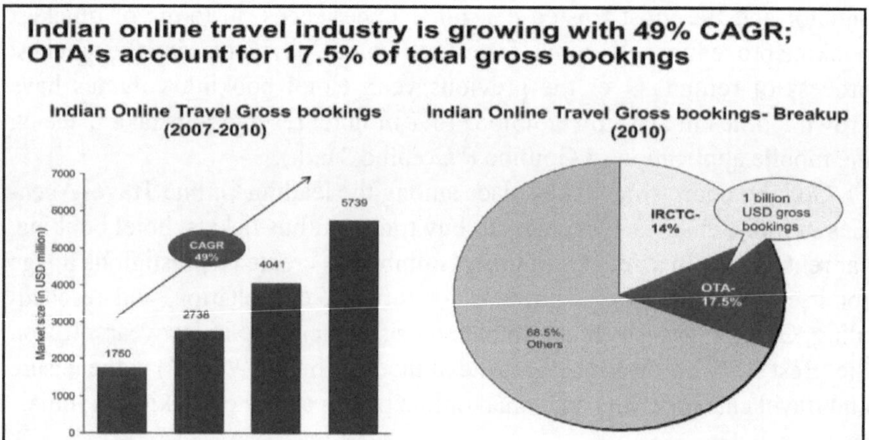

FIGURE 17.1 **(See color insert.)** Indian online travel industry.

to 2010. In 2010, the contribution of OTAs in Online Travel Gross Bookings was 17.5% while IRCTC's was 14% and from others, it was 68.5%.

17.3 PROBLEM DISCUSSION

In India, you can see numerous travelers waiting in a long queue for a long time at the ticket counter just to purchase tickets or book travel-related items. One of the major reasons that they decline to utilize online travel is that they are worried about payment security. Another reason that keeps customers away from purchasing on the web is a privacy issue (Kolsaker, Lee-Kelley, & Choy, 2004). Or they are not aware of its uses. This is not the same as the developed nations, for example, the USA where the vast majority search and book travel products online since it is more helpful, and the types of travel items are exceptionally beneficial. The price is transparent, and it is easy to compare prices for price-sensitive consumers. Consumers can book the travel products online 24 hours seven days and verify instantly. Travelers can get customized travel items and services by way of consulting customer service representatives (CSRs) both online or by means of calling the service center. The internet has a great impact on social, organizational and public life. The web has changed the distribution channels about travel and tourism related information (Bubals & Law, 2008). Online travel is web-based. So IT plays a significant role in the travel and tourism industry. Although some of the studies have been completed in this sector, the marketing condition of online travel has changed to a great extent in the past few years; the online travel operators confront unexpected difficulties than before. So there is a need of updated research. At the present time, Indian travelers would like to travel in free time or on holidays. But now there are few issues coming up in the travel sector, for example, price competition, increasing travelers' complaints, web security and diminishing trust issue and heavy traffic jam. Additionally, the historical background of the online travel booking industry in India is not long. So it has numerous issues which need to be resolved despite everything it has a long distance comparison with America and European travel industry.

17.4 METHODOLOGY

The data and information used in this paper are secondary data, while the method used to collect the data is academic literature, websites, magazines, and other relevant data. The objective is to find effective solutions to cope with challenges and uphold the competitiveness and competitive advantage.

Setting up a marketing strategy in a vibrant environment is not easy because there are many variables influencing the strategy; the chapter focuses on Information Technology.

17.5 MARKETING STRATEGIES: SCENARIO OF OTAs IN INDIA

17.5.1 WORD OF MOUTH IN TOURISM AND TRAVEL INDUSTRY

The meaning of WOM is to the impact or transmission that travelers communicate with others about a travel product, service or an organization. (Litvin, Goldsmith, & Pan, 2008). After the improvement of the web, WOM has a more extensive definition. Goldsmith (2006), describes that the electronic word-of-mouth (eWOM) refers to a type of casual communication. The communication is about travel items, services, or the sellers. The communication has an existence between travelers and producers who produce travel products and as well as travelers themselves. Electronic word-of-mouth is a type of WOM. Sources of word-of-mouth majorly originate from two aspects, first is from mass media in which opinion pioneers utilize mass media to impact opinion seekers. As per Piirto (1992), they give proficient learning or proposals to searchers in a particular field and are trusted by searchers. For instance, we can have a look on official ads of Momomdo posted on the social sites, i.e., Facebook. The second aspect is from travelers' own experiences. Travelers have a wish to share their experiences with others about elements identified with feelings, for example, happiness or anger. In the modern world, individuals would rather trust customer review sites rather than mass media since they assume that personal experiences are straightforward and more reliable.

17.5.2 BRAND AWARENESS: BRAND RECALL AND BRAND RECOGNITION

The role of technology in OTAs is very crucial, and it also changes the competitiveness in the economic-related field; it also makes brand popular among consumers. A decent brand can give an intangible value that is difficult for competitors to copy. Hence, a strong corporate brand gives a sustainable competitive advantage for the organization and lead to client loyalty (Jo Hatch & Schultz, 2003). With regards to the tourism sector, branding strategy is important to enhance corporate reputation, maintain customer loyalty, and charge a premium price as well. Brand awareness gives a benefit that the brand of an organization remains in the client's mind (Pappu, Quester, & Cooksey, 2005). Brand awareness alludes to the capacity of customers that they recall and recognize the details about the brand within several items or services and then make a choice (Rossiter & Percy, 1987). Brand awareness comprises of two structures: brand recognition performance and brand recall (Keller, 1993). Brand recognition alludes to the capacity of customers who can confirm that they have seen the brand before when the brand is given as a hint. Brand recall alludes to the capacity of consumers who can tell a brand name from the memory when given the item classification as a sign (Hutchinson & Raman, 1994). Brand awareness has an impact on the word-of-mouth (WOM) in the travel and tourism industry.

17.5.3 SERVICE MARKETING

In the modern society, tourists have more needs and demands for high-quality travel products and services so OTAs must use ITs to improve tourists' experiences. Service is a procedure that one individual or body provides for another individual or body (Zeithaml, Bitner, & Gremler, 2006). The vast meaning of service is that the outputs are intangible items rather than physical items. The service additionally offers some benefit included items, for example, enjoyment or satisfaction. There are four fundamental components of service that are not quite the same as goods: intangibility, simultaneous production, and consumption, heterogeneity and perishability. Service sector or enterprises are segments which core item is service. Service can occur in a practical place, such as giving foods to clients in the hotel. It can likewise occur via telephone or on the web, for example, giving booking service by OTAs (Zeithaml, Bitner, & Gremler, 2006).

17.5.4 COLLABORATION

As per Wernerfelt's (1995) resource-based theory, firms can obtain and keep an aggressive benefit through core competence. The core competency is the potential to coordinate various production competencies and combine many distinct technologies (Allred, Fawcett, Wallin, & Magnan, 2011). From the viewpoint of competitiveness, Collaboration can assist firms in a way that it can increase the return on the investment which is useful for an organization to get a competitive advantage and grab the new chance and reap the goal. Day (2006) proposes that incorporating business resources can assist the organization in getting a sustainable competitive advantage. For instance, as indicated by Tsaur, Yung and Lin (2006), travel agencies enhance their item advancement in the market by working together with accomplices. Airlines enhanced the client benefit by connecting its online reservation operation to different Airlines, inns, and travel and tourism industry (Duliba, Kauffman, and Lucas, 2001).

17.5.5 PERSONALIZATION

As per Chellappa and Sin (2005), personalization alludes to a marketing technique. Before seeking web-based, as indicated by attributes of customers, for example, browsing records or buying records, suppliers prescribe items and services to particular consumers. From the point of view of consumers, the advantages of personalization are exceedingly proficient, advantageous, customized, amicable and efficient. When they make choices for travel, they will pick organizations which give personalized items and services (Ho & Kwok, 2003). So personalization will result in a positive attitude toward service and product and enlarge purchasing intention (Sheng & Nah, 2008). From the viewpoint of the organization, it is more proficient for the organizations to give the strategic production of items and services, execute the consumer relationship management and direct marketing (Milne, 1997).

17.5.6 PRODUCT INNOVATION

Innovation is progressively essential with regards to the fierce business environment. Product innovation can bring benefits (Lokshin, Gils, & Bauer, 2009). For instance, organizations grow new items by means of innovation to expand the market share. Furthermore, product innovation can make strong perceived desire of a product or service of customers and as a consequence limit the client gap (Lee, 2010). For example, in order to give better worth to customers, the organization needs to adjust to the dynamic condition and create new products to fulfill the necessities of customers. The main challenge of new product development (NPD) is to develop new products with regards to understand consumer needs (Martínez-Sánchez, Vela-Jiménez, Pérez-Pérez, & De-Luis-Carnicer, 2009).

17.5.7 BIG DATA AND THE TOURISM INDUSTRY

There are a large number of data created every day; the travel and tourism industry should endeavor to utilize the big data to enhance the consumer experience. The meaning of big data is making efficient use of data measured over megabyte and gigabyte. The reason for using the information that customers already have generated to furnish better, extra focused and profitable items and services to them. The business venture should incorporate helpful and consumer-centric data stream. However, the venture should not only control and utilize the data stream but also take care of the details about lifestyle and preference of consumers. The web-based organizations can get the resources by gathering cookies, log documents and different other sources. Thus, only those travel organizations that deliver consumer travel experience in a very excellent way can succeed in the future.

17.6 CONCLUSION

The study shows that IT has an impact on the marketing strategies of online tourism agencies. Travelers can get dependable and exact information on the basis of information technologies. Information Technology empowers travelers to book travel products on the road when the traditional method no longer suitable. IT leads to higher satisfaction and make service quality advantageous. Traveler's satisfaction relies upon accurate information and has an impact on service quality. There is tough competition and travelers are demanding. Through collaborating and incorporating marketing strategies, the OTAs can give better items and specialized services to tourists and can enhance their tour experiences. The major services of OTAs are not restricted to ticket booking or hotel booking since travelers can reserve the tickets online by themselves. Consequently, OTAs should concentrate on giving consultancy or improved service through participation to maintain competitive advantage and preserve consumer loyalty.

17.7 RECOMMENDATIONS

This section outlines recommendations to help OTAs in working towards better integration of marketing strategies and use of information technology to flourish online travel business.

- Marketing strategy of OTAs should combine all the marketing objectives into one inclusive plan.
- Spreading awareness about modern online travel booking websites in simple language so that people can understand it properly.
- Campaign to encourage tourists to book any of the travel products through online travel agencies.
- Increasing internet speeds so that none of the transaction would fail.

- Should be given the option of self-service or FAQ page.
- OTAs should ensure that the website loads rapidly, whether on a PC or a mobile.

KEYWORDS

- **information technology**
- **marketing strategies**
- **online travel agencies**
- **tourism industry**

REFERENCES

Allred, C. R., Fawcett, S. E., Wallin, C., & Magnan, G. M. (2011). A dynamic collaboration capability as a source of competitive advantage. *Decision Sciences. 42*(1), 129–161.

Buhalis, D., & Law, R. (2008). Progress in information technology and tourism management: 20 years on and 10 years after the Internet—The state of eTourism research. *Tourism Management. 29*(4), 609–623.

Chellappa, R. K., & Sin, R. G. (2005). Personalization versus privacy: An empirical examination of the online consumer's dilemma. *Information Technology and Management, 6*(2), 181–202.

Dichter, E. (1966). How word-of-mouth advertising works. *Harvard Business Review*, November, 131–144.

Duliba, K. A., Kauffman, R. J., & Lucas, H. C. (2001). Appropriating value from computerized reservation system ownership in the airline industry. Orga*nization Science, 12*(6), 702–728.

Goibibo.(n.d.). Retrieved from https://www.goibibo.com/info/aboutus/.

Hajra Problems and prospects of online travel agencies in India (power point slides). (2013). Retrieved from https://www.slideshare.net/VanditaHajra/problems-and-prospects-of-online-travel-agencies-in-india1.

Ho, S. Y., & Kwok, S. H., (2003). The attraction of personalized service for users in mobile commerce: An empirical study. *ACM SIGecom Exchanges. 3*(4), 10–18.

Hutchinson, J. W., Raman, K., & Mantrala, M. K. (1994). Finding choice alternatives in memory: Probability models of brand name recall. *Journal of Marketing Research*, 441–461.

Jo Hatch, M., & Schultz, M. (2003). Bringing the corporation into corporate branding. *European Journal of Marketing. 37*(7/8), 1041–1064.

Keller, K. L. (1993). Conceptualizing, measuring, and managing customer-based brand equity. *The Journal of Marketing*, 1–22.

Kolsaker, A., Lee-Kelley, L., & Choy, P. C. (2004). The reluctant Hong Kong consumer: purchasing travel online. *International Journal of Consumer Studies, 28*(3), 295–304.

Lee, C. H., & Cranage, D. A. (2011). Personalisation–privacy paradox: The effects of personalisation and privacy assurance on customer responses to travel Web sites. *Tourism Management, 32*(5), 987–994.

Litvin, S. W., Goldsmith, R. E., & Pan, B. (2008). Electronic word-of-mouth in hospitality and tourism management. *Tourism Management, 29*(3), 458–468.

Lokshin, B., Van Gils, A., & Bauer, E. (2009). Crafting firm competencies to improve innovative performance. *European Management Journal, 27*(3), 187–196.

MakeMyTrip (n.d.). Retrieved from https://www.MakeMyTrip.com/about-us/company_profile.php. 2016.

Martínez-Sánchez, A., Vela-Jiménez, M. J., Pérez-Pérez, M., & De-Luis-Carnicer, P. (2009). Inter-organizational cooperation and environmental change: moderating effects between flexibility and innovation performance. *British Journal of Management, 20*(4), 537–561.

Milne, G. R., & Culnan, M. J. (2004). Strategies for reducing online privacy risks: Why consumers read (or don't read) online privacy notices. *Journal of Interactive Marketing, 18*(3), 15–29.

Pappu, R., Quester, P. G., & Cooksey, R. W. (2005). Consumer-based brand equity: improving the measurement–empirical evidence. *Journal of Product & Brand Management, 14*(3), 143–154.

Rossiter, J. R., & Percy, L. (1987). *Advertising and Promotion Management.* McGraw-Hill Book Company.

Sheng, H., Nah, F. F. H., & Siau, K. (2008). An experimental study on ubiquitous commerce adoption: Impact of personalization and privacy concerns. *Journal of the Association for Information Systems, 9*(6), 344.

Tsaur, S. H., Yung, C. Y., & Lin, J. H. (2006). The relational behavior between wholesaler and retailer travel agencies: evidence from Taiwan. *Journal of Hospitality & Tourism Research, 30*(3), 333–353.

Zeithaml, V. A., Bitner, M. J., & Gremler, D. (2006). *Services Marketing: Integrating Customer Focus Across the Firm*. Boston, MA: McGraw-Hill/Irwin. http://library.wur.nl/WebQuery/clc/1809666.

WEBLINKS

- www.business-standard.com
- www.dnaindia.com
- www.goibibo.com
- www.google.co.in
- www.indiadigitalreview.com
- www.indiastat.com
- www.livemint.com
- www.MakeMyTrip.com
- www.newmediatrendwatch.com
- www.redseer.com
- www.reportlinker.com
- www.researchandmarkets.com
- www.thehindubusinessline.com
- www.travel.finacialexpress.com
- www.travelbizmonitor.com
- www.tripadvisor.in

CHAPTER 18

VIDEO GAMES GENERATING TOURIST DEMAND: ITALY AND THE ASSASSIN'S CREED SERIES

BRUNO H. M. CARVALHO,[1,a] GABRIELLA G. C. BERTOZZI,[1,b] and CYNTHIA CORREA[1,c]

[1]*University of São Paulo, Rua Arlindo Bettio 1000, São Paulo, 03828–000, Brazil*

ABSTRACT

In a world in which information plays a vital role in human activities, everything becomes connected, and a new sort of creative economy rises. This scenario provides the development of new communication and entertainment means, new behavior patterns, cultural values, and emerging habits. Consequently, there is an intensification of the flow of individuals in search of new tourist products throughout the globe. Several cities and their attractions have their structures projected into virtual realms, for instance, through video games, an electronic game played on a computing device (personal computer, gaming console, tablet or mobile phone). In this work, the term video game will be used in the broadest possible sense, including any forms of computer-based entertainment software, either text or image-based, using whichever electronic platform and involving one or multiple players in a

[a]*Graduated in Business Management by The European College of Management Dublin in 2012 and in Leisure and Tourism by the University of São Paulo in 2017,*
E-mail brunohmcarvalho@gmail.com

[b]*Technically certified in Hospitality by Fundação Escola de Comércio Álvares Penteado in 2012 and graduated in Leisure and Tourism by The University of São Paulo in 2017,*
E-mail: gcbertozzi@gmail.com

[c]*Associate Professor of the Graduate Program in Cultural Studies, University of São Paulo,*
E-mail: cynthia.correa@outlook.com

physical or networked environment. This study observes the tourist demand for destinations and locations due to playing video games, specifically the Assassin's Creed Series in Italy. Video game players visit cities, villages, fortresses, castles virtually, consuming information about these places. These people look for cultural products regarding cultural activities, which own cultural and social importance, feeding their interests and hobbies. Focused on user-generated content (UGC) produced on Assassin's Creed II and Assassin's Creed Brotherhood, this research aims to identify a demand from players who, either have visited or intend to visit Italy due to the game, and to verify different types of promotion, as well as official and unofficial support to these travelers. The results of this analysis show that through playing those games, people are becoming aware of some destinations and locations, by combining their desire to visit these places grew after playing the games. There is a demand for services related to the Assassin's Creed players and the sites presented in the video games. However, the lack of official support and a wider range of services offered to these tourists make them create their information, displaying it online on blogs, virtual galleries, customized tour guides, and video tours. Private companies are offering guided tours in some of the locations; but very little official support from Italian tourism entities could be found.

18.1 INTRODUCTION

The current economy is grounded on the connection and interconnection of networks, and this scenario offers the development of new communication apparatuses, broadening and blending the boundaries between the field of entertainment and some economic activities such as tourism. A variety of cities and countries began to promote their attractions and structures by means of technological resources and projected into virtual realms. These domains turn internet users and gamers into travelers, who are in search of information based on other experience they have had or for the desire of experiencing something new. This type of experience is also created by the modeling of subjective elements, amongst them the historical interest.

This piece of study aims to verify the tourist demand for destinations and locations due to playing video games, the presence of services for this emerging public, the participation of private and public organs in supporting these travelers. Still, the evolution of the Assassin's Creed series in helping directly or indirectly the tourist to become more aware of numerous places is investigated. Video game players visit cities, villages, fortresses, castles virtually, and consume information about these places. These people are far from being passive or traditional in their behavior as tourists. They look for cultural products regarding cultural activities, which own cultural and social importance, feeding their interests and hobbies.

The cinema and television industries have been involved with tourism since their early stages of production. Famous landmarks are present in films and TV shows on a constant basis boosting tourism even more. These industries also feature unknown locations in their productions attracting hordes of tourists and turning these places into tourist spots. Video games have become present in popular culture over the last decades, and they have helped to shape several industries ever since their creation: cinema, clothing, literature, events, and television. The video game industry has also caused impacts in tourism when hosting conventions, championships and promotional events in different cities of the globe as well as when making use of real-life locations and famous landmarks in open-world games.

Open-world games allow players to explore in-game locations with no boundaries and few rules to be followed. This type of gameplay is present in many modern, well-known games with their vast, detailed open-world scenarios. Some of these in-game locations are completely fictional, like in The Elder Scrolls V: Skyrim by Bethesda Game Studios (2011) or in The Witcher 3: Wild Hunt by CD Projekt Red (2015). Certain is partially fictional, as Grand Theft Auto IV by Rockstar Games (2008), which take players to Liberty City inspired by New York City; the case of Red Dead Redemption (2010), whose world is a fictional version of the western United States and northwest Mexico, or Batman: Arkham Knight (2015) by Rocksteady Studios that allows players to explore Gotham City as they please.

There are those with a focus on delivering highly detailed replicas of the real-world locations in their games and a better exploration experience through providing information about famous landmarks, historical background, and facts on the cities. The popular Assassin's Creed series is famous for recreating real-life locations with an impressive amount of details and has become a reference when the subject is virtual environments in games. Several locations are covered in the series set in different eras, from the age of the Crusades in Jerusalem, to Victorian London during the industrial revolution, the series not only takes players back to the past but also feeds these players with information on sites and some events that happened in real life. Assassin's Creed II and Assassin's Creed Brotherhood are set during the Italian Renaissance and for the purposes of this research; these two games and their influence on post-gaming visitation to Italy will be considered. There will be no discussion on the storyline of the game, nor on the characters, but in the cities, themselves as virtual replicas of real-world locations.

18.2 VIDEO GAMES MARKET

Firstly, to begin a discussion on video games, it is necessary to demystify some of the common sense of gaming, as their target audiences. In 2017, the average age of players around the globe was 35, households that have a device for playing video games was 65%. About gender and purchases, 59% are male gamers that represent 60% of video game purchasers. Hence, 41% are female players, being 40% of game purchasers. The act of playing video games can be evaluated as a social outlet since 54% of the most frequent gamers play with others, and 53% of them feels video games help keep friends connected. Another 42% of the most frequent players feel that video games make them spend time with the family. Each more people are becoming active players with the advance of computer games and console games, associated with the spread of gaming culture itself. Video games have come a long way since the first games emerged in the 1970s. Today's video games offer photorealistic images, graphics and simulation of reality by means of virtual reality (VR) techniques to a degree which is overwhelming in uncountable cases.

Moreover, in recent years, with the emergence of social networks, smartphones, and tablets, new categories of casual games were introduced, as mobile and social games. The advance of casual games (especially on smartphones) as a revolution, a reinvention of what video games can be and who the players are. These games usually are played in short periods, and they do not require deep knowledge and experience, attracting people who were not consistent video game players. This ascension of casual gaming is a rediscovery of what was considered the simplicity of early video games, as new flexible designs in gaming are letting video games fit into the lives of more people. It is a reinvention of the video games, and the image people have about them and their usage.

The growing popularity of games translates into amazing amounts of money. The gaming industry is fast becoming a financial giant. The total revenue of the video game industry in the United States in 2017 amounted to 36 billion U.S. dollars, up from 30.4 billion recorded a year earlier. Looking into specific segments, in terms of hardware unit sales, the tendencies point to popularity of new generation video console platforms. A forecast of the value of the video game industry in the United States from 2011 to 2020, indicates the market will expect revenues to reach 20.28 billion U.S. dollars by 2020.

Video games demand attention due to their cultural and aesthetic elements; the aesthetic developments of game design are more intense, persistent, and exciting. This explosive evolution of creativity in gaming is influencing meaningfully distinct types of expression, like cinema. After the Matrix trilogy, after the Grand Theft Auto, Uncharted, and Red Dead Redemption games, it is evident that movies and games are working in partnership, appropriating from each other's arsenals. This is exactly the case of the game series analyzed in this chapter, which featured the release of the movie Assassin's Creed Syndicate in 2016. A Franco-American production based on the eponymous franchise created by Ubisoft and directed by Justin Kurzel.

Given the advancement of economic interest in video games and the spread of gaming culture itself, video games' theme has been thoroughly studied by different academic fields; then, game researchers show a multidisciplinary background. In general, from academic perspectives and by utilizing diverse methodologies, studies on games concentrate in five crucial aspects: the game (the idea is to look at games in themselves, considering their structure, techniques and other elements of game design); the players (linked to behaviors and how players use games as a type of social space); the culture (games as cultural objects); ontology (philosophical foundations of games), and metrics (quantitative measures of player behavior).

The emerging field of studies on video games faces an assortment of definitions, as the following:

> Video game theory is itself a convergence of a wide variety of approaches including film and television theory, semiotics, performance theory, game studies, literary theory, computer science, theories of hypertext, cybertext, interactivity, identity, postmodernism, ludology, media theory, narratology, aesthetics and art theory, psychology, theories of simulacra, and others.

Since the video games' characteristics, such as narrative, simulation, art, remediation, performance and due to their educational value or the behavioral psychology behind the act of playing video games, some authors contemplate video games as ergodic tools (work). Others accept video games as ludic tools (play), as toys and entertainment means.

The interactive aspect and the impossibility of players do not simply surrender to a game with a proper story are central features of video games. This is what differs video games from other media like television, movies or books. Video games are created for people actively take part with their systems and, in turn, for these systems to react to the players' behaviors. They can be played in cooperative or competitive modes, alone, with physically present players or even with millions of other online players throughout the globe. Due to the huge diversity of themes and goals, a comprehensive taxonomy of current games is of extreme difficulty.

Besides the complexity of the electronic game structure, for generations of gamers, video games are crucial to how they express themselves artistically and even in the way they conceive the world. Also, community ties and values between players cross the boundaries of the game, with the interactivity made possible by online channels.

18.3 TOURISM MARKETING

The process of progressive incorporation of information and communication technologies (ICT) into tourism has transformed the sector. The technology enables consumers and organizations to communicate and interact, as well as improving basic business roles, such as commerce and marketing. Due to the intangible nature of tourism products and services, the tourist will probably not have the chance to try them before a purchase order. In the case of localities, it is necessary to travel to the destination to evaluate the place. From this point of view, ICT are strategic tools to promote the sector, enhancing the competitiveness of the sector.

The dissemination of tourism information is commonly shaped by the production of graphic material such as brochures, folders, posters, which are advertised in travel agencies and trade fairs, as well as the so-called familiarization trips, incentive campaigns, billboards, and traditional media, like television, radio, and press. With the advent of digital technologies, the distribution is happening through several channels and supports. Among the first actions of digital marketing in the sector, there was the creation of web pages, blogs, and portals of companies of various sizes and segments, highlighting the promotion of destinations, known as e-Destinations. Therefore, the use of social media and applications for mobile devices has acquired relevance in the tourist trade.

Other arrangements on digital marketing embrace advertisements, sponsored posts in profiles of opinion makers and online influencers (YouTubers, Bloggers, and Instagrammers), production of web series, panoramic videos, enabling a 360-degree view, and the use of gadgets correlated to more traditional presentations. This material should provide basic and complete information, assisting potential consumers in decision making, by the use of technology innovation in the marketing mix. In addition, marketing strategies grounded on Advergames (advertisement and games) and on gamification in tourism have been investigated. As well as actions grounded on technologies evaluated as more advanced, for example, those based on the applications of augmented reality – AR, and the use of VR.

With respect to definitions on gamification, the most popular saying that gamification denotes the idea of using game design elements in contexts, not related to the game to motivate and increase engagement, and also keep up the user loyalty. In a practical way, the reasoning behind the structure of the games is used in the business field, especially in digital marketing. Gamification can be described as the application of thought and mechanics of games to involve users to solve problems or to practice activities collectively, establishing a sense of commitment. In a commercial foresight, the involvement designates the link between a consumer and a product or service. However, this notion can be amplified, and the engagement considered a type of association that founds a loyal relationship between people and companies or brands.

About the idea of working with gamification in tourism, in fact, the application of game design elements is common, but it may not have been recognized or named according to the concept of gamification. For instance, each loyalty card or frequent flyer program indicates a gamification approach; meanwhile, points are collected to be exchanged for a bonus. Among the applications currently in service in tourism, some destination marketing organizations (DMO), entities responsible for the promotion of tourism destination globally, have designed games to release themselves, as in the cases of Thailand, Cape Town, Ireland and China.

The introduction of video games in association with more innovative features of VR represents a new mechanism for communication and the promotion of experiences. The creation of virtual worlds where the user can walk, see, hear and manipulate objects, can constitute an important resource for the discovery of tourism since, in this simulated reality, tourists can get to know, in an active way, some details about the destinations and services before the trip. Consequently, the employment of VR and games has seemed as strategic for DMO. VR is a quality human-machine interface that involves simulation and real-time interaction through multiple sensory channels. These sensory modalities are visual, auditory, tactile, aromatic and palatable. The main features of VR involve interaction, immersion, and imagination.

After decades of development, VR is closer to consumers; sales of devices powered by this technology are expected to grow 84.5% by 2020. Thus, VR provides many possibilities for different areas, such as sports, medicine, law, education, and tourism, the field focused on this study. Indeed, the tourism sector has appropriated VR for the sale and dissemination of products and services, both for professionals working with tourism marketing and for consumers to make decisions concerning travel. VR has the potential to revolutionize the promotion and sale of tourism and can serve as a marketing tool for hotels, travel agencies, and attractions.

In the context of cities and regions, VR can be defined as, a computer-mediated sensory experience which facilitates access to visual and audio dimensions of a destination. This explanation delivers a contextual significance to the necessities appreciated exclusively by the travel endeavors. The idea is to provide a sample of the kind of experience that can be experienced by the consumer, assisting in the decision-making process. The traveler depends solely on the quality of the information offered regarding the tourism product or service, and the more complete the information, the greater the chances of the product or service to be acquired. In this way, tourism marketing has become more dependent on technologies in order to expand the global market.

18.4 METHODOLOGICAL APPROACHES

For this qualitative research, the interpretative method was the chosen methodology as it aims to enable insights on the experiences of the personal significance of people in a given context. The research was based on the analysis of specialized literature as support of the change in the video game industry over the last decades and the relevancy of video games in current society.

To demonstrate the desire of players in visiting the locations, or for having based their choice in visiting them due to playing the games, a content analysis of the UGC produced on YouTube via Loomer's Assassin's Creed Channel and on online forums designed by Ubisoft for discussion on trips based on its video game series were selected.

Besides this, to identify and examine the offer of services to tourists, information was verified on four websites: Wandering Free, Eco Green Tours, Guided Tours of Florence, and The Four Seasons Hotel Florence with a brief analysis on TripAdvisor for their reviews.

For the availability of official support, the official Italian tourism website, the official tourism websites of the cities of Venice, Rome, Florence and Monteriggioni were also verified. For unofficial support, a website called USA Today, and its virtual picture galleries were analyzed, and two comments were extracted from them. A Google Maps tool was found with its route based on the game's storyline. All the information mentioned above was collected from March to July 2016.

The Assassin's Creed series was the elected video game due to its high details in reconstructing real cities, for having a wide range of locations, and for allowing players to explore them freely. Assassin's Creed II and Assassin's Creed Brotherhood, focused on this research, portray the 15th century Italy. In the action game series, Abstergo is a multinational technology company, which created a virtual reality machine – Animus, which allows people to experience their ancestors' lives through their DNA memories. For centuries, the Creed of the Assassins and the Templars have fought to obtain information about a long-gone civilization that created artifacts capable to give power to those who possess them. The assassins wish to hide them from humanity until it matures enough to hold such power, whereas the Templars wish to control those artifacts to save humankind from its imminent ruin their own way. Abstergo uses the Animus to experience past lives of assassins and Templars to get to know about the hidden location of the objects and to obtain more information on the ancient civilization.

18.5 RESULTS AND DISCUSSION

Assassin's Creed is a franchise of action-adventure games from Ubisoft Montreal. The first game of the Assassin's Creed series was released on November 13th, 2007, based on an Islamic Assassin cult and set during the Crusades, allowing players to explore key locations of the Holy Land, such as Acre, Damascus, Jerusalem, and the Assassin stronghold of Masyaf during the 12th century. The game was an instant success and reviews from the specialized media have praised the cities in the game; Kevin Van Ord wrote for Gamespot in 2007:

> Not enough can be said about the living, breathing world that you'll inhabit in Assassin's Creed... Each city is beautifully rendered from top to bottom and features meticulously crafted towers that reach for the sky, bustling market squares, and quiet corners where citizens converse, and drunks lie in wait to accost you. As you wander the streets (and rooftops), you will push your way through crowds of women carrying jars on their heads, hear orators shout political and religious

> wisdom, and watch town guards harass innocent victims. Altaïr has a profound effect on this world, but the cities are entities all their own, with their own flows and personalities [sic].

More than 93 million Assassin's Creed games were sold between 2007 and 2015. Furthermore, the series has made appearances on the following video game platforms and gadgets: iPhone, iPod Touch, Android, Windows Phone, iPad, Microsoft Windows, Apple Macintosh, Nintendo DS, PlayStation 3, PlayStation 4, PlayStation Portable, PlayStation Vita, Xbox 360, Xbox One, and Wii U. Since the release of the first game, the franchise has spawned sequels and prequels (currently consisting of nine main games and seventeen spin-off games), graphic novels, comic books, animated films, figurines, weapon replicas, board games, clothing, a movie, and other merchandise.

On November 17th, 2009, Ubisoft released the first major sequel, Assassin's Creed II, set during the 15thcentury Italian Renaissance. In this game, players can explore the streets of Florence, Monteriggioni and San Gimignano in Tuscany, Forli and the Apennine Mountains in the Romagna region, the cities of Venice (Figure 18.1) and Rome (The Vatican only).

The game is superior, presents better graphics and improved mechanics while fixing some of the problems with the previous version. Common Sense, an organization dedicated to helping kids thrive in a world of media and technology, rated Assassin's Creed II as a 5 out of 5, whereas the first game is rated 4 out of 5, both age 18+.

FIGURE 18.1 (See color insert.) Saint Mark's Basilica in Venice in Assassin's Creed II. *Source*: Figure adapted from an external source (https://www.moddb.com/groups/assassins-creed-fans/images/piazza-san-marco).

A community member on BuzzFeed posted in 2013 a list of 20 historically precise things in Assassin's Creed II, one of the topics being the accuracy of the cities in the game:

> The cities in Assassin's Creed II are a true testimony to the determined and unintimidated work ethic of the makers of the Assassin's Creed franchise. Florence. San Gimignano. Monteriggioni. Venice. Besides recreating everyone's favorite tourist spots (like the Duomo in Florence, the Piazza San Marco in Venice, the towers in San Gimignano, etc.) with astounding details, this team really did their research. Whether looking at them from a distance or exploring them on foot or by roof, the cities of Assassin's Creed II really give off the vibe that this is what they actually might have looked like in the 1400s and 1500s. It's unbelievable [sic].

Assassin's Creed Brotherhood was released a year later, November 16th in 2010, and it continues the storyline from the previous game. Players are taken to a 15th century Rome (Figure 18.2), which was the largest map the producers had made for the franchise at the time.

In this game, players can wander through the crowded streets of Rome, they can visit the Colosseum, The Roman Forum, The Pantheon, several churches, tombs, the Pyramid of Cestius, the Roman Aqueducts, the building site of the Sistine Chapel, among other famous locations in the city. They can also visit the cities of Florence, Venice, Monteriggioni, Naples and the regions of the Alban Hills, Mount Circeo and Valnerina but without the same freedom, as

FIGURE 18.2 (See color insert.) The city of Rome in Assassin's Creed Brotherhood. Source: Figure adapted from an external source (http://www.push-start.co.uk, 2016).

these locations can be visited only during certain missions with limited access of the map.

As commented, these virtual environments, where users can stroll and interact, constitute a significant resource for the discovery of tourism. Consequently, Ubisoft's sales numbers of the Assassin's Creed franchise show that more than 90 million people are exploring the virtual streets of Venice, Florence, Istanbul, New York, Rome, Paris, and other cities from their homes. These people are not doing it in silence; they are posting their reviews of the games through online forums, their experience in playing them on social media, in-game tour videos on their YouTube channels, and more people can get to know the in-game locations.

A YouTube user called Loomer posted a series of six videos on his channel shot during his trip to Italy and Turkey in July 2014. The videos are entitled "Assassin's Creed Pilgrimage." In these videos, the user explores the streets and key locations of Rome, Florence, Monteriggioni, Siena, Venice (present in Assassin's Creed II and Assassin's Creed Brotherhood), and Istanbul (Assassin's Creed Revelations) while comparing real locations to in-game locations.

Loomer also provides a Google Maps Tour Guide entitled "Ezio Trilogy Tour" about the main character in the games set in Italy and Turkey. He pinned key in-game landmarks, or places, which are related to the story and characters of the game on the map so that people can visit these real locations. There are thirteen pinned locations in the city of Rome, nine for the city of Florence and twenty-one from the cities of Venice, Forli, Venice, and Istanbul. Loomer has 28,531 followers in his channel (April 2016), some subscribers left comments showing their interest in getting to know the places after playing the games, for instance:

> Hey Loomer, thanks for sharing this with us! All the locations in the Ezio Trilogy are absolutely gorgeous, and I hope to visit them one day. Even though Assassin's Creed is just a video game franchise, fans from all over have a personal connection with these places (User 1).

> I've booked a trip to Rome next year to do the exact same thing! [sic] (User 2).

> Words can't describe how happy this video makes me feel. Ever since playing through the Ezio series, it has been my dream to travel to Italy and visit the locations featured in the game [sic] (User 3).

> I'm such a fan of this series and had plans for doing something similar in the future [sic] (User 4).

> Hi, Loomer, this series is the best possible tribute to Ezio absolutely fantastic, this is something I would have done myself, but I don't have time and money to do so [sic] (User 5).

Still, in response to Loomer's videos, others prefer to share their experience of visiting those locations after having played the video game series, being the reason that made these followers know the places. The following excerpts from UGC illustrate this behavior well.

> Thank you so much for sharing this with us! Really interesting to see all the locations in real life. I've been to Forli myself to see the city and take some AC-ish photos and was so impressed by how real the AC version looks [sic] (User 6).

> Incredible, Loomer. Great work. This was amazing to watch and took me back in the series! Really want to go play the old AC games AND travel to Rome again [sic] (User 7).

> Thanks, Loomer, it reminded my trip in Florence (Firenze) two years ago, I went for the same reason, knowing better my country and discovering Ezio's native city [sic] (User 8).

> I lived in Rome two years ago I did the same thing, went to Florence just for my love for the Ezio's AC chapters…I love how a game gave that feeling to a lot of people that want to come here just for that [sic] (User 9).

> This is great. I went to Florence in April, always wanted to go since Assassin's Creed 2 [sic] (User 10).

> Great work! I was also inspired by the series and followed Ezio's steps in all these cities in 2013 (User 11).

Particularly, User 12 feels motivated to explore the place to see if there are similarities between the projection in the game and the location in real life when he wrote:

> Such a great video, Loomer, I hope to visit some of those same locations someday, 'Especially Florence' the hometown of the great 'Ezio Auditore Da Firenze!!!!!' Always wondered what his home or what the building in real-life that's in the same area of his home in the game looks like [sic].

User 13 went further by expressing not only his admiration for the series, but the fact that has been visited Italy with the purpose of comparing real-world Italian locations of the games. Additionally, other subscriber showed the intention of visiting different countries due to the games of the series, like Turkey. User 13 stated:

> I loved this series so much! I also made a trip to Italy (because I love traveling and to compare it to the game) and it was amazing! I have been to the same places except for Forlí, but I have also been to San Gimignano. I have also been to Jerusalem and Akko in Israel but could not visit lots of places that were in the game because it was a school trip. I really wish I could go to Turkey someday, and I really want to visit Istanbul and Cappadocia [sic].

Ubisoft has an official forum in which people can comment and talk about their trips inspired by the Assassin's Creed games. User 14 spent a week in Italy, and he/she makes it very clear that he traveled to the country because of the game when he/she said:

> Hi guys, a few months ago, I posted for ideas on places to visit for my upcoming 'Assassin's Creed Tour of Italy.' Well, I did it in the last week of July, and it was awesome. I was in Italy for a week, and I visited Florence, Rome and also Monteriggioni.

The User 14 also declared that: "One of the things I did not mention in my previous posts is that at every location I visited, I would mention to the staff at the different landmarks that I was visiting Italy based on Assassin's Creed, and in every location, they would say, 'oh really? You're not the only one!'" [sic]. This demonstrates that according to the local staff, people are visiting these locations and they are doing so because they played the games and they want to see them in real-life after seeing them in virtual reality.

As an outcome, there are websites in which users publish photo demonstrating how precise the recreation of the locations is and build virtual galleries of their trips to make a comparison of real-life images to in-game locations. An example is the USA Today website that displayed a photo tour gallery called "Photo tour: Around the world in Assassin's Creed video games." In this gallery, viewers can browse through 22 pictures of different cities of the Assassin's Creed video game series with a brief description of each one of the photos. User 15 commented, "Playing Assassin's Creed II made fall in love with Italian Architecture."

User 16 emphasizes not only her/his interest in visiting the cities in real-life, but also the difficulties in finding official information on tours grounded on the games: "hi there. There does not seem to be much information on the web about assassins Italy... I am planning a trip myself with my son... surely there is a wider interest in a trip to see the town these games are based on... I was convinced I would find tour info online, but they seem rare…" [sic].

Concerning tourist service, the website Wandering Free provides a tour called Assassins Creed Tour of Italy. It consists of a 9 nights/10 days self-guided tour in Florence, Venice, and Rome, with highlights from the games, this tour could not be found on TripAdvisor in April 2016. Another website named Guided Tours of Florence offers the Assassin's Creed Tour; in which tourists are guided through the streets and landmarks of the city while learning about the history of the city and its landmarks, and the Assassin's Creed storyline. This tour has received 117 reviews on TripAdvisor, being 107 as excellent, 8 as very good, 1 as average, nine as poor, and 1 as terrible in April 2016.

For tourists with disabilities or who are not willing to walk, the website Eco Green Tours proposals two hours, guided itinerary on a golf cart through the streets of Florence named Assassin Tour of Florence. According to the website, the monitor presents the differences and similarities between the game and reality. The tour is available in English, Russian, Spanish, Portuguese, and Polish; and its slogan is "Enjoy our Assassin Creed Tour and keep in touch with the reality of virtuality." The tour received 83 reviews, being 69 as excellent, 11 as very good, 2 as average, 1 as poor, and 0 as terrible.

The Four Seasons Hotel in Florence offers an Assassin's Creed Tour on their official homepage with no information other than this brief description: "A tour of the city center on the historical action-adventure video game 'Assassin's Creed' can be arranged with a private English-speaking guide. Highlights include visiting the same sights that Ezio Auditore, the protagonist of the game, frequented."

In Assassin's Creed II and in Assassin's Creed Brotherhood, players can visit Florence, Monteriggioni, San Gimignano in Tuscany, Forli, the Apennine Mountains in the Romagna region, Naples, the regions of the Alban Hills, Mount Circeo and Valnerina, the cities of Venice and Rome. Nevertheless, Florence, Venice, Rome, and Monteriggioni are the four most important locations in the game, being the cities in which players can explore more freely and with higher details. Venice's official tourism website brought no results for the terms "Assassin's Creed" on its homepage, and there is no information about guided tours based on the series. The official tourism website for the city of Florence has no reference to the games and no information on tours and locations for tourist visit although there are several in the games. The official tourism website of Rome showed no result or any sort of information regarding the series.

The exception is Monteriggioni's official tourism website that advertises the city as a location of Assassin's Creed on its homepage where users can find this short description of the game but without any extra information about the availability of the guided tour:

> Are you a fan of the famous video game? Come and discover for yourself the castle of Monteriggioni, the place where the magnificent feats of Ezio Auditore are set. You will recognize the powerful outer walls that for centuries resisted the attacks of the Florentines, and which you can walk along just like a warrior of old. Entering the museum of arms and armor, you'll feel like you're in the armory of Ezio's villa, and you can feel what it's like to be in the 'shoes' of a legendary knight of old.

The official Italian tourism website separates the country into four categories: North, Centre, South, and Islands. Each of these categories includes the main cities and attractions belonging to their region. A brief search for the terms "Assassin's Creed" at the home page and browsing through the four categories brought no immediate results.

In fact, whether the interaction between video games and tourism can be considered quite new, the interaction amid the cinema industry and tourism has been happening since the early stages of the film industry, which could become a role model for that new demand for services. It is possible to find specialized tours based on famous films in different countries and locations where there are companies offering these tours and governmental entities supporting this type of tourism. A positive example of different types of support to the combination of these two industries is the government of New Zealand with Hobbiton Movie Set Tours, a filming location for The Lord of The Rings movie trilogy and The Hobbit films.

18.6 CONCLUSION

As video games technology becomes more available and present in the daily routine, current society is changing its perception on how we play these games and the reason why we play them. Video games are not only seen as an entertainment mean, but as a new way of socializing, a facilitator of daily tasks, a tool for formal and informal education, skills training and even as trendsetters. The list of fields in which video games become more active grows with the advance of this technology and their integration with other industries. The well-consolidated cinema industry has affected tourism since its early stages by increasing the numbers of tourists worldwide after having direct and indirect promotions of cities, natural landscapes, and attractions in films.

The strongly rising of video game industry has demonstrated the same trend over the last decade. As good graphics engines are created and the capacity of virtual recreation increases, video game producers are being able to recreate locations with much higher details encouraging the public's curiosity about these places. The Italian cities reproduced in Assassin's Creed II, and Assassin's Creed Brotherhood have aroused players' intentions on visiting them in real-life. There are websites in which travelers post their comments, pictures, and hints about these visits based on the games, virtual galleries comparing the game to real-life locations, tour guides and Google Maps made by players, which will take travelers to places that are relevant to the game's plot. Channels on YouTube are also available with videos of trips, videos showing the similarities between the cities in the games and real cities.

There is a lack of support from the private sector as very few private companies provide tours or guidance to those travelers. These services can be found in Florence without difficulties, they are present in hotels, tour operators and travel agents, and local tour guides' websites, receiving good ratings on TripAdvisor through users' comments and evaluations. There is an absence of information on these services in other cities featured in the games except for the city of Monteriggioni, which invites tourists to visit it with the slogan "Discover the original settings for Assassin's Creed" right on its homepage. Official Italian entities regarding tourism make no mention of the games or any services related to them on their official websites, and the public can feel that when looking for information on guided trips and virtual guides as it was mentioned by several internet users.

It is important to emphasize that this matter will not be restricted to Italy, as Ubisoft has put effort into recreating different locations with the closest accuracy to the reality that current technology allows the company to do in its recent releases. The process of making new games of the Assassin's Creed series involves actual visits to the places that are going to be considered for the adaptations, proper historical research, studies on the local architecture, culture studies, cults and religions and even clothing related matters are considered.

These concerted efforts could be used for purposes other than selling games and private entertainment. Not only the company could make use of tourism in advertising its products, but also cities could advertise themselves through Ubisoft's most famous video game series. Ubisoft could create a type of "tourism mode" in which players could simply wander around the virtual cities without having to worry about guards or enemies attacking them. These games could be available for the public in tourism, conventions worldwide for mutual promotion. Local entities regarding tourism could link their city image in the series, as did the town of Monteriggioni with Assassin's Creed and New Zealand with The Lord of the Rings movie trilogy and The Hobbit films.

The list of locations covered in the main Assassin's Creed games, up to this moment, is vast. These games have taken players to the Holy Land during the age of the crusades, to Renaissance in Italy, to the Ottoman Empire in Turkey, the Colonial period in the United States, the Caribbean Islands during the Golden Age of Piracy, the French Revolution, and the Industrial Revolution in London.

In an era of intense technology usage and in a moment where video games are becoming an important part of human activity and actual trendsetters, it becomes vital for both public and private sectors to be aware of new trends in the entertainment industry, to be able to understand this emerging public and to learn how to encourage this type of tourism. Some cities do pay attention to the cinema industry related to the tourism field. However, they tend to ignore these video games for tourists. This new market could be a differential in already saturated destinations, bringing the spotlight to unknown locations. Also, as mentioned, it can prepare incoming tourists for cultural differences and even help with historical background and previous tour guidance.

KEYWORDS

- **destination marketing organizations**
- **information and communication technologies**
- **tourism marketing**
- **virtual reality**

REFERENCES

Abi Research (2016). Abi Research Anticipates More Than 50 Million Mobile VR Devices to Ship by 2020 as Virtual Reality Takes Step Forward at the 2016 GSMA Mobile World Congress, 2016. Retrieved from https://www.abiresearch.com/press/abi-research-anticipates-more-50-million-mobile-vr/.

Assassin's Creed Tour – Forums. Ubisoft. Retrieved from https://forums.ubi.com/showthread.php/267627-Assassin-s-Creed-%20Tour/page2?s=5b57c9383b601740310fcae7a371f7fe.

Azazello. (2007). Historically Accurate Things in Assassin's Creed 2 and brotherhood.2007. Retrieved from https://www.buzzfeed.com/azazello/20-historically-accurate-things-in-assassins-cree-cqjf.

Borrero, F., Sanjuán, P., & Ramírez-González, G. (2015). "Gamification techniques in tourism, application test, Casa Mosquera Museum." In *Sistemas & Telemática, 13*(33), 63–76.

Buhalis, D., & Wagner, R. (2013). E-destinations: Global best practice in tourism technologies and applications." *Tourism Tribune, 28*(1), 3–6.

Bujari, A., Ciman, M., Gaggi, O., Palazzi, C. E. (2016). "Using gamification to discover cultural heritage locations from geo-tagged photos." *Personal and Ubiquitous Computing*, 4–18.

Burdea, G. C., & Coiffet, P. (2003). *Virtual Reality Technology.* John Wiley & Sons, Inc., New Jersey.

Cheong, R. (1995). "The virtual threat to travel and tourism." *Tourism Management, 16*(6), 417–422.

Crogan, P. (2018). Indie dreams: Video games, creative economy, and the hyperindustrial epoch. Games and Culture. ISSN 1555–4120. Retrieved from http://eprints.uwe.ac.uk/34123.

Deterding, S., Dixon, D., Khaled, R., & Nacke, L. (2011). From game design elements to gamefulness: Defining "gamification." In: Proceedings of the 15th International Academic MindTrek Conference: Envisioning Future Media Environments. Tampere, Finland, September 28–30.

Di Pietro, L., Di Virgilio, F., & Pantano, E. (2012). Social network for the choice of tourist destination: attitude and behavioral intention. *Journal of Hospitality and Tourism Technology, 3*(1), 60–76.

Eco Greens Tours (Florence). Assassin Tour of Florence. 2015. Retrieved from http://www.ecogreentours.eu/assassin-tour-of-florence.html.

Egenfeldt-Nielsen, J., Smith, J. H., & Tosca, S. P. (2016). Understanding Video Games: The Essential Introduction. New York, NY: Routledge.

Gamespot. Assassin's Creed Review. Retrieved from https://www.gamespot.com/reviews/assassins-creed-review/1900–6182795/.

Garg, S. (2017, December 29). Understanding and Analyzing Video Game Sales. Video Game Sales Project. Retrieved from https://rpubs.com/sahilgarg30/videoGamesProject2.

Gavalas, D., Konstantopoulos, C., Mastakas, K., & Pantziou, G. (2014). "Mobile recommender systems in tourism." *Journal of Network and Computer Applications, 39*, 319–333.

Granic, I., Label, A., & Rutger, C. M. E. (2014). The Benefits of Playing Video Games. *American Psychological Association, 69*(1), 66–78.

Guided Tours of Florence. Retrieved from http://www.guidedtoursofflorence.com.

Guttentag, D. A. (2010): "Virtual reality: Applications and implications for tourism." *Tourism Management, 31*(5), 637–651.

Hays, S., Page, S. J., & Buhalis, D. (2013). Social media as a destination marketing tool: its use by national tourism organizations. *Current Issues in Tourism, 16*(3), 211–239.

Hobbiton Movie Set Tours. Retrieved from https://www.newzealand.com/br/plan/business/hobbiton-movie-set-tours/.

Horan, P. (1996). "Virtual Reality Applications in the Hospitality/Tourism Industry." Hospitality Information Technology Association – Electronic Journal, pp. 1–7.

Huang, Y. C., Backman, K. F., Backman, S. J., & Chang, L. L. (2016): "Exploring the Implications of Virtual Reality Technology in Tourism Marketing: An Integrated Research Framework." *International Journal of Tourism Research, 18*, 116–128.

Italy. Agenzia Nazionale del Turismo – Enit. Italian Tourism Official Site. 21 June 2016. Retrieved from http://www.italia.it/en/home.html.

Italy. Direzione Turismo. Dipartimento Attività Culturali e-Turismo. Roma, Sito Turitico Ufficiale. 21 June 2016. Retrieved from http://www.turismoroma.it/.

Italy. Monteriggioni tourist office. Monteriggioni Turismo. 21 June 2016. Retrieved from http://www.monteriggioniturismo.it/en/.

Italy. Tourism departments of Comune di Firenze. Firenze Turismo. 21 June 2016. Retrieved from http://www.firenzeturismo.it/en/.

Italy. Turismo Venezia – Azienda di PromozioneTuristicadellaProvíncia di Venezia. Turismo Venezia. 21 June 2016. Retrieved from http://en.turismovenezia.it/.

Juul, J. (2010). A Casual Revolution: Reinventing Video Games and their Players. Cambridge, London, USA, UK: The MIT Press, 252 p.

Kwon, J. M., Bae, J.-I., & Blum, S. C. (2013). Mobile applications in the hospitality industry. *Journal of Hospitality and Tourism Technology, 4*(1), 81–92.

Lofgren, K. (2017, April 5). Video Game Trends and Statistics – Who's Playing What and Why? Trends and Statistics. Retrieved from https://www.bigfishgames.com/blog/2017-video-game-trends-and-statistics-whos-playing-what-and-why/.

Loomer. Assassin's Creed Pilgrimage – Tour of Rome. 2014. Retrieved from https://www.youtube.com/watch?v=uIC1HCl0lxE&list= PL3izZvY3LWNcnxrWZWQGBVB8uGOs8PnE.

Loomer's Assassin's Creed Channel! Retrieved from https://www.youtube.com/watch?v=QrZ3WDq6wHQ.

Molina, S. O. (2003). Opós-Turismo. São Paulo: Aleph.

Popesku, J. (2014). Social media as a tool of destination marketing organizations. E-Business in the tourism and hospitality industry, 715–721.

Putro, H. T. (2015). "Immersive Virtual Reality for Tourism and Creative Industry Development." In: *Proceedings of the International Conference on Creative Industry, 2015*. Academic Press, Bali, 1–6.

Salen, K., & Zimmerman, E. (2004). Rules of Play – Game Design Fundamentals. Cambridge, Massachusetts, London: Massachusetts Institute of Technology.

Sambhanthan, A., & Good, A. (2013). "A Virtual World Model to Enhance Tourism Destination Accessibility in Developing Countries." Cornell University, New York.

Sillis, B. (2016, November 14). 10 Amazing Assassin's Creed Facts. Retrieved from https://www.redbull.com/nz-en/assassins-creed-facts.

The Four Seasons Hotel Florence. Retrieved from https://www.fourseasons.com/florence/services_and_amenities/family_at_four_seasons/family_activities/.

Total and segment revenue of the U.S. video game industry from 2016 to 2017 (in billion U.S. dollars). Statista Inc., 2018.Retrieved from https://www.statista.com/statistics/249996/annual-revenue-of-the-us-video-game-industry-by-segment/.

Trip Advisor. Assassin's Creed Tour Florence. 2013. Retrieved from https://www.tripadvisor.com.br/ShowUserReviews-g187895-d2181488- r181628319-Guided_Tours_of_Florence-Florence_Tuscany.html.

Ubisoft Montreal. Assassin's Creed series. Retrieved from https://www.ubisoft.com/en-us/game/assassins-creed/.

USA Today (United States of America). Photo tour: Around the world in Assassin's Creed video games. 2015. Retrieved from http://www.usatoday.com/picture-gallery/travel/destinations/2015/10/16/photo-tour-around-the-world-in-assassins-creed-video-games/74059358/.

Value of the video game market in the United States from 2011 to 2020 (in million U.S. dollars. Statista Inc., 2018. Retrieved from https://www.statista.com/statistics/246892/value-of-the-video-game-market-in-the-us/.

Wandering Free (South Africa). Assassins Creed Tour of Italy. 2015. Retrieved from http://www.wanderingfree.co.za/geek-tours/assassins-creed-tour-of-italy/.

Williams, P., & Hobson, J. S. P. (1995). "Virtual reality and tourism: fact or fantasy"? *Tourism Management, 16*(6), 423–427.

Wolf, M., & Perron, B. (2003). An introduction to the video game theory. Revista de Comunicació Audiovisual, 5, 1–25.

Xu, F., Tian, F., Buhalis, D., & Weber, J. (2013). "Marketing Tourism via Electronic Games: Understanding the Motivation of Tourist Players." In: Proceedings of the 5th International Conference on Games and Virtual Worlds for Serious Applications. Poole, United Kingdom, September 11–13.

Xu, F., Weber, J., & Buhalis, D. (2014). "Gamification in Tourism." Zheng Xiang & Iis Tussyadiah (Eds.), Information and Communication Technologies in Tourism 2014. In: Proceedings of the International Conference. Dublin, Ireland, January 21–24.

Zichermann, G., & Cunningham, C. (2011). Gamification by Design: Implementing Game Mechanics in Web and Mobile Apps. Sebastopol, CA: O'Reilly Media.

CHAPTER 19

DEMOGRAPHIC ANALYSIS OF TOURISTS' PERCEPTION REGARDING VACATION DESTINATION

MANVINDER SINGH PAHWA[1] and ANUKRATI SHARMA[2]

[1]*Associate Professor, Department of Business Administration, Manipal University Jaipur, India VPO Dehmi Kalan, Sanganer Jaipur, (Rajasthan) India, E-mail: manvindersingh.pahwa@jaipur.manipal.edu*

[2]*Associate Professor, Department of Commerce and Management, University of Kota, Kota (Rajasthan) India, E-mail: dr.anukratisharma@gmail.com*

ABSTRACT

Tourism is one of the important industry of any country. It attracts the people of different countries and plays a role of abridging the sociocultural exchange of states and countries around the world and helps in maintaining political balances amongst the places. Tourist move from their native place to explore and exploit the nature, culture, and society across the world and provides revenue to the destination, thereby strengthening the economy. Therefore, the facilities and amenities at the destination induce the tourist to visit and stay. When a tourist makes a plan to visit a destination, he has his mind set to determine the vacation destination. His decision of choosing a destination depends on various demographic and other factors.

The chapter is an attempt to study the association of certain demographic attributes like gender, education level, marital status and income level on his perception towards self-image/prestige, value for money and leisure. The research design used is exploratory as well as descriptive. A convenience sample of 112 tourists has been considered. Cross-tabulation and Chi-Square test has been used. The study is helpful for understanding the packaging of the tourism services and product specific to changing the demographic profile of the tourists and may help the local administration to develop appropriate attractions for tourists.

19.1 INTRODUCTION

The literal meaning of vacation as per Cambridge English Dictionary is "a time when someone does not go to work…." It is necessary for all of us to do "nothing" for some time. This is doing "nothing" signals towards taking a break from the routine work or job as it helps in reducing stress and facilitates refreshing and rejuvenating self to regain energy to perform better in life. People, especially children wait for vacations not to do "nothing" but to do what they want to. Most people plan their vacation somewhere outside their hometowns. Some people try multiple destinations as one enjoys doing his own things in his own way. But what induces him to choose a destination for a vacation. Variety of tourist plans their vacation with a variety of purpose. Some want adventure, but some need peace. Very commonly hear from the peer groups and people in our surroundings and often seen on Social Media proclamation like; "I desire to visit a peaceful place to come out from stress. I need a break from routine life, want to go to a beautiful place where I can find lots of vegetation." Even sometimes for Shopping, for enjoying the weather, for self-image/status building and many more reasons. Tourist is searching and going to different destinations. Doesn't matter it is an explored or unexplored destination, National or International, small or large. That's untimely increased the competition. From every nook and corner, one can see the competition in Tourist Destinations. The competitiveness is so profound among the destinations to attract and grab more tourists. It is understandable that the reasons are several behind pushing the destination marketing by Destination Marketing Organizations, Government and Private Players in the tourism industry. This study aims to find out the main factors which influence the choice of tourists to select and visit a destination. An attempt has been made to study vacation destination tourist behavior in the present study. The authors supposed that the satisfaction of tourists is very much indomitable by the image of a destination. To attract more tourists and to enjoy a competitive advantage, it is crucial to address the diverse requirements of diverse tourists segments. Meantime effective marketing and promotion to make a distinct destination image are mandated. The empirical results of this study show the important factors which impact the destination choice. The outcomes of the study will be beneficial for policymakers, marketers, DMOs and other stakeholders of the tourism industry.

19.2 LITERATURE REVIEW

The importance of tourist's perspective regarding the choice of vacation destination can never be denied as they play a very important role in the development of the tourism industry around the world. The tourist is the consumer as well as a customer of the product and services of the tourism industry and is directly responsible for the generation of revenue. Therefore, it is foremost important to study and analyze the demographics of tourists so that the impact of the change in demographics may be addressed from the point of view of the provisioning of services. The image of a destination makes an impact on the selection of the destination a lot (Crompton & Ankomah, 1993, Goodalll, 1988). Different authors studied the impact of the image on the process of destination selection. There are several factors which influence the selection of a destination. The factor could be related with age, budget, mental or physical requirements, time viability, societal symbol are the factors which impacts the visitors' choices. According to Leisen (2001), economical and time constraints are associated with the selection of destination. Although Fesenmaier & Jeng (2000) argued that selection of destination is an assessment making process which comes across from different channels.

Abderrahim Daoudi and Tanja Mihalič, (1999) in their study stated that there is a strategic importance of tourism for a country which varies with economic weight and political actuality. Tourism is an unpredictable industry and weakened by inconsistent factors such as currency exchange rates, environmental disasters and political causes (Roe & Urquhart, 2001). Thus, safety issues are another major reason which influences the choice of the destination selection. It has been also observed when visitors recognize the risk of visiting a destination they hold up their decisions (Dhebar, 1996). According to Gartner (1993), the selection of destination is highly depending on the availability of the information of the destination on different sources. Visitors are seeing information of safety and security these days prior to visiting any destination. Several reasons work behind the feeling of their insecurity to visit any other destination. Especially the females seem more conscious about this factor while selecting a destination for their visit. Murphy et al. (2007) present their views in a different way according to their study those visitors who love taking a risk and need adventure are not much focused on collecting information about a destination.

Alegre & Cladera (2012) identified tourist profiles and analyzed their demographics for accessing their propensity to spend on shopping. The study facilitates the identification of the type of tourist who will be more interested to participate in shopping at the destination. Correia, Kozak, & Ferradeira (2013) used a unidimensional measure of satisfaction and attempted to explore about the push and pull satisfaction on 323 tourists of Lisbon and concluded that the tourist satisfaction increased as they were provided exposure to sociocultural activities of the destination. Correia et al. (2013) in their research work tried to identify the sector-specific destination attributes which induce to satisfaction of tourist visiting a destination. Seven dimensions were identified by them in this context comprising of basic facilities like boarding and lodging, internal and external accessibility, safety, on-site and nearby attraction, and information.

Chiu, Zeng, & Cheng (2016) revealed that cognitive image directly affects the effective image and Therefore, the destination image. Aziz, Hussin, Nezakati, Nerina, & Yusof (2018) conducted a survey of Muslim families in Malaysia using convenience sampling and elucidated that change in socio-demographic attributes directly impacts the motivation of tourists except for gender and income. The study seems to be of great help to the packages of the services in product differentiation according to demographics.

Xiao & Huang (2014) presented composition and demographic profile of leisure-based tourist market at Changchun, Northeast China's Jilin Province. It was revealed that the study of demographics has helped in understanding tourist behavior and accordingly became instrumental in the modification of city, enhancement of the image of the destination, it's management and marketing.

In the light of the above literature review, the researchers made an attempt to study the demographics of tourist who are planning their vacation destination. The paper tries to address the question, "What demographic attributes associate with prestige, economy, and leisure of a tourist in India?"

19.3 METHODOLOGY

This study is about analyzing the demographics of tourist perception regarding their choice of vacation destination. The study was conducted across the city of Jaipur just before the summer vacation declared in the schools and colleges. This time was deliberately taken in order to get an intimate response from the respondents.

19.3.1 RESEARCH DESIGN

The present chapter uses descriptive research. All the data prepared on MS Excel Sheet was exported to SPSS 21.0 Software for the purpose of analysis. Initially, mean scores, standard deviation, and normality of distribution were analyzed using SPSS 21 software. Thereafter Cross Tabulation was used for ascertaining the demographic profile of the respondents and to access their perception regarding prestige, value for money and leisure using the statements given in Table 19.1.

19.3.2 MEASURING VARIABLES

For the purpose of this chapter, three demographic attributes have been used, i.e., gender, marital status, education level, and annual household income. The respondents were asked to choose the demographic variables which best suits them. The gender of respondents was measured using "male" and "female" while marital status was measured using "unmarried," "married but single," and "married." But as the response under the category "married but single" was significantly low; Therefore, it was not considered for the purpose of analyzing the gender. The education level was measured by asking the respondents about the highest qualifications earned by them out of "Graduate," "Post Graduate," and "Professional/Doctorate." For the purpose of measuring income, the respondents were asked to mention their estimated household income in Indian Rupees (INR). Household income instead of personal income was asked because the decision of vacation is done mostly with family members and friends; hence it is assumed that respondents will be going for vacations with their family members and friends. Moreover, social status is depicted by household income instead of personal income. The slabs of income tax for the financial year 2018–19 with a slight modification have been used for measuring estimated annual household income ranging from "Less than INR 500,000" to "More than INR 1,000,000."

The three aspects namely, Prestige, Economic and Leisure were measures using certain statements as shown in Table 19.1.

TABLE 19.1 Aspects of the Survey

S. No	Aspect of Destination	Question asked	References
1.	Prestige	The destination should be such that my friends should think high of me if I visit this destination	(Lassar, Mittal, & Sharma, 1995)
2.	Value for Money	The destination should be economical to visit	(Sweeney & Soutar, 2001) and (Grace & O'Cass, 2005)
3.	Leisure	The destination should provide a quality experience.	(Aaker, 1991) and (Sweeney & Soutar, 2001)

Besides mentioning their demographic characteristics, the respondents were asked to record their perception of 7-point Likert scale for the statements above depicting prestige, value for money and leisure.

19.3.3 DATA COLLECTION

For the purpose of the collection of primary data, a structured questionnaire was prepared in both online modes using Google Forms and offline printed on the paper. For those willing to fill online, the link was shared with the respondents through a message on their mobile number requesting them to fill the same online. Those who could not access the internet and willing to give information were requested to fill printed paper copy. The data so collected from the printed paper was merged with the online data generated from filling of the questionnaire. The data for this research was collected from 148 respondents through convenience sampling. Of which 112 valid questionnaires were finalized for the purpose of analysis, filtering missing responses and incomplete questionnaire, recording usable response rate of 75.67.

19.4 RESULTS AND INTERPRETATION

The present study aims at examining the demographic characteristics of tourist with respect to their perception towards the choice of vacation destination. The research was conducted in the city of Jaipur, Rajasthan, India which itself is the destination for heritage tourism in India. Table 19.2 provides for the basic descriptive statistics about the respondent tourist demographic profile depicting gender, education, marital status and income of tourist respondents.

TABLE 19.2 Descriptive Statistics

Demographics	Attributes	Value		Mean	SD	Skewness	Kurtosis
Gender	Male	60	53.6	1.46	0.501	0.145	–2.015
	Female	52	46.4				
	Total	112	100				
Education	Grad.	32	28.6	3.13	0.833	–0.257	–1.514
	Post Grad.	33	29.5				
	Prof./Doc.	47	42.0				
	Total	112	100				
Marital Status	Unmarried	55	49.1	1.51	0.502	–0.036	–2,035
	Married	57	50.9				
	Total	112	100				
Income	> 5	44	39.3	2.97	0.875	0.52	–1.701
(In Rs Lakhs)	5 to 10	27	24.1				
	10<	41	36.6				
	Total	112	100				

The analysis shows that there were 112 valid respondents in all of which 60 (53.6) were males and 52 (46.4) were females, with a mean score of 1.46 with a standard deviation of 0.501 with skewness and kurtosis of 0.145 and –2.015, respectively. Of these respondents, all are literate with 32 (28.6) are graduates, almost equal 33(29.5) were postgraduates, and 47(42) were having a professional qualification. The number of married 57(50.9) is as equal as that of unmarried 55(49.1). There are 44 (39.3) respondent tourist having a household income less than INR 5 lakhs whereas almost equal number persists in higher income group with 41 (36.6) earn more than INR 10 lakhs as household income. There is also a middle-income group ranging from INR 5 lakhs and INR 10 lakhs to the extent of 27 (24.1) respondents with a higher standard deviation of 0.875 with skewness of 0.52 and kurtosis –1.701.

The cross-tabulation of demographic attributes gender, education, marital status and household income with personal attributes – prestige, value for money and leisure are presented in Tables 19.3, 19.4, and 19.5, respectively.

From Table 19.3 it can be seen that females disagree than males on the point that vacation destination should be consistent with their own image. They do not take pride in linking their vacation destination with that of their self-image. The χ^2 statistics for cross-tabulation between gender and prestige

TABLE 19.3 Prestige

			Strongly Disagree	Disagree	Neutral	Agree	Strongly Agree	Total	χ^2 Value	χ^2 Sig.	Symmetric Measures Phi	Symmetric Measures Sig.
Gender	Male	Count	13	8	7	16	16	60	**9.582**	**.048**	**.029**	**.048**
		%	41.9%	57.1%	50.0%	84.2%	47.1%	53.6%				
	Female	Count	18	6	7	3	18	52				
		%	58.1%	42.9%	50.0%	15.8%	52.9%	46.4%				
	Total	Count	31	14	14	19	34	112				
		%	100.0%	100.0%	100.0%	100.0%	100.0%	100.0%				
Education	Graduate	Count	5	8	4	5	10	32	14.86	.062	.364	.062
		%	16.1%	57.1%	28.6%	26.3%	29.4%	28.6%				
	Post Graduate	Count	13	2	7	4	7	33				
		%	41.9%	14.3%	50.0%	21.1%	20.6%	29.5%				
	Prof / Doctorate	Count	13	4	3	10	17	47				
		%	41.9%	28.6%	21.4%	52.6%	50.0%	42.0%				
	Total	Count	31	14	14	19	34	112				
		%	100.0%	100.0%	100.0%	100.0%	100.0%	100.0%				
Marital Status	Single	Count	14	9	6	9	17	55	1.736	.784	.125	.784
		%	45.2%	64.3%	42.9%	47.4%	50.0%	49.1%				
	Married	Count	17	5	8	10	17	57				
		%	54.8%	35.7%	57.1%	52.6%	50.0%	50.9%				
	Total	Count	31	14	14	19	34	112				
		%	100.0%	100.0%	100.0%	100.0%	100.0%	100.0%				
Household Income (INR)	Below 500000	Count	12	6	5	9	12	44	3.172	.923	.168	.923
		%	38.7%	42.9%	35.7%	47.4%	35.3%	39.3%				
	From 500000 to 1000000	Count	8	4	3	2	10	27				
		%	25.8%	28.6%	21.4%	10.5%	29.4%	24.1%				
	More than 1000000	Count	11	4	6	8	12	41				
		%	35.5%	28.6%	42.9%	42.1%	35.3%	36.6%				
	Total	Count	31	14	14	19	34	112				
		%	100.0%	100.0%	100.0%	100.0%	100.0%	100.0%				

The image of the destination should be consistent with my own self-image

9.582 is significant (p = 0.048); it can be concluded that the null hypothesis failed to be accepted. Further Phi value is also very low (0.029) and is also significant (p = 0.048). This shows that males and females perceive differently regarding prestige. On further analysis, it can be seen that the level of agreement for males is 84.2% as compared to females at 15.8% which shows

TABLE 19.4 Value for Money

			Strongly Disagree	Disagree	Neutral	Agree	Strongly Agree	Total	χ^2		Symmetric Measures	
									Value	Sig.	Phi	Sig.
Gender	Male	Count	20	6	9	16	9	60	9.352	.053	.364	.062
		%	48.8%	54.5%	60.0%	80.0%	36.0%	53.6%				
	Female	Count	21	5	6	4	16	52				
		%	51.2%	45.5%	40.0%	20.0%	64.0%	46.4%				
	Total	Count	41	11	15	20	25	112				
		%	100.0%	100.0%	100.0%	100.0%	100.0%	100.0%				
Education	Graduate	Count	14	3	4	5	6	32	5.963	.651	.231	.651
		%	34.1%	27.3%	26.7%	25.0%	24.0%	28.6%				
	Post Graduate	Count	14	3	3	8	5	33				
		%	34.1%	27.3%	20.0%	40.0%	20.0%	29.5%				
	Prof / Doctorate	Count	13	5	8	7	14	47				
		%	31.7%	45.5%	53.3%	35.0%	56.0%	42.0%				
	Total	Count	41	11	15	20	25	112				
		%	100.0%	100.0%	100.0%	100.0%	100.0%	100.0%				
Marital Status	Single	Count	19	5	8	9	14	55	.902	.924	.090	.924
		%	46.3%	45.5%	53.3%	45.0%	56.0%	49.1%				
	Married	Count	22	6	7	11	11	57				
		%	53.7%	54.5%	46.7%	55.0%	44.0%	50.9%				
	Total	Count	41	11	15	20	25	112				
		%	100.0%	100.0%	100.0%	100.0%	100.0%	100.0%				
Household Income (INR)	Below 500000	Count	15	4	8	5	12	44	10.69	.219	.309	.219
		%	36.6%	36.4%	53.3%	25.0%	48.0%	39.3%				
	From 500000 to 1000000	Count	13	4	1	3	6	27				
		%	31.7%	36.4%	6.7%	15.0%	24.0%	24.1%				
	More than 1000000	Count	13	3	6	12	7	41				
		%	31.7%	27.3%	40.0%	60.0%	28.0%	36.6%				
	Total	Count	41	11	15	20	25	112				
		%	100.0%	100.0%	100.0%	100.0%	100.0%	100.0%				

The destination should be economical to visit

that males consider it prestigious to go at a destination which they want to reflect or show off for the status.

On analyzing the education level with that of prestige, it was found that respondent tourist of all education level was indifferent in feeling pride to visit a destination. In other words, the level of education among the respondents does not impact the prestige point of vacation destination as χ^2

TABLE 19.5 Leisure

			Strongly Disagree	Disagree	Neutral	Agree	Strongly Agree	Total	$\chi 2$		Symmetric Measures	
									Value	Sig.	Phi	Sig.
Gender	Male	Count	16	8	9	10	17	60	4.399	.355	.198	.355
		%	51.6%	50.0%	69.2%	71.4%	44.7%	53.6%				
	Female	Count	15	8	4	4	21	52				
		%	48.4%	50.0%	30.8%	28.6%	55.3%	46.4%				
	Total	Count	31	16	13	14	38	112				
		%	100.0%	100.0%	100.0%	100.0%	100.0%	100.0%				
Education	Graduate	Count	9	4	3	7	9	32	8.306	.404	.272	.404
		%	29.0%	25.0%	23.1%	50.0%	23.7%	28.6%				
	Post Graduate	Count	12	4	2	2	13	33				
		%	38.7%	25.0%	15.4%	14.3%	34.2%	29.5%				
	Prof / Doctorate	Count	10	8	8	5	16	47				
		%	32.3%	50.0%	61.5%	35.7%	42.1%	42.0%				
	Total	Count	31	16	13	14	38	112				
		%	100.0%	100.0%	100.0%	100.0%	100.0%	100.0%				
Marital Status	Single	Count	17	6	5	8	19	55	2.233	.693	.141	.693
		%	54.8%	37.5%	38.5%	57.1%	50.0%	49.1%				
	Married	Count	14	10	8	6	19	57				
		%	45.2%	62.5%	61.5%	42.9%	50.0%	50.9%				
	Total	Count	31	16	13	14	38	112				
		%	100.0%	100.0%	100.0%	100.0%	100.0%	100.0%				
Household Income (INR)	Below 500000	Count	7	9	2	6	20	44	**16.02**	**.042**	**.378**	**.042**
		%	22.6%	56.3%	15.4%	42.9%	52.6%	39.3%				
	From 500000 to 1000000	Count	11	3	3	1	9	27				
		%	35.5%	18.8%	23.1%	7.1%	23.7%	24.1%				
	More than 1000000	Count	13	4	8	7	9	41				
		%	41.9%	25.0%	61.5%	50.0%	23.7%	36.6%				
	Total	Count	31	16	13	14	38	112				
		%	100.0%	100.0%	100.0%	100.0%	100.0%	100.0%				

This destination should provide quality experience

statistics; in this case, being 14.86 is insignificant (p = 0.062). But if tested at 90% confidence level it will come out to be significant. Therefore, the hypothesis at $\alpha = 0.05$ is accepted.

Marital status also does not significantly impact the prestige of visiting a tourist destination for vacations as χ^2 statistics reported is 1.736 which was also found insignificant (p = 0.784) with Phi 0.125 (p = 0.784).

On analyzing household income, it can be concluded that income also does not impact the feeling pride in visiting a vacation destination by tourists as it reports χ^2 statistics of 3.172 which was found insignificant ($p = 0.923$) with insignificant Phi 0.168 ($p = 0.923$).

Table 19.4 presents the association of demographic attributes with consideration of "Value of Money" for the vacation destination tourists.

On cross-tabulating Gender with vacation destination providing value for money, χ^2 statistics reported a value of 9.352 which is found to be marginally significant as $p = 0.053$. The further analysis gives interesting results that where males always think of saving their pockets, but while going for vacation, they behaved differently than females. Though both have almost equally strong disagreement (males 48.8% and females 51.2%) that the vacation destination should be economical. Though the null hypothesis is failed to be accepted but

The education level and value for money for vacation destination also have no association with each other as χ^2 statistics, in this case, has been reported to be 5.963 which is insignificant ($p = 0.651$).

Marital status also could not report any association with value for money for choosing a vacation destination as χ^2 statistics is 0.902 which is also insignificant as $p = 0.924$ with Phi 0.090 with insignificant p-value = 0.924.

Another interesting result depicted here is that even income level also has no association with an economy of vacation destination as χ^2 statistics were found to be 10.69 which is insignificant as $p = 0.219$ with insignificant Phi value 0.309 ($p = 0.219$).

Table 19.5 presents the results of cross-tabulation between demographics and tourist's perception regarding leisure and quality of experience.

It can be seen that the demographics gender, education, and marital status have no association with the leisure and quality of experience as the χ^2 statistics reported from the cross tabulation of these attributes were insignificant as reported 4.399 ($p = 0.355$), 8.306 ($p = 0.404$) and 2.233 ($p = 0.693$), respectively.

The household income found to be associated with the leisure and quality of experience from a vacation destination. Those people having higher income expect a higher level of quality of experience than the people of lower income as they satisfy with marginally lower satisfaction. The χ^2 statistics reported here is 16.02 which is significant as $p = 0.042$ with Phi 0.378 found significant ($p = 0.042$).

19.5 CONCLUSION

The study of demographics is very important for the marketers who are offering tourism packages to the tourist going to choose the vacation destination. Moreover, the vacation destination administration or the department of tourism of various places must consider these demographics while offering and demographic-specific tourist attraction.

For this study, the 12 hypotheses were framed from the attributed identified from the review of the literature. The data were also collected at a very relevant time when the respondents were preparing for booking the tour for spending good time with family and friends and have certain things in mind while choosing a vacation destination.

The results of cross-tabulation showed that the males were significantly sensitive toward a vacation destination which increases their self-image. They were found to have a "show-off" attitude regarding choosing and enjoying a vacation destination. At the same time, where males are generally money savers but for a vacation destination, they were found more "ready to spend" than females. Further even those having low income tend to have quality experience in vacations.

Any study is incomplete without mentioning limitations so as this is too. This study used convenience sampling method which affects the generalization beyond the geographical area. The study may not equally replicate its findings for tourists of other places. The results may be biased as the majority of people were regional.

The study also signals a lot of scope of future study as the interaction and intersection of demographics can give some eye-opening results. Further, the demographics like occupation, age, etc. can also be examined. The findings are based on categories of demographic profiles and not on the profile of individuals.

KEYWORDS

- **demographic analysis**
- **destination marketing**
- **destination tourism**
- **tourist perception**
- **vacation destination**

REFERENCES

Aaker, D. (1991). *Managing Brand Equity*. New York: The Free Press.

Abderrahim Daoudi, & Tanja Mihalič, (1999). "Strategic importance of tourism as a part of an integrated development strategy of countries and places," *The Tourist Review, 54*(3), 18–18, https://doi.org/10.1108/eb058309.

Alegre, J., & Cladera, M. (2012). Tourist characteristics that influence shopping participation and expenditures. *International Journal of Culture, Tourism and Hospitality Research, 6*(3), 223–237. https://doi.org/10.1108/17506181211246375.

Aziz, Y. A., Hussin, S. R., Nezakati, H., Nerina, R., & Yusof, R. (2018). The effect of socio-demographic of Muslim family tourists in Malaysia. https://doi.org/10.1108/JIMA-03-2016–0016.

Chiu, W., Zeng, S., & Cheng, P. S.-T. (2016). The influence of destination image and tourist satisfaction on tourist loyalty: a case study of Chinese tourists in Korea. *International Journal of Culture, Tourism and Hospitality Research, 10*(2), 223–234. https://doi.org/10.1108/IJCTHR-07–2015–0080.

Correia, A., Kozak, M., & Ferradeira, J. (2013). From tourist motivations to tourist satisfaction. *International Journal of Culture, Tourism, and Hospitality Research, 7*(4), 411–424. https://doi.org/10.1108/IJCTHR-05–2012–0022.

Dhebar, A. (1996). "Speeding High-Tech Producer, Meet the Balking Consumer." *Sloan Management Review 37*(2), 37–49.

Fesenmaier, D. R., & Jeng, J., (2000). "Assessing structure in the pleasure trip planning process." *Tourism Analysis, 5,* 13–27.

Gartner, W. C., (1993). "Image formation process." *Journal of Travel and Tourism Marketing, 2,* 191–215.

Grace, D., & O'Cass, A. (2005). Service branding: Consumer verdicts on service brands. *Journal of Retailing and Consumer Services, 12*(2), 125–139. https://doi.org/10.1016/j.jretconser.2004.05.002.

Lassar, W., Mittal, B., & Sharma, A. (1995). Measuring customer-based brand equity. *Journal of Consumer Marketing / International Journal of Retail and Distribution Management / Journal of Product and Brand Management, 12*(3), 11–19. https://doi.org/http://dx.doi.org/10.1108/07363769510095270.

Leisen, B., (2001). "Image segmentation: the case of a tourism destination." *Journal of Services Marketing, 15*(1), 49–66.

Murphy, L., Mascardo, G., & Benckendorff, P., (2007). "Exploring word-of-mouth influences on travel decisions: friends and relatives vs. other travelers." *International Journal of Consumer Studies*, ISSN 1470–6423.

Roe, D., & Urquhart, P., (2001). "Pro-poor tourism: Harnessing the world's largest industry for the world's poor." Opinion paper for World Summit on Sustainable Development, International Institute for Environment and Development, London.

Sweeney, J., & Soutar, G. (2001). Consumer Perceived Value: The development of Multiple Item Scale. *Journal of Retailing, 77*(2), 203–220.

Xiao, H., & Huang, A. (2014). Research in brief Leisure-based tourist behavior : a case study of Changchun. *International Journal of Contemporary Hospitality Management, 12*(March), 210–214.

INDEX

D

T

U